Rediscovering Optics

Nahum Kipnis

Rediscovering Optics

BENA Press
Minneapolis
1993

First edition

Send all inquiries to:
BENA Press, 3200 Virginia Ave S, Suite 304, Minneapolis, MN 55426

Illustrations, layout, and design by Berta Kipnis

Cover illustration from C. Shott,
Magia universalis naturae et artis... Würzburg, 1657, v. 1

ISBN 0-9636784-0-X
Library of Congress Catalog Card No. 93-090416

Printed in the United States of America

About the Author

Nahum Kipnis was born and educated in the former U.S.S.R. He received master's degrees in physics and mathematics and taught physics at high school and college levels for 19 years. In 1979, he emigrated together with his family to the United States, and five years later he received his Ph.D. in the history of science and technology from the University of Minnesota. Since 1985, he has been a Science Educator at The Bakken: A Library and Museum of Electricity in Life in Minneapolis. Every summer, he instructs teachers in how to make learning science more interesting and effective by combining the history of science with an investigative experimentation. And students of these teachers discovered that repeating historical experiments using materials at hand is both enjoyable and instructive. These experiments constitute the core of *Rediscovering Optics.*

Nahum selected optics for his first book for teachers (there are plans for others) because it is his favorite topic. His scholarly monograph *History of the Principle of Interference of Light* (Basel; Boston: Birkhaüser, 1991), based on his Ph.D. dissertation, is also devoted to the history of optics, and so are a number of articles

TABLE OF CONTENTS

ACKNOWLEDGEMENTS

This material is based in part upon work supported by the National Science Foundation under Grant No. TPE-8855532. Any opinions, findings, and conclusions or recommendations expressed in this material are those of the author and do not necessarily reflect the views of the National Science Foundation.

I am indebted to Samuel Devons for the idea of recreating historical experiments in the classroom. I thank participants of summer institutes at The Bakken for stimulating my research on the historico-investigative method, the staff of The Bakken for various help, and Joe Cain and Karl Amo for their assistance in designing optical apparatus.

My deep gratitude goes to my wife Berta for the idea of this book and for her outstanding job on its illustrations and layout, and to my son Constantine for copy editing.

For the permission to reproduce diagrams I thank:

Blackwell Scientific Publications, Boston:
R. S. Snell & M. A. Lemp, *Clinical Anatomy of the Eye*, 1989; and
Sinauer Associates Inc., Sunderland, MA and L. M. Hurvich:
L. M. Hurvich, *Color Vision*, 1981.

INTRODUCTION

This book is written primarily for science teachers, although it can also be used by motivated high school students. It will help teachers deepen their understanding of optical concepts and improve their teaching techniques in a whole new way, namely, through studying the discovery of these concepts and recreating this process in the classroom.

This approach, called *historico-investigative* method, has been taught to science teachers at The Bakken, a Library and Museum of Electricity in Life in Minneapolis, MN for the past seven years. The participants of the Bakken programs tested this method in their schools and found that it increased students' interest in science and improved their knowledge.

The historico-investigative method evolved from a search of new ways to adapt the teaching of science to the needs of modern life. The time has gone when studying science was necessary only for future scientists and engineers. Nowadays, every one needs to know some science (even physics!) simply to cope with his/her work and daily life. This "knowledge of science" should include not only memorizing scientific facts but also learning how to **think** scientifically. Scientific thinking is a special approach to solving problems, which can be used not only for making scientific discoveries or inventing new technologies, but also for solving unexpected problems or improving performance in any job. Frequently, students view science as a compendium of answers to all possible questions, and whenever they have a question they want to memorize a ready explanation. We need to change this attitude by transforming a student from a passive recipient of knowledge into an active participant in its creation. The motto should be **learning science is studying nature,** where "studying nature" means observing phenomena and deducing their laws **before** trying to explain them through a general theory.

One can learn the art of scientific thinking in the process of solving problems. In a science class a "problem" means an experiment with an unknown answer (*open-end* experiment). In this book, I focus on a specific sub-class of these experiments, which I call *investigations*. An investigation serves not only to teach students correct experimental procedures but also to obtain a correct result, i.e. as close an approximation to the complete solution of the problem as possible given the circumstances. Students determine whether their results are correct by comparing them among themselves, but the final verdict is that of the teacher. It is better to start with qualitative experiments, which allow students to focus on understanding the nature of the phenomenon. Without such an understanding, quantitative experiments are reduced to mathematical exercises with physical data. Such a situation can be prevented by offering quantitative experiments only to students who had already mastered the qualitative ones This book focuses on qualitative investigations.

Of all possible topics for investigations, repeating a **historical discovery** has a special appeal. First, students like all things "real," and reproducing an important event in the history of science sounds better to them than doing an exercise conceived for the sake of exercising. Secondly, students are exhilarated with the notion that they can be as smart as Galileo, for instance, or Newton, when some of their ideas turn out to resemble the arguments of these famous scientists. This will boost students' self-confidence and interest in learning, and also play an important role in some students' choice of their future profession.

Thirdly, the scientific thinking is much faster developed when students do the experiments than when they watch their teacher to do them. To a teacher, shifting the emphasis from demonstrations to labs may imply a much greater expense. Fortunately, the history of science can help with this, too, because many historical instruments can be replicated, with some modifications, from readily available materials. With simple and inexpensive replicas of historical apparatus students can repeat a number of important historical experiments described in this book. Fourthly, by going through a sequence of experiments students can recreate the development of a new scientific concept (for instance, the rectilinearity of light) and see for themselves how interaction of theory with experiment produces new knowledge. This will improve students' understanding of the nature of science as an intellectual endeavor, which can help them thereafter in choosing a profession and in making responsible decisions as citizens. Modern science is too complicated for this purpose.

Fifthly, some old theories are very suitable for teaching purposes because of their simplicity and ease of modeling. The history of science legitimizes their use because it teaches us that there are no "final" theories in science, and each theory sufficiently supported by experiment retains its explanatory value **within its range**, whatever new theories will come to replace it. This means, for instance, that if some phenomena are easier to explain on the basis of the emission theory of light, teachers should not hesitate to apply it instead of the more modern wave theory. Sixthly, an excursion into the history of science would show teachers that some of their priorities and requirements concerning experiments are groundless and arbitrary. They would learn, for instance, that until the 19th century physics was primarily a qualitative science, and that a number of famous quantitative physical laws were derived from rather crude measurements. Finally, the history of science provides us with a number of interesting stories that show the human side of science.

By blending optics with elements of its history and suggestions on performing investigative experiments, this book shows a teacher how to present optical concepts to students in the process of their discovery and make students participate in it. Chapters 1 to 6 (geometrical optics) can be used in both physical science and physics. Chapter 7 (on colors) is only for junior high school: it deals with some phenomena of physical optics without explaining them, with the idea that the theory will be supplied in senior high school. Chapters 8 through 12 are recommended only for physics class. Some materials can be used in other classes, such as life science, anatomy, astronomy, geometry, and arts. Although only a small portion of this book can be used in a physical science or physics class in any given year, eventually the teacher should learn everything in it. The practical way to do it is by varying the topics and experiments offered in the same class in different years. The investigative technique described in this book is universal: having learned it from optical experiments, teachers will be able to use it in other parts of physics and in other sciences as well.

Teachers are to be ready for certain difficulties in using the historico-investigative method. One is that many students, especially in junior high school, lack interest in history. Since the appreciation of history depends on students' age and the level of their intellectual development, the teacher should adjust the historical presentation to the audience. In its full extent, the historical aspect may enchant only a few students, but for these few it can be of great importance. Another difficulty is that students come to study physical science or even physics unprepared to do investigations. Teachers have no choice but to train students in this art, which requires teachers to learn the investigation technique for themselves. To help with this, the book provides samples of investigative experiments from simple (called *demos* and *labs*) to quite complex (labeled *investigations*), which follow a specific plan derived from various historical experiments. Some points in this plan are frequently neglected by teachers, for instance, the origin of the problem, the role of preliminary observations, and the selection of variables. To introduce students to the new type of experiments, the teacher can start with simple demos, discussing with students the general plan of the experiment and then involving them in every part of the procedure, including selecting variables, formulating hypotheses, suggesting verifications, and making conclusions. Next, it is advisable to do two-three labs together with students where they discuss only the major steps of the procedure. Eventually, the teacher limits his/her interference in the students' work to helping them formulate the problem, summarizing the results, and suggesting what to do when different groups obtain different results.

It is strongly recommended that investigative experiments be conducted not only in the classroom but also **at home**. This applies to both experiments unfinished in the classroom and the new ones. Home experiments allow more time for a given topic (and the investigative experiments do require more time than the ordinary ones). They also provide students with greater opportunities for developing their creativity, allow them to proceed at their own pace, and give shy students a chance to do what they could only watch in the classroom. Finally, there is an issue of evaluating students' performance. It is recommended to take into consideration both the effort and the result with the general idea of encouraging achievements rather than punishing mistakes. For instance, a greater number of variables investigated or an ingenious experimental idea deserve extra points.

Being a multi-purpose book affects its style: some sections resemble a physics text, others, the history of science resource, and still others, lesson plans. The latter give examples of how to alternate lecturing with short experiments and discussions (with history incorporated throughout) so as to make a student a participant in discovering a new phenomenon or introducing a new concept. The history of science is presented as a "drama of ideas" (A. Einstein) to show that new knowledge does not come ready-made but is created in the process of scientific controversy. Most experiments are described as in the process of being performed. In some sections the presentation is dramatized, which helps to emphasize the role of different components of the investigation plan, a possibility of different points of view, discrepancy of results, difficulties in the procedure, etc., in short, to prepare the teacher for the ups and downs of a real situation in the laboratory. Signs on the margin help in locating different activities and attract attention to specific points. Typesetting the historical materials in a different font serves the same purpose.

Remember that the **descriptions of experiments are *not* for reading**: take the book into your lab and perform each experiment comparing your results to those described in the book. With a cardboard sheet to cover a window any room can be transformed into a "lab," and, in fact, at one time or another I repeated all these experiments at home or in the back yard. Except for a laser, all the equipment necessary can be found at home or made from readily available materials, or bought very cheaply. This makes investigative experiments feasible as home assignments.

Developing creativity is one of the major objectives of this book. It is impossible to provide teachers with ready-to-use student assignments, because no assignment works equally well with students of different ages and different degree of preparedness. However, a creative teacher will be able to adjust examples labeled "student instruction sheets" to a specific student audience, to available materials, and to other circumstances. Among other things, the teacher should include into student's instructions certain **safety** requirements. Some precautions are mentioned in this book, but they are limited to experiments described: for instance, warnings not to look into a laser beam or at the sun, or to have adults' supervision when working with burning mirrors (or lenses) and candles at home. The teacher should think of the necessity of additional precautions, especially for home experiments, such as avoiding inflammable liquids, keeping the staff away from small children, etc. Besides, some schools have their specific safety requirements, which may affect the choice of apparatus or materials.

I hope that using this book will be as much fun as writing it. Good luck! Your comments and suggestions are welcome.

Chapter 1

LIGHT and VISION

I. HOW DO WE SEE?

history and discussion

Teacher. Let's imagine a prosecution witness testifying in a criminal court about seeing from his window someone committing a murder in the street. One of the strategies of the defense attorney could be to question the ability of the witness to see the event properly. What should he ask the witness to discredit his story?

John. To see at a distance, one needs **healthy eyes**. Thus, the first question should be how good is the witness' vision.

Teacher. Very good! The eye is at the root of the vision process. The problem of visual perception has preoccupied scientists since antiquity. Actually, there are two parts to this problem. One deals with the intermediary agent which brings information from an object to the eye, and the other with the way the eye interacts with this agent. First, we will study how the problem was resolved in the past, and at the end of this unit we'll discuss the modern views. There had been several theories of vision in antiquity. According to one theory (we will call it *extramission* theory), originated by the ***Pythagorean*** school, the eye has an internal "fire" which it sends towards an object to discover the object's shape and color. ***Empedocles*** (ca. 492- ca. 432 B.C.), for instance, compared the action of the eye to a lantern making things visible at night. The other theory (the *intromission* theory), initiated by ***Democritus*** (b. ca. 460 B.C.), asserted that it was the object that sends its image (*eidola*) into the eye. According to ***Leucippus*** (ca. 500-440 B.C.), "*we do not see the objects coming nearer to us when we perceive them, therefore, they must send to our soul 'something' which represents them, some image,...some kind of shadow or some material simulacrum which envelopes the bodies, quivers on the surface and can detach itself from them in order to bring to our soul the shape, the colours and all the other qualities of the bodies from which they emanate.*" (Ronchi, 7) ***Epicurus*** (ca. 341- 270 B.C.) made the corpuscular nature of the *simulacra* obvious: "*For particles are continually streaming off from the surface of bodies, though no diminution of the bodies is observed, because other particles take their place. And those given off for a long time retain the position and arrangement which their atoms had when they formed part of the solid bodies.*" (Lindberg, 1976, 2) Neither theory could explain the nature of visual perception, and ***Plato*** (ca. 427-347 B.C.) supposed that the key to it was in the interaction between the internal "fire" and the external agent.

Aristotle (384-322 B.C.) rejected both the intromission and extramission theories. In his view, vision was due to changes in the medium between the eye and an object. He said, for instance, that the color of a body "*moves the transparent medium,... and this, being continuous, acts upon the sense organ.*" (*On the Soul, 107*) Eventually, Aristotle's idea gave rise to wave theories of light, but in the ancient world it had few followers. It is worth noting that although the rudiments of both the wave and corpuscular theories of light had already existed in antiquity, the main focus in optics until the seventeenth century was on the nature of vision rather than on the nature of light. For several centuries the debate on vision did not rely on anatomical evidence. ***Galen*** (ca. 129 - ca.199 A.D.), gave the first anatomical description of the eye. He mentioned three coats of the eye: the *Cornea* being the outermost, the *Uvea,* the middle one, and the *Retina,* the innermost. He also described three transparent media: *aqueous* humor, *crystalline* humor, and *vitreous* humor and compared the retina to an extension of the brain. According to Galen a visual fluid (*pneuma*) originates in the brain and passes through a hollow optic nerve into the eye. After being diffused in the vitreous humor it reaches the crystalline and makes the eye and the adjacent air sensitive to light coming in from outside.

Thus, the main distinction between ancient theories of vision was on **whether vision is initiated from within or without the eye**. What is your opinion about this?

Mary. Since we cannot see in the dark when there is no light coming into the eye, it is obvious that the extramission theory is wrong. Why did Greek philosophers conceive so strange an idea?

Teacher. Your argument is similar to that of Aristotle: "***if vision were produced by means of a fire emitted by the eye, like the light emitted by a lantern, why then are we not able to see in the dark?***" (*On Sense,* 230) However, the proponents of the extramission theory had their reasons. Have you ever read in fairy tales of a fire or lightning coming out of the eyes of wild beasts? This is a very ancient belief, perhaps based on the fact that the eyes of some animals glow in the dark.

Ruth. Indeed, our cat certainly supports this theory, and it seems that an expression about a fire issued from the eyes has become quite common in describing people's emotions too.

Teacher. Galen's objection to the intromission theory was that if the eye receives imprints of objects reproducing their shape and size, an imprint of a mountain could not enter the eye through a small pupil. Another objection referred to the difference between someone's eye serving as an object and its image in a mirror: if this image were an imprint (or a mask) of the eye which left the eye and reached the mirror, one would see this mask from behind rather than from the front. Now, could you give me an argument in favor of the intromission?

Michael. An obstacle between the eye and a body prevents it from sending light into the eye.

David. This won't work, for it could be used by the rival theory as well.

Teacher. Democritus had a better point than Michael. He said that if the imprints of the surrounding bodies did not reach the eye, we wouldn't see their images in the eyes of people. Now, check his statement by looking into the eyes of one another. Can you see there an image of a bright object, such as a light or a window?

Dorothy. I can see it very well.

Teacher. It is important to note that despite the lack of understanding of the nature of light and vision, optics became one of the most advanced sciences of antiquity. This happened because of a special approach to optics developed by ***Euclid*** (ca. 300 B.C.). Its nature will become transparent after the following experiment.

II. RECTILINEAR PROPAGATION OF LIGHT

1. Is the line of vision straight?

lab

Teacher. We have already concluded that to see something clearly one needs healthy eyes. Is that sufficient?

Mary. No. Since light travels rectilinearly, there should be **no obstacles** between the eye and the body. Thus, the attorney should ask the witness whether at the critical moment his line of view was not blocked by something (a passerby or a car, for instance).

Teacher. Right. However, how certain are you that light always travels along a straight line?

Mary. Meaning how do I prove it? I would use three pins and a ruler. First, draw a straight line on paper using a ruler and place two pins at its ends. Second, move a third pin between the two until all three appear to coincide. Finally, check whether the third pin is on the same straight line with the others,

lab

Teacher. This sounds like an easy experiment. Let us do it following Mary's instructions. Since you have pins with flat heads, place them on their heads rather than stick them into the table. Repeat the experiment three times. What are the results?

John. It's true: all the pins make the same straight line.

Teacher. Incidentally, do you take it for granted that a ruler's edge is straight?

John. No, I can prove it. For instance, if I look along the edge of a ruler and see all its points on a straight line, the edge is straight.

David. Wait a second! We've just proved experimentally that light moves rectilinearly on the basis of the coincidence of the line of vision with the line drawn with a ruler. Now, you are saying that the ruler's edge **is** straight because it coincides with the line of vision. This is circular reasoning!

good point!

2. Visual rays vs. light rays

Teacher. Indeed! One way to resolve this paradox is by assuming that the visual line is straight in the geometrical sense. That was Euclid's idea, which he realized in his book *Optics*. Although the famous geometer favored an emission from the eye, he abandoned all speculations on its physical nature and left only one property for this emission: to travel along a straight line. That was the concept of the *visual ray*. While for Euclid the rectilinearity of the visual rays was an **axiom**, some scientists tried to justify it mechanically (see Ch.2) without resorting, however, to direct experiments. Nonetheless, the theory of visual rays dominated optics for many centuries and even affected a major change in the rival theory.

history

In the early Middle Ages the center of learning moved from Europe to the Muslim world. Arab scientists translated works of Greek philosophers, commented upon them, and also did original investigations. In optics, two scientists were especially important, ***Abu Yusuf Ibn Ishaq al-Kindi*** (813-866 A.D.), and ***Abu Ali al-Hasan Ibn al-Haytham*** (ca.965-1039 A.D.), also known as Alhazen. Al-Kindi aimed at improving the extramission theory by rectifying some flaws of Euclid. In particular, he wanted to prove experimentally the rectilinearity of visual rays. Instead, he accomplished this for **luminous** rays, by studying the shape of shadows and images created by apertures. Perhaps he believed in the similarity of the properties of the two sorts of rays. Al-Kindi brought forth new arguments against the intromission theory. If external objects sent their images to the eye, he said, we would see all of these objects simultaneously, which is impossible. Actually, when reading a book we see only one word at a time rather than the whole page. On the other hand, it is easy

history

to grasp that an internal agent could scan the surrounding objects sideways selecting them one at a time. He also noted that while the concave shape of the ears serves for receiving sound, the spherical shape of the eye and its mobility are necessary for emission rather than reception. In Al-Kindi's view, the fact that a ring seen sideways appears to be a straight line contradicts the idea that the eye receives the ring's form, which is a circle. He assumed that a visual perception can be created only by a solid body striking the eye's sensor, which implied that rays must be three-dimensional. This concept, as applied to rays of light, was later transformed first into the *physical ray* and then into the *wave front*.

Like al-Kindi, Ibn al-Haytham was preoccupied with demonstrating the rectilinearity of light rays. He had different devices for light of different origin. As concerning sunlight,

> *when it enters through a hole into a dark chamber the air of which is cloudy with dust or smoke, the light will appear to extend rectilinearly from the hole through which the light enters to the place on the chamber's floor or walls which that light reaches. If the air in the chamber is clear and pure and the extension of the light through it is not visible, and if an experimenter wishes to examine the interval through which the interval through which light extends, then let him take an opaque body and, approaching the rectilinear interval between the hole and the place on the chamber's floor or walls where the light is, let him intercept it by the opaque body: he will find that the light will appear on that opaque body and vanish from the place where it showed on the chamber's floor or walls. (The straightness of this interval can be tested with a straight rod.) ...Let an experimenter take an opaque body and, having made a minute hole in it, let him hold it opposite the body of the sun: he will find that the light goes through the hole, extending on a straight line. If he tests the interval on which the light just described has extended by applying a ruler to it, he will find it to be perfectly straight.* (Sabra, 13)

To test the light of a flame, Ibn al-Haytham employed a tube with a pinhole. To extend his proof to the light scattered in the atmosphere, he let light into a room through an one-foot hole into the outer wall and two identical holes in the inner wall. He found that a luminous spot **A** (Fig.1.1) on the floor appeared only when this spot and the holes **B** and **C** in both walls were on the same straight line as verified by a stretched thread. Ibn al-Haytham supported the intromission theory. He found a way to explain its greatest difficulties: how to reconcile a large imprint of a body with a small pupil,and why ten thousand people can simultaneously receive the same imprint. He supposed that the whole image is formed in the eye by means of many rays coming from different points of a body instead of a single imprint of the body's size and form.

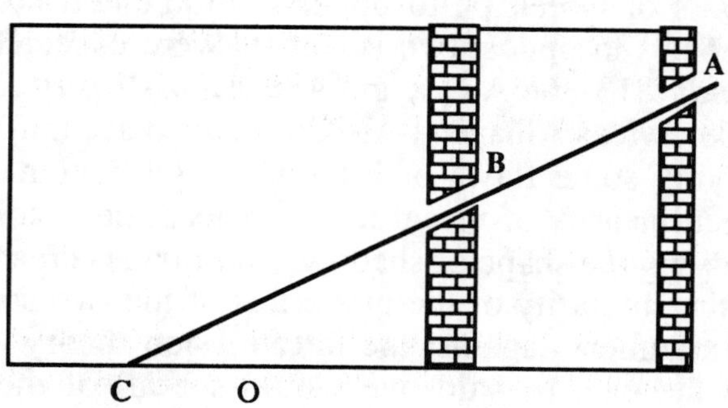

Fig. 1.1. Alhazen's demonstration of the rectilinear propagation of light

In Ibn al-Haytham's view, the eye's sensor is located on the front surface of the crystalline lens, where each point of an object has its image. To preserve an one-to-one correspondence between point objects and their images, he ignored refraction and all rays coming into the eye from the body but those which fall perpendicularly to the cornea and pass through the lens' center (Fig.1.2). Like Galen, he believed that the image is produced directly in the brain. He was not sure that a camera obscura was a good model of the eye because he could not explain how two eyes produce a single direct image while two holes create two inverted images.

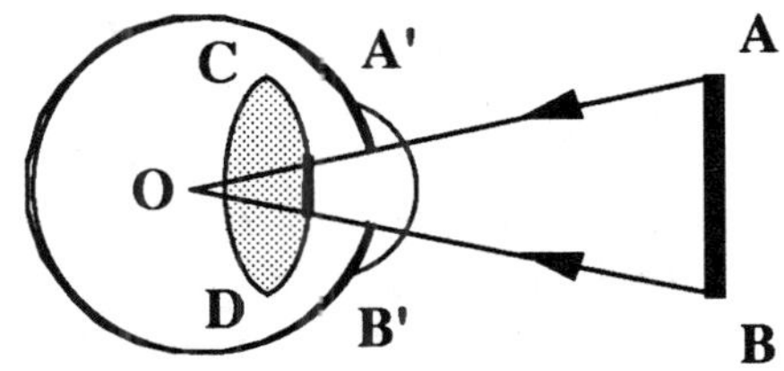

Fig. 1.2. Alhasen's model of the image formation in the eye.

discussion

Teacher. Incidentally, do you find Ibn al-Haytham's proof that light travels along a straight line convincing?

Ruth. Only when referring to a light **beam**. However, it proves nothing about the one-dimensional Euclid' ray. If we compare the beam of light to a ruler that Ibn al-Haytham used to test the direction of light and rays of light to individual fibers of wood within it, it is easy to imagine that the ruler can be straight despite some of its fibers being curved.

David. What he actually proved is that the concept of straight light rays is compatible with experiments but not that these rays **must** be straight. Can a better method be devised?

Teacher. No, if you mean a direct proof. However, there is an indirect method involving a comparison of the shapes of a luminous source, of an aperture, and of the image they produce.

Mary. But how can one reconcile a circular disk of the sun with a square aperture in a screen? Will the image be round or square?

Teacher. Let us start with an experiment.

3. What should a projected image resemble?
(Aristotle's problem)

demo

Teacher. I cut two holes in an index card, one circular and the other rectangular. I will shine light on them from a desk lamp, and you watch their images on a white screen. What do you see?

John. Both images are round. But this is impossible!

Teacher. What would you expect?

John. A circle and a rectangle.

Teacher. That was what Aristotle thought too:

> *Why is it that when the sun passes through quadrilaterals, as for instance in wickerwork, it does not produce a figure rectangular in shape but circular? Is it because the sun's rays fall in the form of a cone and the base of a cone is a circle, so that no matter what object they fall upon the rays of the sun*

must appear circular? For if the rays were straight the figure formed by the sun would necessarily be bounded by straight lines. For when the rays fall straight on to a straight line they do produce a rectilinear figure (*Problems*, 334-5).

Here Aristotle is referring to the geometrical theorem that if the straight lines originating from a single point touch the extremities of a figure, its projection on a parallel plane must be similar to the original (Fig.1.3). The discrepancy between this theory and experiment made him suggest that perhaps the sun's rays are cones with their apexes on the sun rather than straight lines. While observing a solar eclipse Aristotle discovered again that the sun's image resembled the sun rather than the opening in a screen:

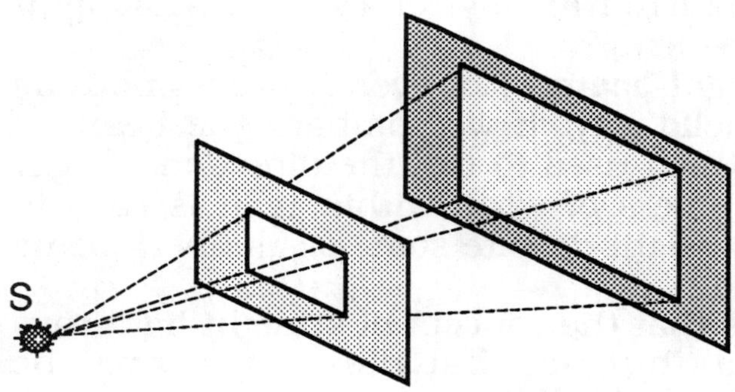

Fig. 1.3. An illustration of Aristotle's argument.

Why is it that in an eclipse of the sun, if one looks at it through a sieve or through leaves, such as a plane-tree or other broad-leaved tree, or if one joins the fingers of one hand over the fingers of the other, the rays are crescent-shaped where they reach the earth. Is it for the same reason as that when light shines through a rectangular peep-hole, it appears circular in the form of a cone? (*Problems*, 341)

The problem raised by Aristotle of why the image made by sunlight passing through an opening is similar to the luminous body and not to the opening, had baffled scientists for almost two thousands years. Let us **investigate** it in some detail.

investigation

Equipment. Index cards, masking tape, razor blade or scissors, paper or cardboard white screen. In each index card make two round holes with a paper-punch and several square holes of different size using a razor blade or by cutting pieces with scissors and pasting them together.

Preliminary part

Background

Teacher. We begin this investigation with a certain problem in mind found in Aristotle: can a rectangular hole produce a round image of the sun? We are curious to know whether Aristotle was right, which prompts us to repeat Aristotle's experiment. This can be done by observing an image produced by sunlight passing through a square hole in a cardboard sheet. Incidentally, in the 16th century astronomers also became interested in this problem, but that came not from reading Aristotle but from the needs of their science.

Experiments

Ruth. Our group discovered that Aristotle's conclusion is correct only when the screen is far away from the index card. He probably never tried to bring the screen close to the hole, for if he did, he would have seen a square image instead of a round one.

John. Distance is not the only thing that matters. We've found that at the same distance between the screen and the card a large square opening produced a square image, and a small square hole projected a circle.

Formulating a problem

Teacher. We see that Aristotle was only partially right. This finding suggests that we examine a more general problem than his: why does the image produced by sunlight coming through the same opening at some circumstances resemble the sun and at others, the opening?

Selecting variables

Mary. We found two factors affecting the shape of the image: 1) the distance from the hole to the screen, and 2) the size of the hole. These could be our variables.

Main part

a. distance

Preliminary experiments

David. I would suggest placing the screen in the shadow and moving the index card to and from it.

Ruth. It looks as if all square holes produce square images when close to the screen.

Hypothesis

John. The closer the hole is to the screen, the more the image resembles the opening.

Mary. Why do you call this conclusion a hypothesis? Haven't we just proved it?

Teacher. We compared only two holes of a specific shape (square). Are we sure that holes of other shapes behave in the same way? Do we know that the result will not change if we make the hole either very large or very small? In both cases, the answer is "No." That is why we need additional experiments. **The preliminary experiments are necessary to advance a hypothesis, but to prove it one has to perform more experiments and of a somewhat different kind.**

important

Test

Ruth. Let us see how holes of different shape but about the same size behave at different distances. Instead of cutting additional holes we changed the shape of the original openings by covering them partially with tape. In our experiments the images of triangles and rectangles closely resembled the apertures when the screen was no further than 15 cm from the card.

John. We tried parallelograms and trapezoids. In our view, the similarity was preserved for distances up to 20 cm.

conclusion ➡ The image resembles a middle size opening only when the latter is close to the screen.

b. size of the opening

Preliminary experiments

John. Let us watch the images of two different squares, first, when the card is close to the screen, and then when the card is far from it. Apparently, in both cases the larger square hole makes a sharper square image.

Hypothesis.

Mary. The larger the opening, the closer its picture resembles the original, whatever its distance from the screen.

Test

John. We can check this in two ways: 1) by making the hole much larger and much smaller; 2) by studying openings of other shape (for instance, triangles of different size).

conclusion ➡ The hypothesis is true, and the conclusion of the previous part is confirmed again.

general conclusion ➡ Sunlight reproduces the shape of any given opening only when the screen is close to it. On the other hand, at a sufficiently large distance any opening produces a round image. Perhaps at large distances we obtain the image of the luminous body. This hypothesis can be tested by experimenting with light sources of different shape.

Explanation

David. You said in the beginning that this experiment has something to do with the rectilinear propagation of light?

Teacher. Yes, I did. Aristotle thought that having a round image of a square opening contradicts the idea of one-dimensional straight rays and requires light cones instead. Do you agree with him?

David. If light rays were cones, the images would have been round at all distances from the opening, which is not true. However, I don't understand why square images become rounded at large distances.

Teacher. Let us assume that light rays **are** straight lines and consider a simplified problem: to find the images **CD** and **C'D'** of a square aperture **ab** (Fig. 1.4a), created by the center **O** of the sun **AA'** and its peripheral point **A'**, respectively. Let **S** and **S'** be the centers of these images. The shift between the images **SS'** depends on the angular radius of the sun α (α= 0.0045) and on the distance **L** between the aperture and the screen

$$SS' = L \tan \alpha$$

One can find from this equation that the shift is 9 mm for L=2 m, and 0.9 mm for L=20 cm. Imagine that each point of the sun creates its own rectangular image of the opening. These partial images overlap, which produces different results close and far from the opening.

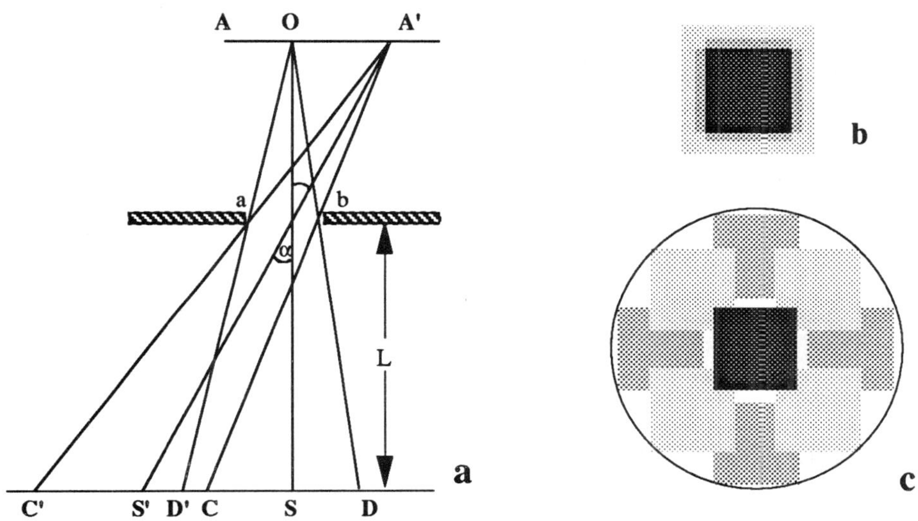

Fig.1.4. An explanation of Aristotle's problem: **a.** images produced by the central and peripheral portions of the Sun; **b.** overlapping of partial images at a small distance from the aperture; **c.** overlapping of the images at a large distance from the aperture.

While the partial images have almost the same size as the aperture whatever the distance **L**, their mutual shift depends on this distance. For instance, if a square aperture is of 8 mm, at L=20 cm the shift is much smaller than the partial image (Fig. 1.4b). As a result, the partial images created by different portions of the sun practically coincide, and the total image reproduces the shape of the hole. However, at L=2 m the shift is comparable with the size of the partial image, and the total image becomes rounded (Fig. 1.4c).

Thus, the shape change of the image with distance is fully consistent with the concept of one-dimensional rectilinear rays of light. ***important***

4. The projection method in astronomy.

history

The uncertainty about the image created by an opening could have been the reason for astronomers not to use the projection method to measure the angular size of the sun. Initially, ***Archimedes*** (287-212 B. C.) and ***Claudius Ptolemy*** (2c A.D.) measured it directly, by looking at the sun behind a proper obstacle. Archimedes, for instance, used a cylinder placed as far from the eye as was necessary to let the rays only to touch its sides (Fig.1.5). Ptolemy preferred Hypparchus' diopter in which the movable plate had two pinholes and was moved so that the eye saw the top of the sun

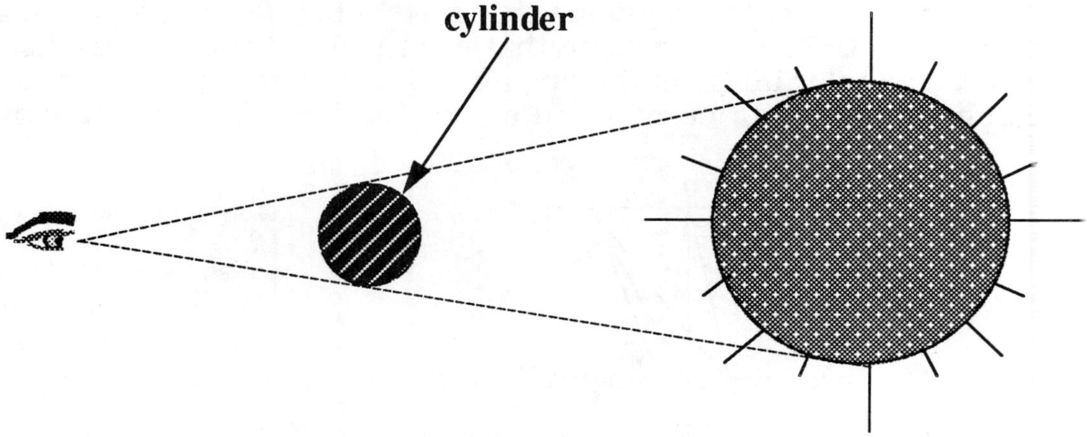

Fig. 1.5. Archimede's method of measuring the solar diameter.

disk through the upper hole and its bottom through the lower hole. The disadvantage of this method was the necessity to look at the sun which is dangerous for the eyes (there were no dark filters available at the time). For this reason, in antiquity the size of the sun was measured only at sunrise or sunset, and when in the 16th century astronomers started measuring the sun's diameter during solar eclipses, they had to do it by projecting it through a small opening on a white screen. In 1598, ***Tycho Brahe*** (1546-1601), a famous Danish astronomer, discovered to his surprise that the image of the moon covering that of the sun was 20% smaller than the accepted size of the full moon. The puzzle was resolved by ***Johannes Kepler*** (1571-1630), and this set him on the way to great discoveries in optics. An interesting part of his solution was using a mechanical model. Kepler set a book, which played the role of the sun, high over a table with a polygonal opening. Then he attached several threads to one corner of the book and stretched them so that they reached the floor after touching each edge of the aperture. He marked each projection on the floor and repeated this procedure with other corners of the book. Since the image obtained resembled the optical ones, Kepler concluded that the rectilinearity of light was upheld.

interesting!

Optical projection found important practical applications. This method provided the basis for the *camera obscura* which became very popular in the 16th century, in part due to ***Giovanni Baptista Porta*** (ca.1538-1615). Although pinholes had been used earlier (by Ibn al-Haytham, for instance), it was not until the 16th century that scientists began to study systematically the **shape** of the image. Porta made a small hole in the shutter of a window and observed the inverted images of all external objects on the opposite wall. He found that a convex lens, placed near the hole, made the images more distinct and suggested using the camera obscura with a lens to make exact pictures of persons or things and for observing solar eclipses (Fig.1.6). Porta also made the first slide projector: a picture drawn on a thin paper was placed outside

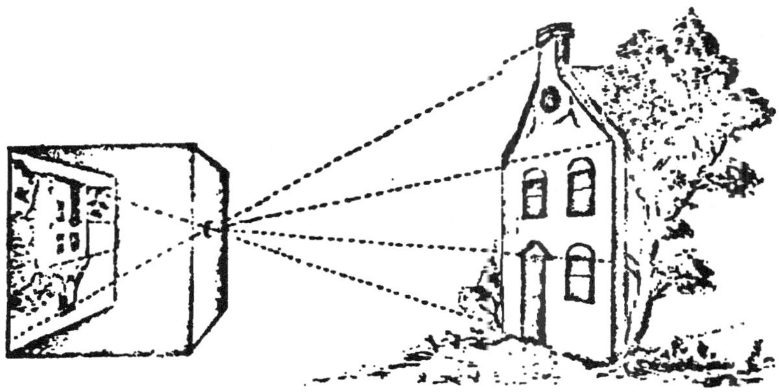

Fig.1.6. Camera obscura with a lens. From J. Priestley, *History of Light* (1772), Fig.14.

the room near the lens and illuminated by the sun. Later, ***Athanasius Kircher*** (1601-1680), a Jesuit Professor of Mathematics, used this idea in his *magic lantern* (Fig. 1.7), which was a more convenient and optically advanced instrument. The luminous source was a candle, the light of which was collected by a lens and sent on the glass

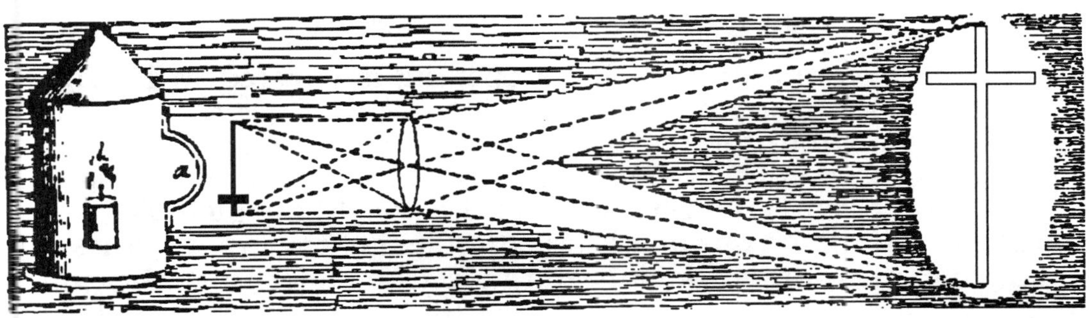

Fig.1.7. Kircher's magic lantern. From J. Priestley, *History of Light* (1772), Fig.48.

slide with a picture drawn on it. This picture was an object for the second lens which projected it on the wall. Kircher also claimed to invent a device to project letters inside the room from outside using sunlight and a mirror, the letters being written on the mirror surface (Fig. 1.8).

? ***can it work ?***

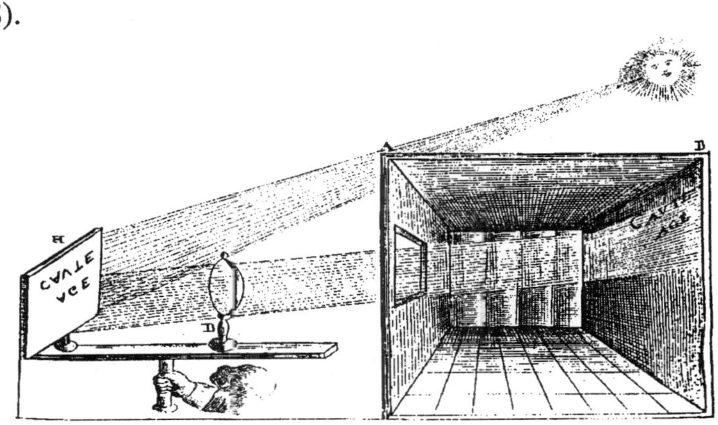

Fig.1.8. Kircher's "slide projector" from C. Shott, *Magia Universalis Naturae et Artis* (1657), v.1, p.429. Courtesy of The Bakken Library.

III. EXPERIMENTS

1. Measuring the sun's diameter

(Ptolemy's experiment)

lab
home

Equipment. This is a modification of Ptolemy's measurement of the sun's angular diameter: a diopter has one large hole instead of two small ones. The instrument consists of a meter stick and two plates attached to it perpendicularly (Fig.1.9). One plate (**A**) is fixed near the eye and has a small opening (1mm). The other plate (**B**) is directed towards the sun and has a circular opening of about 7-9 mm. This plate has a sliding

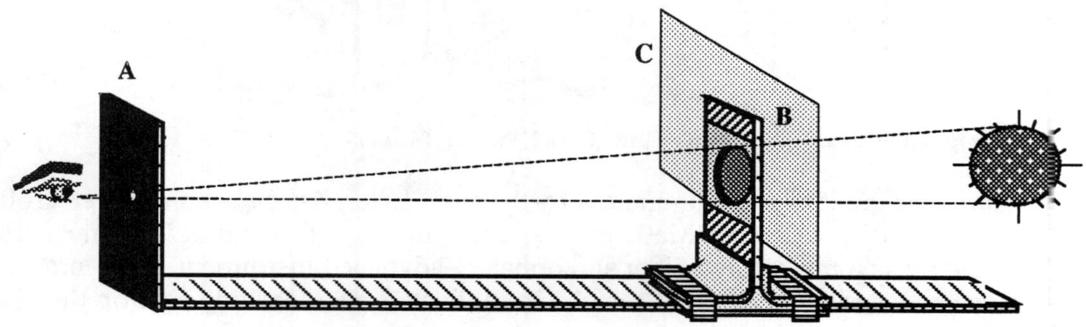

Fig. 1.9. Measuring the angular diameter of the sun.

base made of aluminum. The large opening is covered with a piece of exposed and developed photographic film. To protect the eye from an excessive glare from the region around the sun, tape to the plate **B** a cardboard sunscreen **C** of 15 cm square. In the middle of this screen make a hole 1x1 cm. and set it against the hole in the plate **B**.

Procedure. Bring the movable plate 20-30 cm of the eye aperture, direct the instrument towards the sun and move the slider until the sun disk will fill the whole opening. If the size of the opening is **d** (Fig. 1.10) and the distance between the plates **L**, the angular diameter of the sun α can be found from the equation $\tan\alpha/2 = d/2L$. Since α is very small, $\tan\alpha \approx \alpha$ and thus $\alpha = d/L$ (rad). Repeat the experiment with the openings of different diameter, compare the results, and calculate the average angular

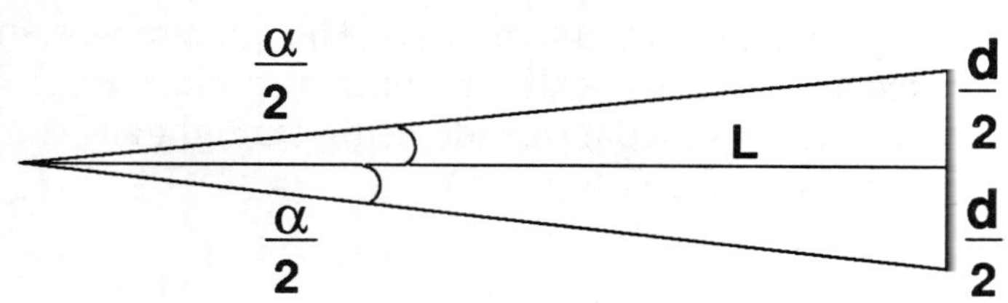

Fig.1.10

diameter. When assigning this as a home experiment suggest to repeat the measurements when the sun is near the horizon and also when it is high in the sky. Do these results square with the visual perception of the sun in these positions?

Warning. Never look at the sun directly. Use dark filters made of an exposed and developed photographic film.

2. Camera obscura
(student instruction sheet)

lab home

Equipment. A cardboard box with a cover or folds to cover it, aluminum foil, waxed paper, scissors, a razor blade, masking tape. Make an aperture about 1x1 cm in the center of a narrow (front) side of the box (Fig. 1.11). Cut a piece of foil 2x2 cm and tape it over the front aperture. In the center of the back side of the box make an aperture for the eye about 1x1 cm. To cut stray light tape to this aperture a cardboard cylinder 2 cm long and 4 cm in diameter (a piece of a paper towel core). Make

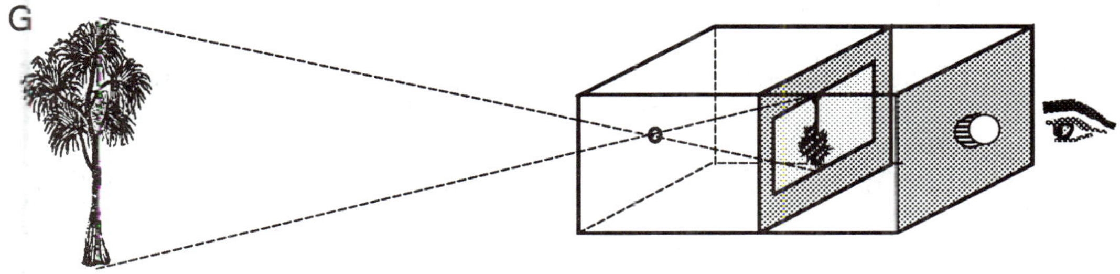

Fig. 1.11. Camera obscura.

a movable screen of about the same dimensions as the back flap of the box and insert it inside the box parallel to both apertures. Cut out a rectangular opening in it so as to leave a 2-3 cm margin on all sides. Cover this opening with waxed paper and fasten it along the margins with masking tape. Each time you move the screen attach it to the box in two points with masking tape. Having a movable screen inside the box reduces stray light and allows to investigate one more variable (which one?). While experimenting outside you may find helpful to tape to the top of the box a dark cloth and drape it over your head.

Procedure. Make a 2 mm hole in the foil. While standing in a dark part of the room point the camera outside through a door or a window and look into the back aperture. You will see images of the outside objects: trees, buildings, etc. Then repeat your observation outside.

Investigate how the picture's size, brightness, sharpness, and coloration depend on:

1. the size of the pin-hole;
2. its shape (replace the original foil window with others which have holes of different size or shape);
3. the distance from the screen to the pin-hole;
4. the distance from the camera to the object.

Conclusion.

Remember: change only one variable at a time.

BIBLIOGRAPHY

Aristotle XV, Problems I, in Loeb Classical Library (Cambridge, MA: Harvard Univ. Press, 1970).

Aristotle, "On the Soul," in *Aristotle VIII* in Loeb Classical Library (Cambridge, MA, 1936), pp. 8-203.

Aristotle, "On Sense and Sensible objects," in *Aristotle VIII*, in Loeb Classical Library (Cambridge, MA, 1936), pp. 214-283.

Morris Cohen & L. Drabkin, *A Source Book in Greek Science* (Harvard Univ. Press, 1958), pp. 257-61

'The Optics of Euclid', *Jour. of the Optic. Soc. of Amer.* 35 (1945), 357-372.

The Optics of Ibn Al-Haytham, trans. by A.Sabra, 2 vols. (London: Warburg Institute, 1989).

Saleh Omar, *Ibn aL-Haytham's Optics* (Minneapolis: Bibliotheca Islamica, 1977), pp. 80-99.

The Photismi de Lumine of Maurolycus (New York: Macmillan, 1940), pp. 105-121.

Edward Grant (ed.), *The Source Book in Medieval Science* (Harvard Univ. Press, 1974), pp. 392-409.

David Lindberg, *Theories of Vision from Al-Kindi to Kepler* (Chicago: Univ. of Chicago Press, 1976).

David Lindberg, *Studies in the History of Medieval Optics* (London: Variorum reprints, 1983).

Johann Kepler, *Paralipomenes a Vitellion* (1604), tr. into French, (Paris: J.Vrin, 1980), pp. 303-377.

Alan E. Shapiro, "Archimedes's Measurement of the Sun's Apparent Diameter," *Jour. for the Hist. of Astronomy* 6 (1975), 75-83.

Stephen Straker, "Kepler's Optics" (Unpublished Ph.D. diss., Indiana Univ., 1970), ch.1

Chapter 2

REFLECTION of LIGHT

I. THEORY

1. When do we see things?

discussion

Teacher. We've established that to see something one needs good eyes and an unobstructed line of view. Anything else?

Mary. If the object does not emit light by itself, there must be a **source of light** shining on it. Since the event took place in the evening, a natural question would be: were there any street lights or other lights nearby?

David. To be seen, an object must **reflect light** towards the eye and in sufficient quantity. Thus, we have to ask how well the place was lighted and what color was the suspect's clothing.

Teacher. Fine. To understand how light is reflected from clothing and other natural bodies we have to discuss first a simpler kind of reflection performed by mirrors.

2. Can you see a mirror?

lab

Teacher. Which of you have ever seen a mirror?

Dorothy. Everyone has.

Teacher. Is that so? Well, examine a mirror now and tell me what do you actually see.

John. I see dust, a scratch, and a window.

Ruth. Wrong, the dust and the scratch are objects **on** the mirror but not the mirror itself. My mirror has neither scratches nor dust on it.

Mary. And the window is only a reflection. Unlike dust, the window is seen only if I look in the mirror at a particular angle. Probably, the mirror itself is invisible.

Teacher. What you see either in or on a mirror is the result of *reflection* of light: a portion of light received by a body from a light source is transferred to the eye. Let us call the light ray **SM** coming from a light source **S** (Fig. 2.1) to the body **AB** the *incident ray* and the light ray **ME** reflected towards the eye, the *reflected ray*. Mirrors reflect light differently from other bodies, and we will call this reflection *regular*.

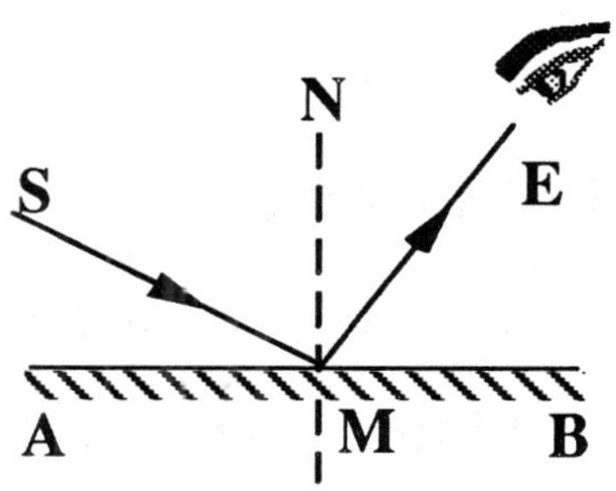

Fig. 2.1

3. Regular reflection

demo

Teacher. There are two ways of studying reflection. One is by observing the actual trajectories of incident and reflected light. For this purpose we will use a laser beam and an empty aquarium filled with smoke. The smoke is necessary to make the beam visible: **light in itself is invisible**, and we only see bodies (for instance, particles of smoke) illuminated by light. I place on the floor of the aquarium a mirror and a piece of aluminum foil. Then I put a strip of masking

important

tape about 10 cm long on the foil, light it, and close the aquarium with a cardboard sheet covered on the bottom with aluminum foil. Now, I shine the beam on the mirror. Do you see both the incident and reflected rays? Imagine now a perpendicular drawn to the mirror at the point of reflection. We will call the angles which this perpendicular makes with the incident and reflected rays the *angle of incidence* (**i**) and the *angle of reflection* (**i'**), respectively (Fig. 2.2). How does the angle of reflection change when I vary the angle of incidence?

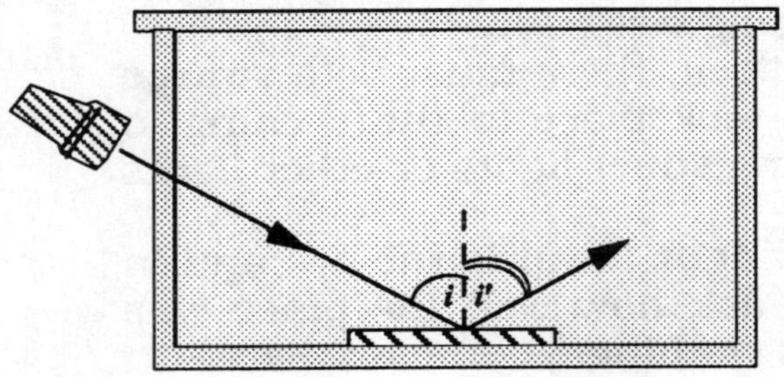

Fig 2.2

John. When the angle of incidence increases so does the angle of reflection. They appear to be equal, don't they?

Teacher. To answer this question one has to measure the angles. We are going to do this in a short while, and now I want you to watch how a reflected ray changes its direction in space when I change the direction of the incident ray.

Ruth. When you point the beam at the northern wall, the reflected ray also points northward; when you move the beam westward, so does the reflected ray. It looks like the reflected ray lies in the same plane as the incident ray and the perpendicular to the reflecting surface at the point of incidence.

Teacher. Correct. Now that you have seen the real trajectories of the incident and reflected rays, we can start working with their traces. When I shine the beam onto a mirror outside the aquarium you see a red spot on the mirror and another one on the wall. The former is the trace of the incident ray, and the latter is the trace of the reflected ray. We can also imagine the perpendicular to the mirror and the angles of incidence and reflection. When I increase the angle of incidence the spot on the wall moves so as to show that the angle of reflection increased too. In this way we can study reflection without using any smoke.

II. THE LAW OF REFLECTION

1. Finding the trajectories of rays

investigation

Equipment. A front-surface plane mirror 6x6 cm or larger, a rectangular block of wood, a letter-size sheet of paper, a flat-head pin, a 3x5 in. index card, a ruler, masking tape, scissors, a marker, and a protractor.

Procedure. Set a mirror vertically by taping it to the rectangular wooden block flush with the table. Take a sheet of paper, fold it in two widthwise, draw the line **AB** along the fold, and tape the paper to the table flush with the edge. Position the mirror along **AB** and tape the block to the paper.

Formulating a problem

Teacher. We have already noted that we see an object in a mirror only if the mirror is properly oriented about the object and the eye. Let us find out more specifically how the incident and reflected rays must be arranged about the mirror.

Selecting a procedure

Teacher. The experiments we have done with a laser have already hinted at the equality of the angles of incidence and reflection, and we could extend those experiments to measuring the angles. However, I find it more instructive to solve the problem in the way that had been in use for many centuries before lasers. First, let me ask you how would you measure the angles of incidence and reflection when you don't see the reflected and incident rays?

John. The angle of reflection can be measured directly with a protractor equipped with a movable sight: we would place the protractor on the table so that its base were perpendicular to the mirror and then sight the image of the object. However, I don't know what to do about the incident ray.

Ruth. Using a device for measuring angles is a clever idea but it looks too expensive for a lab. If we were able to trace both rays on paper, it would be easy to measure the angles.

Mary. I know how to trace the reflected ray: if we place one pin near the eye and another next to the mirror where we see the object, a straight line connecting the two pins will be the reflected ray and another straight line connecting the object and the pin near the mirror will be the incident ray.

David. The idea is fine but there is a catch in it: you **presume** that a light ray is a straight line. In fact, even if this were true for direct light, that does not necessarily apply to reflected light. Of course, we saw a reflected laser beam, but how did people prove it before lasers?

good point!

Teacher. David is right, before getting around to measuring the angles we have to prove that incident and reflected rays exist and are straight lines. If we want to talk of single rays, we need a very small object, such as the point of a pin. The idea of tracing rays is excellent but we don't have to presume anything about their true trajectories. We can determine them experimentally by marking several points belonging to the same ray and connecting them. How do we find these points?

important

Dorothy. This can be done by blocking light with a very narrow obstacle, such as a pencil.

Preliminary experiments

Teacher. Keep the object (a pin) and the eye steady and move a pencil (vertically) before the mirror (Fig. 2.3). At some positions of the pencil the

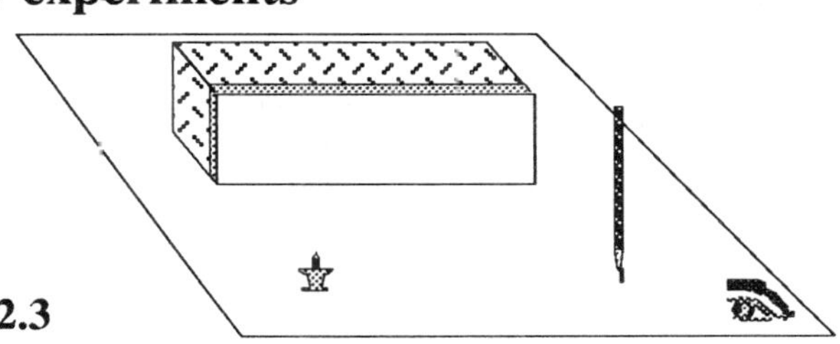

Fig.2.3

image will disappear . How do you explain this?

Hypothesis 1

John. The image vanishes when the pencil intercepts the rays coming from the image to the eye.

Test 1

Dorothy. We've found that when the pencil was between the image and the eye at any given distance from the mirror, the image disappeared. Thus the hypothesis is confirmed.

Michael. I disagree: we observed a disappearance of the image even when the pencil was not between the eye and the image.

Hypothesis 2

John. The mirror sends light to the eye in several directions.

good point!

David. This is nonsense. If several reflected rays entered the eye from different directions, we would see several images of the same object since we believe that the image is located in the direction of light coming into the eye.

Test 2

Ruth. This may be true, but why don't we check it experimentally. I would repeat the previous experiment in the following way: move the pencil parallel to the mirror from one edge of the paper to the other and mark the points where the pencil produces a disappearance of the image. If we do it four-five times, at different distances from the mirror, we will have enough data to determine the trajectory of light.

Dorothy. We've done this and found two sets of points which appear to make two straight lines coming from the same point at the mirror. One line points to the eye, while another is apparently directed to the object. The latter one cannot be a reflected ray because the eye is on the other side. Thus, we have enough evidence to advance the following hypothesis.

Hypothesis 3

The image also disappears when we block the light coming from the object to the mirror, or the incident ray.

Test 3

good point!

Mary. To make sure that we actually block the incident ray and not a reflected ray bent twice, I would move the object further back until it is behind you. Ask your partner to move the pencil before the object but behind you.

Michael. We've done as Mary suggested and the image disappeared again. It is extremely unlikely that a reflected ray could make such a great detour.

Test 4

David. There is another way of checking whether the second ray comes from the object. If we reduce the amount of light coming from the object (or, in other words, reflected by the object) to the mirror, the visibility of the image must decrease.

John. It works. I blocked the light coming to the pin from the window, and the image became darker.

Ruth. The same thing happened when we replaced a shiny pin with a black one.

conclusion ➡ | We see an image of an object when the latter reflects light to the mirror, and the mirror reflects some of this light into the eye.

2. How large a mirror do we need?

Preliminary experiments

David. We have traced one direction of incident light, but isn't it true that light is coming from the object to the mirror in many different directions? These rays are to be reflected in different points, but only one of them reaches the eye because we see only one image. If this is true, we may suggest the following

investigation

Hypothesis 1

It is possible to obtain an image with a very small mirror.

Dorothy. I don't believe that one can see much of anything in a tiny mirror.

Test 1

Ruth. Don't forget that our object is tiny too. Instead of breaking a mirror into small pieces I will cover most of it with paper.

David. Remember, however, that covering the point where we see the image is equivalent to intercepting reflected light. Thus, we had better mark the image on the mirror with a marker and cover the mirror so as to leave the mark open.

Teacher. This is going to be one of those cases where the eye must be absolutely steady throughout the experiment. To be sure that you don't move the eye, you can look at the mirror through a hole made in an index card attached to the edge of the table more to one side of the mirror.

John. I covered three quarters of the mirror and found no difference in the image.

Mary. We gradually covered the whole mirror except for a small spot around the mark. The image is still there.

conclusion ➡ | Even a very small mirror can produce the image of a point object.

David. Well, what if we keep the same object and the same small open part of the mirror but move the object, say, 10 cm to the left? Will we see the image?

Test 2

Dorothy. No, the image disappeared.

Hypothesis 2

For the given positions of the eye and a very small mirror there is a single position for a small object to be seen in the mirror.

Test 3

Mary. When I moved my eye to the right I had to move the pin to the left to see its image again. This confirms the hypothesis.

conclusion ➡ There is a connection between the positions of the eye, a small object, and a small mirror: if you move one, you have to move one of the others.

Michael. It is nice that a tiny mirror is of some use. However, in real life most objects are larger than a small mirror. Does that mean that a small mirror is of no **practical** value?

David. In my view, the previous experiments shown that a small mirror can be utilized with large objects too. However, let us prove this by direct experiments. For instance, let one student from each group seats steady and have the other try to see her face in a tiny mirror.

Dorothy. I could see only the nose.

Mary. I moved my eye and was able to see different parts of my partner's face. Of course, a great deal depends on the position of the face and that of the mirror.

John Right. For this reason, we decided to hold the mirror in hand. By turning the mirror and moving the eye, we were able to scan the whole face.

important

conclusion ➡ Even a very small mirror can be used to see a big object by moving both the mirror and the eye. Naturally, it is not a convenient method; however, if a bigger mirror is unavailable, a small one will do.

Discussion

Teacher. Have we shown the existence of incident and reflected rays?

good point!

David. Apparently, we have. I see one difficulty, however, in our procedure: the lines we draw on paper are not the rays but rather their projections on the table. The difference was not really important in the experiment we just completed, but it will become significant if we measure the angles, for the angle between two lines in space is not the same as the angle between their projections on a plane.

Teacher. Your remark is very relevant. In order to have the angles between projections equal the angles between the light rays and the perpendicular to the mirror we have to make these rays parallel to the table. How do we do this?

John. I guess if the eye is at the same level as the object relative to the table, both the incident and reflected rays will be parallel to the table. To achieve this we have to make sure that the distance of the hole from the table equals the pin's length.

3. The law of reflection

Formulating problem 1

Dorothy. Let us compare the angles of incidence and reflection.

Hypothesis 1

Michael. Apparently, the angles of reflection and incidence are equal.

Test 1

David. Tape an index card to the edge of the table along **CD** (Fig. 2.4). Make two holes in it 3 mm in diameter at the distance from the table equal to the length of the pin **OO'**. Look in the mirror through the hole $\mathbf{E_1}$ and mark on the mirror the reflection point $\mathbf{M_1}$. Then project the points $\mathbf{M_1}$ and $\mathbf{E_1}$ on the paper and mark their projections $\mathbf{M_1'}$ and $\mathbf{E_1'}$. Between the mirror and the eye draw 4-5 lines on paper parallel to the mirror. Holding a pencil vertically move it along one of the lines. When the image disappears, press the pencil to paper and mark its position. You will get two such marks, one for the incident ray and another for the reflected ray. Repeat the experiment by moving the pencil along other lines. Check with a ruler whether all points belonging to the same ray make a straight line, if in doubt repeat the experiment and obtain several more points. Connect the points: you have obtained the projections of the incident and reflected rays $\mathbf{O'M_1'}$ and $\mathbf{M_1'E_1'}$ (**O'** is the center of the pinhead). Since in our case

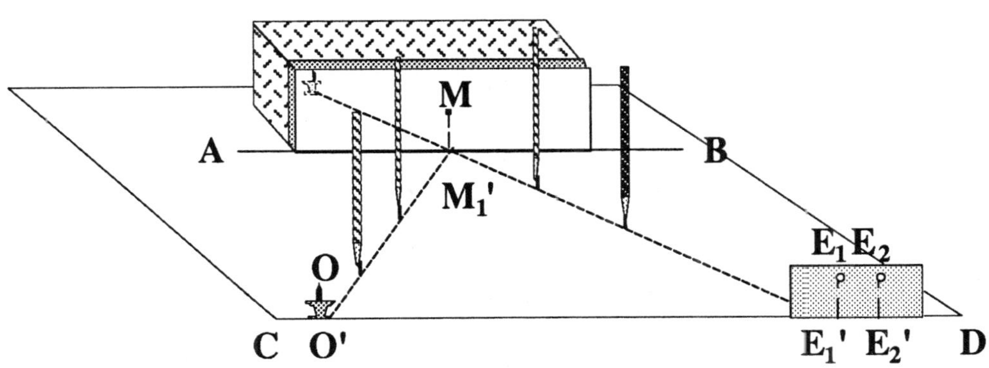

Fig. 2.4

the plane of reflection is parallel to the table, the angles formed by reflected and incident rays with the normal to the mirror equal the respective angles between the projections of these lines.

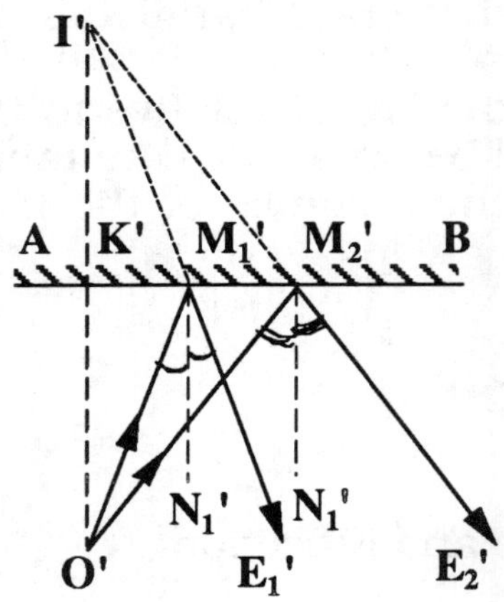

Fig. 2.5.

Thus, draw **$M_1'N_1'$** perpendicular to **AB** and measure the angles **$O'M_1'N_1'$** and **$E_1'M_1'N_1'$** (Fig. 2.5) with a protractor. Repeat the same experiment with the eye looking through the hole **E_2**. You will see the image at the point **M_2**. Now, the projections of the incident and reflected ray will be, respectively, **$O'M_2'$** and **$M_2'E_2'$**. Measure the angles **$O'M_2'N_2'$** and **$E_2'M_2'N_2'$**.

Michael. We have found the two angles to differ by one degree, which is good enough to make the following

conclusion ➡ **The angles of incidence and reflection are equal.**

Problem 2

Teacher. Let us find the location of the image. How can we do it?

Ruth. Since we believe that the image lies on the continuation of the ray entering the eye, its projection **I'** must be the intersection of the lines **$M_1'E_1'$** and **$M_2'E_2'$**.

David. If **K'** be the intersection of **O'I'** and **AB**, geometry requires that **O'K'=I'K'** and that **O'I'** be perpendicular to **AB**.

Dorothy. That what I found experimentally.

Teacher. The image can be also located through the phenomenon of *parallax*. Place a marker behind the mirror. If you move your eye parallel to the mirror, the marker appears to be shifting relative to the image: when you move your head to the right, the marker appears to the left of the image, and when you move your head to the left the marker shifts to the right of the image. Place the marker so that its apparent shift will be minimal if not zero. If you encircle the base of the marker and find its center you will see it coinciding with the image point **I'**.

general conclusion ➡
a. the angle of reflection is equal to the angle of incidence;
b. the object and its image lie on the same perpendicular to the mirror at equal distances from it.

4. Irregular reflection

Teacher. You know that only some surfaces work as mirrors. If you want to make a mirror, how would you go about it?

John. A surface must be made smooth, by polishing, for instance.

Ruth. This is correct, but it is difficult to say by touch whether the surface is sufficiently smooth. For instance, this paper is silky, but it does not make any images.

discussion

Teacher. Right. Only a microscope can make a proper distinction. Every surface looks under a microscope like a hillside. When the "hills" are high, our fingers feel it; when they are lower, only the eye can tell them. Let us think of how light is reflected by the "hills" when we don't see any mirror images.

Mary. Perhaps we can view the sides of these hills as a number of small mirrors randomly oriented in space (Fig. 2.6). Light coming from the same direction is reflected in all possible directions, which makes the surface visible everywhere.

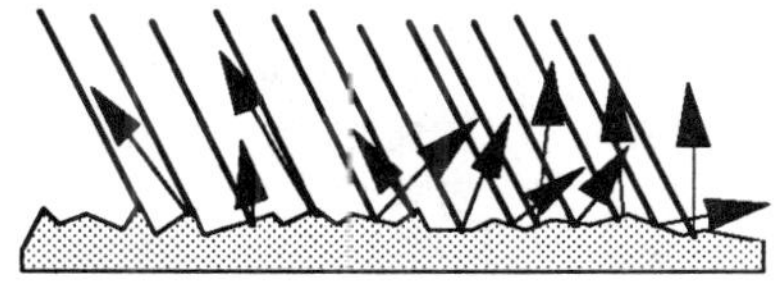

Fig. 2.6

David. O.K. I now see why high "hills" reflect in different directions, but I don't understand why low "hills" (a smooth surface) must reflect light in one direction.

Teacher. The emission model of light that we are now using cannot explain this. The wave model can, but we have to postpone this discussion until Ch.11.

David. Can we prove directly, perhaps with a laser, that some surfaces actually reflect light in all directions?

demo

Teacher. Let us try. Here I have various materials: sand paper, glass, writing paper, and wood. Since the amount of reflected light may be low, I will keep a white screen close to the reflecting surface and darken the room. What do you observe?

Dorothy. We see the whole screen lighted near the table when the screen is set parallel to the beam. It looks like the sky after a sundown.

Ruth. This means that some light is reflected at various angles and in all directions, although most light is still reflected in one direction.

Mary. This is probably true for all surfaces, although we see that the intensity of irregular reflection is higher on rough surfaces.

III. PROVING THE LAW OF REFLECTION

1. Archimedes

***history** and discussion*

Teacher. The ancient philosophers treated reflection of light somewhat differently from the modern textbooks. It is not easy to date various approaches to reflection since many early optical treatises, including Euclid's *Catoptrics*, have not reached us. [Optics consisted then of three major parts: "Optics" proper was devoted to rectilinear propagation of light and perspective, "Catoptrics" dealt with reflection, and "Dioptrics," with refraction.] The earliest known demonstration of the law of reflection belongs to Archimedes. Here is his proof (paraphrased): "Let **B** be the eye (Fig. 2.7), Γ the mirror, and Δ the object. The angle of incidence **Z** is either larger, or smaller, or equal to that of the angle of reflection **E**. Suppose, Z > E. If we switch the eye and the object, **E** will become the angle of incidence and **Z** the angle of reflection, which means that E > Z. Since this contradicts our

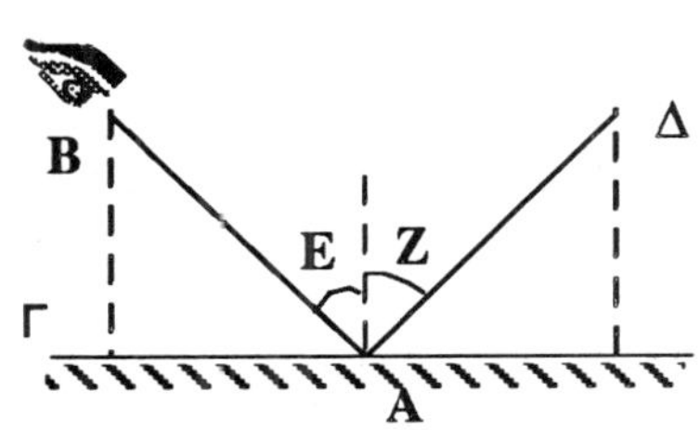

Fig. 2.7 Archimedes' demonstration

assumption, the only choice left is that the angle of incidence equals the angle of reflection."

David. Wait a second! How do we know that when the eye and the object exchange places, the path of the reflected ray will not change?

Teacher. We don't! In fact, such a property of light, called *reversibility*, is not a consequence of other postulates and must be proven experimentally. For this reason, Hero chose a different approach.

2. Hero

history *and discussion*

Teacher. In his book *Catoptrics*, the earliest extant treatise on reflection of light, ***Hero of Alexandria*** (ca. 1c A.D.) asserted that "**practically all who have written of Dioptrics and of optics have been in doubt as to why the reflections are at right angles.**" (M. Cohen, 263) He begins his demonstration with proving the rectilinearity of visual rays. One of his arguments sounds like a rudimentary law of inertia: "**whatever moves with unchanging velocity moves in a straight line.**" (ibid.) He compares the motion of visual rays to that of an arrow shot by a bow. Another argument of his is that all fast-moving bodies tend to follow the shortest path between two points, which is a straight line. In Hero's view, the principle of the shortest path is applicable to visual rays because their velocity is infinite: indeed, after opening one's eyes one immediately sees the most distant objects such as the sun or the stars. Next, Hero shows the existence of reflection. He compares reflection of light to that of a stone thrown against hard or soft surfaces. Unpolished bodies behave like soft surfaces because they have porosities which trap the rays and do not reflect them back. Polishing, however, fills these porosities and makes the surface similar to that of a hard body. Finally, Hero gives a mathematical proof that both plane and spherical mirrors reflect rays at the same angle as their angle of incidence. Here is a part of his proof (paraphrased). "Let **MN** (Fig. 2.8) be a plane mirror, **G** the eye, and **D** the object. Let **GA** be an incident ray, **AD** reflected ray, ∠NAG=∠MAD, and **GB** another incident ray. I say that GA + AD < GB + BD. Draw GE⊥MN and prolong **GE** and **AD** until they meet at **Z**. Now, ∠MAD=∠EAG and ∠ZAE=∠MAD, thus ∠ZAE=∠EAG. Since the angles at **E** are right angles, ZA = AG and ZB = BG. But ZD < ZB + BD. Thus, GA + AD < GB + BD, or the ray reflected at the angle of incidence is following the shortest path between the two given points.

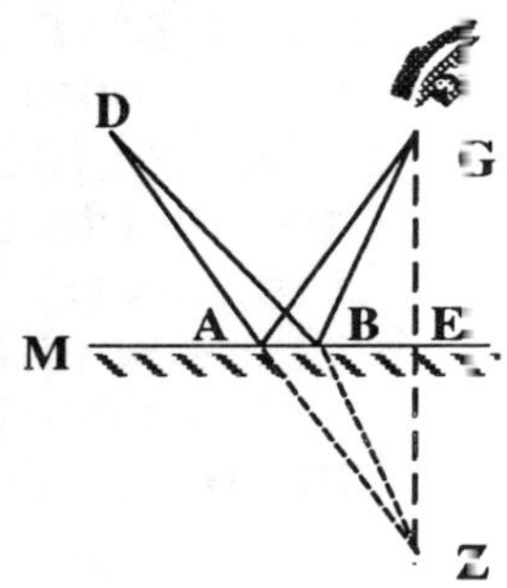

Fig. 2.8 Hero's demonstration

Mary. But this does not prove the law of reflection.

Teacher. Correct. However, following these ideas you can easily prove the opposite: if light takes the shortest path, which means GA + AD < GB + BD for any point **B**, the angles of reflection and incidence are equal. The following passage implies that Hero did some experiments on reflection: "Let **AG** be a plane mirror (Fig. 2.9), **B** the eye, and **D** the visible object. Draw **AD** and **BG** perpendicular to the mirror, and divide **AG** at **H** in such a way that

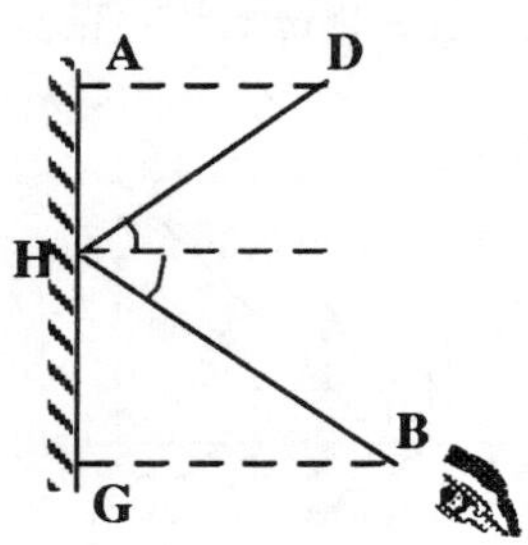

Fig. 2.9

$$AD : BG = AH : HG \tag{2.1}$$

I say, then, that if **H** is covered, the image of **D** is no longer seen." Hero proceeds to show that the angle of reflection at the point **H** equals the angle of incidence. In other words, eq. (2.1) is another formulation of the law of reflection.

3. Ptolemy

history and discussion

Teacher. While Hero discussed neither the position of the image nor the position of the plane of reflection, Ptolemy thought of the law of reflection as consisting of the following propositions: 1) the image lies on the continuation of the visual ray (to Ptolemy, the incident ray was the visual ray, while to us, it is the light ray); 2) the image lies on the perpendicular to the mirror drawn through the object and at the same distance from it; and 3) the reflected and incident rays make equal angles with the perpendicular to the mirror at the point of incidence. His demonstration of the first proposition is based on the disappearance of the image when the point of reflection is covered, for the image must lie on the straight line drawn through the eye and the point of reflection. To prove that the image and the object are on the same perpendicular to the mirror, Ptolemy referred to the fact that a long ruler makes a single line with its image in a plane mirror whatever the position of the eye, only if the ruler is perpendicular to the mirror. Prolemy's proof of the third proposition is more obscure. According to A. Lejeune, he relies on the fact that when one looks in a mirror, each eye sees the other one.

Ruth. This does not mean, however, that the ray coming to the mirror from the right eye is reflected towards the left eye at the same point at which the ray coming from the left eye is reflected to the right eye, does it?

good point!

Teacher. Good point! No, it doesn't. Lejeune suggests that Ptolemy meant the effect of the so-called "cyclopic eye": when one is staring at one's own face in the mirror concentrating on the point between the eyes, one can see between them a third eye. This is the result of superposition of the images of two eyes as seen by each of them: the image of the right eye seen by the left eye coincides with the image of the left eye as seen by the right eye, and we see three eyes instead of four. In this case, indeed, the rays coming from both eyes are reflected at the same point. And in such a case the angles of reflection and incidence are equal, as shown in Fig. 2.10. Indeed, let **ACE** be the mirror (AC=CE), **O** and **O'** the eyes, **C** the point of reflection, **OI, O'I'** and **FG** perpendiculars to **AE** . Then **I** and **I'** will be the images and **O'C** and **OC** the incident and reflected rays. Since O'G=OG, ∠OCG=∠O'CG.

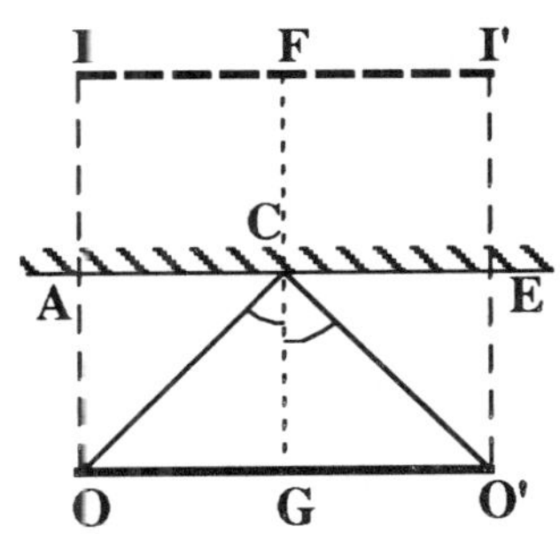

Fig. 2.10

Thus, Ptolemy's formulation of the law of reflection is more complete than the others since it deals not only with the angles but also with the position of the image and (implicitly) with the position of the plane of reflection. It is important to note that in demonstrating the law of reflection Greek philosophers put more emphasis on **reasoning** than experiment. Perhaps, the reason for this was their distrust in the correctness of visual perception based on various optical illusions. Incidentally, they never used the idea that the image of a point can be found at the intersection of two reflected rays coming from the same object. This came much later, with Kepler.

IV. EXPERIMENTS

1. The Law of Reflection

(student instruction sheet)

Part I (qualitative)

lab

Equipment. A front-surface plane mirror 6x6 cm or larger, a rectangular block of wood, a letter-size sheet of paper, a flat-head pin, a free-standing narrow object at least 15 cm long (a flat-head screw or a bolt, a pencil), a 3x5 in. index card, a ruler, masking tape, scissors, a marker, and a protractor.

Procedure. Set a mirror vertically by taping it to the rectangular wooden block flush with the table. Take a sheet of paper, draw a line bisecting it, and tape it to the table flush with the edge. Place the mirror along its middle line and tape it to the paper. Make a 3 mm hole in the index card and tape the card to the edge of the table to one side of the mirror (to the right, for instance). Set a pin on its head to the left side of the mirror, and watch its image in the mirror through the hole.

Investigate.
1. Does a mirror reflect light towards the eye?
Tip. Block light entering the eye.
2. Does a mirror receive light from an object?
Tips. a) Block light coming from the pin to the mirror. b) Reduce the amount of light reflected by the pin.
3. What are the trajectories of the incident and reflected rays?
Tip. Block light with a pencil held vertically. Mark on paper several locations of the pencil and connect them.
4. How big a mirror you need to see a very small object.
Tip. Cover as much of the mirror as possible to retain the ability to see the image of the pin's tip.

Report (a sample)

1. First, I placed my finger close to the hole and everything vanished. Then I moved the finger closer to the mirror and found a position where I saw the mirror but not the image. Thus, some light comes from the image to the eye.
2. When my partner placed his finger between the pin and the mirror the image disappeared. This means that the pin sends light to the mirror. When my partner blocked the pin from the window but not from the mirror, the image became dim. Thus, light comes from the window to the pin, and the latter reflects some of it to the mirror.
3. I had five points marked for the reflected ray and five points for the incident ray. When I connected the points, I obtained two straight lines (a drawing is attached).

4. I marked the image on the mirror with a marker, and my partner began covering the mirror with paper. When the paper covered the mark, I couldn't see the image. When the paper covered most of the mirror except for the mark, I did see it. Thus, to see a small object even a very small mirror is sufficient.

conclusion ➡ We see an image of a body because the body reflects light to the mirror and the mirror reflects it to the eye. The incident and reflected light travel along straight lines. A small body can be seen in a very small mirror.

Part II (quantitative)

1. Prove that the angle of incidence equals the angle of reflection.
Tips. a) Make a hole in the index card at the same height from the table as the tip of the pin; b) mark the location of the pin; c) while looking through this hole, trace the reflected and incident rays on paper; d) draw a perpendicular to the mirror through the intersection of the projections of the two rays; a) measure the angles of incidence and reflection (are they equal?); f) make another hole in the index card on the same level relative to the table as the first one and repeat the whole procedure. Keep the object in the same location.

2. Find the true location of the image.
Tips. a) Place a screw (or a marker) behind the mirror. Move your eye left and right and watch how the image shifts about the screw. Find a position for the screw in which the shift will be minimal if not zero. This is the image position, mark it on the paper.
b) Remove the mirror along with its holder and extend the reflected rays until they intersect. Mark the intersection point: this is the image position. Do the image locations obtained by both methods coincide?
c) Check whether the projections of locations of the object **O'** and of its image **I'** are on the same perpendicular to the mirror **AB**.
d) Check if **O'** and **I'** are equidistant from **AB**.

Report.
In this part, the report includes: 1) the drawing obtained in the lab with all lines, letters, and angles; 2) the answers to questions 1d, 2c, and 2d; 3) the conclusion about the law of reflection; and 4) suggestions about possible sources of errors and methods of their reduction.

Teachers note. A demo should precede this lab, in which the teacher discusses together with students the plan of this investigation and comes up with problems related to reflection and methods to solve them, as described in the instruction sheet. Thus the "tips" are simply the conclusions obtained from the demo. The demo should give students the ideas of such methods as blocking, tracing, steadying the eye, the need to have the eye and the object at the same distance from the table, but it should leave the determination of the light trajectories or the size of the mirror to the lab.

2. Mirrors

home

Investigate which objects in your household show images, and how the brightness and clarity of images depend on the method of processing or treating the surface of these objects.

3. A wise bird

(for students who love mathematics)

home

A bird is sitting on the top of a tree and pondering how to solve the following problem. There is seed on the ground here and there between two trees but there is a hawk nearby. The bird is afraid of the hawk while in flight but feels safe while on a tree. Her plan is to land, pick a seed and immediately fly to the other tree. How should she select the landing point for picking seed in order to shorten her path and minimize the risk?

Teacher's note. According to Hero's principle, light follows the shortest path between two points in the same medium.

BIBLIOGRAPHY

J. Priestley, *The History and Present State of Discoveries Relating to Vision, Light, and Colours* (London, 1772), pp. 1-4, 34-39.
M. Cohen & L.Drabkin, *Source Book*, pp. 257-71.
"The Optics of Euclid," 357-372.
A. Lejeune, "Recherches sur la Catoptrique Grecque," *Mém. Acad.Roy. de Belgique* 52, no. 2 (1957), 47-67.

Chapter 3

MIRRORS

I. PLANE MIRROR

1. Reading and writing with a mirror

Equipment. India ink, pens, typing paper, toilet paper.

Teacher. Are you sure that when you look in a mirror you see yourself? For instance, check whether your image is right-handed or left handed.

John. I am right-handed, but when I raise my right hand my mirror image raises his left hand, and when I close my left eye he closes his right eye.

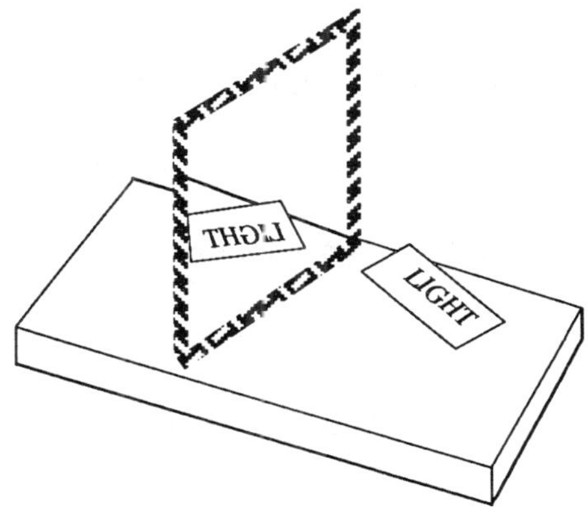

Fig. 3.1. Reading in a mirror.

Teacher. Thus, although the person you see is very much like you, he is **not** you. Let us call this person a *mirror-twin*. It is easy to see that your mirror-twin's language is different from the English: just try to read what he's written (Fig. 3.1).

Ruth. It's impossible!

Teacher. No, it's possible: there is a simple way of translating that person's language into the English. Copy this word on paper and bring it to a mirror! Incidentally, a mirror is very handy if you want to read an ink blotter. Write a word with India ink on a sheet of typing paper and immediately blot it out with a piece of a toilet paper. Can you read the imprint? If you cannot, bring it to a mirror.

Dorothy. It works just fine. I understand now why in old detective stories spies were so much interested in secretaries' blotters.

2. How big a mirror do you need?

Preliminary part

Background

Teacher. The question is practical: the larger the mirror the more expensive it is. It is natural therefore to try to buy a shorter mirror which still allows you to see yourself full-size (Fig. 3.2). What length would you choose?

investigation

Fig. 3.2. How big a mirror is enough?

Discussion

John. I would buy a mirror of my height.

Ruth. My mirror is shorter than I am but I don't remember how much.

Dorothy. If you stand farther from the mirror you can do with a smaller one.

Mary. You can use a smaller mirror if it is tilted to the wall.

Teacher. Since your past experience is insufficient for a positive conclusion, we have to do a full investigation. What are your suggestions about formulating the problem and selecting variables? We also need to devise a proper experimental procedure.

Formulating a problem

Michael. To determine the least length of a mirror sufficient to see oneself full size.

Selecting variables

John. As has been suggested, two variables could be important: 1) the **distance** between the mirror and the person, and 2) the **angle** which the mirror makes with the vertical.

Selecting a procedure

good point!

David. The mirror must not be longer than the image in it, or, more exactly, the length of line drawn on the mirror, which covers the image (the *visible length*). Thus, to save money, we have to reduce the visible length as much as possible. To measure this length I would mark on the mirror (with pieces of masking tape) the head and feet. If only a smaller mirror is available, we can mark, for instance, the hair and the chin. The experiment will consist in checking whether the distance between the marks changes with our variables and comparing the result with the size of the object. With a small mirror, we will use proportions: if, for instance, the head's visible length is 25% shorter than the head, then the necessary length of the mirror will be 25% less than one's height.

Teacher. Very good. We are now ready to begin the investigation.

Main part

a. distance

Hypothesis

Mary. The hypothesis has already been suggested in the preliminary part: when the distance between an object and a mirror increases the visible length decreases.

Test

Dorothy. When I was 1 m from the mirror, my partner placed two pieces of masking tape on the mirror marking my eyes and chin. Then, keeping straight, I walked away from it and saw at the distances of 2 m and 3 m that both my eyes and chin projected on the previous marks.

conclusion ➡ The hypothesis is false: the visible length of an object is independent of its distance from the mirror

b. angle

Hypothesis

Ruth. If I remember it correctly, the greater the angle that a mirror makes with a wall the shorter the visible length of an object.

Test

John. While standing 1 m from the mirror I told my partner where to mark my hair and chin. Then he tilted the mirror about 5° but I found no noticeable change in the visible length of my head. When he increased the angle to about 15° the result was the same.

conclusion ➡ The hypothesis is false, for the visible length does not depend on the angle a mirror makes with the vertical.

Teacher. Good. We have found that the visible length is independent of the distance and the angle. Can you conceive of other variables?
David. I can't. It looks as if it is constant. Let's explore this possibility.

Hypothesis

The visible length of an object depends only on this object.

Test

Mary. Let us measure the actual size of some part of the body and compare it with its visible length.
Dorothy My partner marked the level of my eyes and chin on the mirror and found that the distance between them was 6cm, while the actual distance between my eyes and chin was 12 cm.
Michael. I repeated this experiment measuring the distance between my hair and chin. The true distance was 16 cm, and the visible distance was 8 cm.
Ruth. I did the same with my total height, which is 160 cm. The visible height turned out 80 cm, and it remained the same when I moved to and from the mirror.

general conclusion ➡ **The visible length is equal to one half of the actual length.**

math is a friend !

Teacher. This is correct. Perhaps some of you would be interested in proving this relation geometrically? The solution is easy!

3. Measuring the image

Preliminary part

Background

Teacher. In many cases you are interested in seeing in a mirror not only yourself but other objects and people as well. For instance, imagine you are a tailor and you want to buy a wall mirror that allows you to see your **customers** full-size (Fig. 3.3). How would you formulate this problem?

Fig. 3.3. A tailor's problem.

Formulating a problem

Mary. What is the minimal length of a mirror that allows one to see oneself and other objects full-size?

Hypothesis

Michael. The answer will be the same as in the previous case: one half of the object's height.

Test

Ruth. We can use the same technique as before: measuring a partner's visible height. Since our group had a small mirror, we used a free standing object instead of a person. For instance, we placed a 12 cm marker at 50 cm from a mirror. When my head was 50 cm from the mirror the visible length of the marker was 6 cm, but when I moved my head farther away from the mirror, the image appeared to be of 8 cm.

John. When I was 1 m from a wall mirror, a pencil held in my extended arm appeared to be of 11 cm, but when near my face it changed its visible length to 8 cm (I stood still).

conclusion ➡ The hypothesis is false: we should look for another solution.

Selecting variables

Michael. Two variables are obvious: 1) the distance from the mirror to the eye, and 2) the distance from the mirror to the object.

Selecting a procedure

David. The procedure can be improved by taping to a mirror (vertically) a ruler or a tape measure. Stand erect and look into the mirror so as to project the image of **AB** onto the ruler (Fig. 3.4). Ask your partner to stand near the mirror and take the readings of the top and bottom of the image as **you** see them. Draw up a table: the visible length versus the distance of the eye from the mirror. When moving the eye keep the object steady. and when moving the object stand still and ask your partner to do the moving.

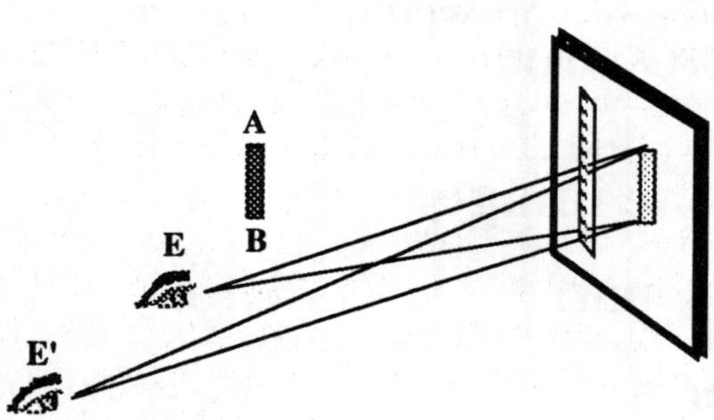

Fig. 3.4.

Main part

a. distance between the mirror and the eye.

Hypothesis

Dorothy. The farther the eye from a mirror, the greater the visible length.

Test

Ruth. We kept an object 0.5 m from the mirror. To the eye placed near the mirror, the image appeared to be 2 cm; at 1 m from the mirror its visible size was 5 cm; and at 2 m away from the mirror, the visible size was 8 cm.

John. We had an object 3 m from the mirror, and an observer moved between 0.5 m and 2 m from the mirror. The results were of the same nature as Ruth's.

conclusion ➡ The visible size of the image increases when the distance between the eye and the mirror increases (the object is steady).

b. distance between the mirror and the object

Hypothesis

Mary. The farther the object from the mirror, the smaller its visible size.

Test

Dorothy. I stood at 1 m from the mirror. When the object was near the mirror it appeared to be 10 cm high; when it was 1 m from the mirror its visible size was 5 cm; and at 2 m it appeared to be 2.5 cm.

Michael. I stood 3 m from the mirror, and my partner moved the object from the mirror to the distance of 2 m: the results were similar to Dorothy's.

conclusion ➡ The visible size of the image diminishes when the distance between the object and the mirror increases (the eye is steady).

Discussion

David. This is strange! Whether I move away the eye or the object, the distance between the eye and the image increases. Thus, in both cases the visible size of the image should decrease: it's a simple geometry!

Teacher. This is a great argument! Indeed, geometry can give us an insight, but let us apply it carefully. Let **E** be the eye (Fig. 3.5), **M** the mirror, **AB** the object, **CD** the image, **PQ** the image's projection on the mirror, **d** the distance from the eye to the mirror, **b** the distance from the object (and the image) to the mirror, **a** the length of the object, and **x** its visible length. Since the triangles ECD and EPQ are similar,

math is a friend !

PQ:CD=EQ:ED, or x:a=d:(d+b). Thus x=a/(1 + b/d). When the object recedes from the mirror, **b** increases and **d** is constant, thus **x** increases. When the eye recedes from the mirror, **d** increases and **b** is constant, and this makes **x** to increase.

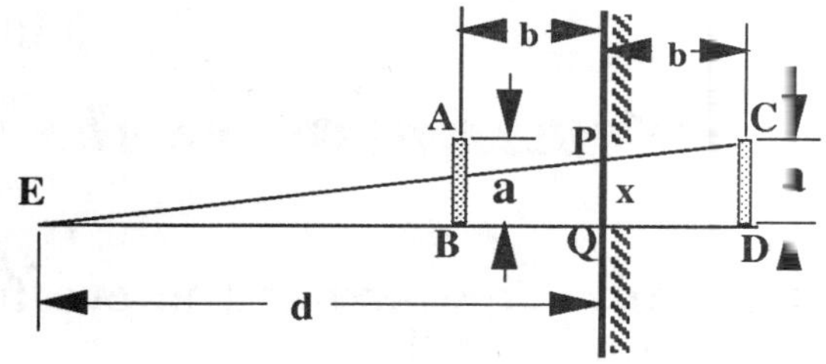

Fig. 3.5.

Michael. It is fine that mathematics agrees with experiment, but why do we need it at all after we've made our conclusion experimentally? Is it not true that experiment is more important than theory?

important

Teacher. Yes, experiment **is** the ultimate judge in science. The problem is that an experiment can be faulty (which is not so rare an event in science), and for this reason experimenters feel more comfortable if their results conform to a theory.

Dorothy. What if they don't?

Teacher. Then scientists do additional experiments, or check the theory, or both.

Ruth. Does it mean that a physicist cannot work without mathematics?

interesting!

Teacher. Theory is not limited to mathematics. There are non-mathematical considerations which help evaluate an experimental result (incidentally, David's argument is based on the **fact** of reducing the apparent angular size of a body moving from the eye, rather than on geometry). However, mathematics can be very helpful.

Michael. Coming back to our poor tailor, what should she do? Is it not true that she has to buy a shorter mirror, stay close to it, and keep customers far away from it?

Mary. This is not very practical, for usually a tailor should be close to a customer, and in such a case she will see the customer in the mirror less than full size.

David. With Michael's arrangement, in the tailor's view, the mirror could be even less than half the customer's height. However, in this case the customer will not see herself in full; besides, at a large distance from a mirror she will not be able to see small details of her outfit.

John. Thus, realistically speaking, the mirror must be no less than one half of the height of the tallest prospective client, as in the previous problem. If the mirror is hung so that its top is at the same level as the head of the client, she can come as close to the mirror as needed.

Teacher. Mary, David, and John, you are all right. Incidentally, the change in the visible size of a virtual image due to movements of the eye is called *parallax*. It is the same phenomenon that we observed while locating the mirror image behind a mirror. Sometimes parallax can be avoided.

Ruth. As in the case when we observed our own image?

Teacher. Yes.

II. SPHERICAL MIRRORS

1. Seeing images

lab

Teacher. Now, we will pass from plane to curved mirrors. Are you familiar with them?

Mary. I use one for make-up, because it makes everything larger.

John. A car side-view mirror allows one to see cars behind at a large distance and makes them look very small.

Teacher. Let us run an experiment to see which mirrors enlarge images and which reduce them. To learn how to identify different mirrors look at this diagram. You see two types of spherical mirrors. One reflects light by its inner surface (Fig. 3.6a) and is called *concave* mirror (the inner surface can be associated for memory sake with a "cave"); another is a *convex* mirror (Fig. 3.6b), where reflection takes place on the outer surface. A concave mirror **converges** light, or narrows a beam; while a convex mirror **diverges** light, or widens a beam. Tell me which of these mirrors you have used.

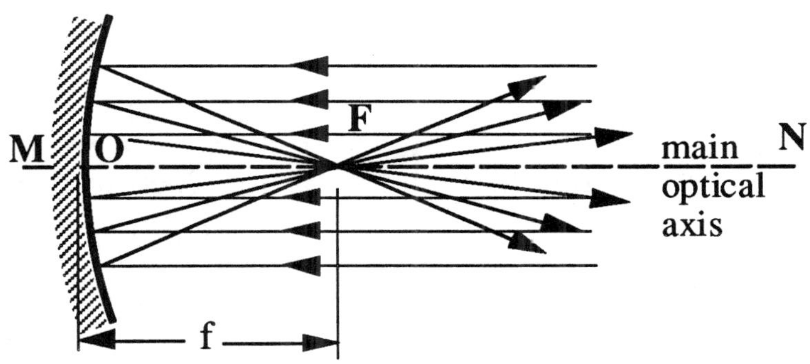

Fig. 3.6a

Ruth. The concave mirror is good for make-up.

Michael. A side-view car mirror is convex.

Teacher. Now try to obtain a reduced image with a concave mirror, and a magnified image, with a convex one.

Dorothy. We have found that a convex mirror does not magnify and a concave one does not reduce.

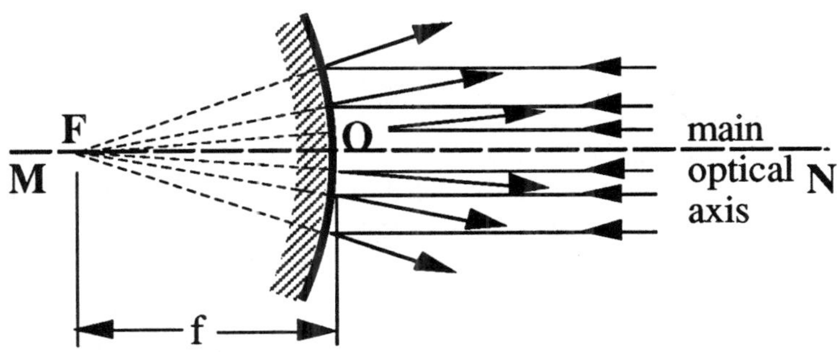

Fig. 3.6b

David. I support the conclusion about convex mirrors and disagree with that on the concave ones. We have seen in a concave mirror a reduced image of an electric light behind us, both direct and inverted.

Teacher. Let us check the controversial point by other means.

2. Projecting images

lab home

Teacher. By directing this mirror towards the sun (or a distant light) I collect the reflected light on a white screen. You see that a bright spot changes its size and clarity depending on the distance between the mirror and the screen. A concave mirror can also project a bright object, such as a candle flame or an electric bulb. Here, for instance, I am producing a very large image of a flame. Now, it is your turn. Can you obtain magnified or reduced images? Are they direct or inverted? Use such objects as a candle flame or a clear electric bulb. First, try a concave mirror and then a convex one. Take notes describing the character of the image versus the object's distance from the mirror.

Ruth. We could not project anything by means of a convex mirror. This is strange because we saw images when looking **in** it.

Mary. By using a concave mirror, we were able to project an enlarged image of a candle flame but not a reduced one.

John. Our group succeeded in getting a reduced image by moving the object far from the mirror. We found a candle flame very good for determining whether the image is direct or inverted. In our case, both the magnified and reduced images were inverted.

David. How come the same mirror produces an inverted image in projection and a direct one when you look into it?

Teacher. To resolve the contradictions you have uncovered we have to turn to the theory of reflection by a spherical mirror.

3. Building a mirror image

theory

Teacher. An important feature of a spherical mirror is its ability to collect a parallel beam into a point. The inverse is also true: if a light source is in a focus, the mirror produces a parallel beam. If the beam is parallel to the mirror's axis of symmetry **MN** (*main optical axis*) (Fig. 3.6a), the refracted rays intersect at the point **F** of the main optical axis called *focus*. The distance between the mirror's apex **O** and the focus **F** is called *focal distance* **f**. The focal distance **f** of a spherical mirror equals half of the radius **OC** of the sphere. Each small part of a spherical surface can be considered as a plane mirror to which we can apply the law of reflection. Since the radius of a sphere is perpendicular to its surface, every ray coming to the mirror through the center **C** will be reflected back along the same line. Thus, incident and reflected rays make equal angles with the radius drawn through the point of reflection (Fig. 3.7a).

On this basis, we can formulate rules for building images in a spherical mirror. To find the shape and location of the image of any given object, it is sufficient to build images of a few points of the object (two points, for instance, if the object is a straight line, three for a plane figure, etc.). To build the image of a given point in a mirror one can apply the law of reflection to **any two** rays, which pass through the object point. However, there are special rays the use of which makes building images easier:

1. **if an incident ray is parallel to the main axis of the mirror, the reflected ray passes through the focus;**
2. **if an incident ray passes through the focus, the reflected ray is parallel to the mirror's main axis; and**
3. **if an incident ray passes through the sphere's center, it comes back the same way.**

In Figs. 3.7a & 3.7b, a concave mirror creates **real** images, or images built by an intersection of real reflected rays. Such images can be projected onto a screen. In Figs. 3.8 & 3.9, both concave and convex mirrors produce **virtual** images; these images are built with the aid of continuations of reflected rays rather than the rays themselves. Since a virtual image is fictitious, it cannot be projected onto a screen, but it can be seen by the eye. A concave mirror can also produce a virtual image, when the object is closer to the mirror than its focus. When the object is at a focus, the image fills the whole mirror. For instance, if the object is a candle flame, the whole mirror will be lit.

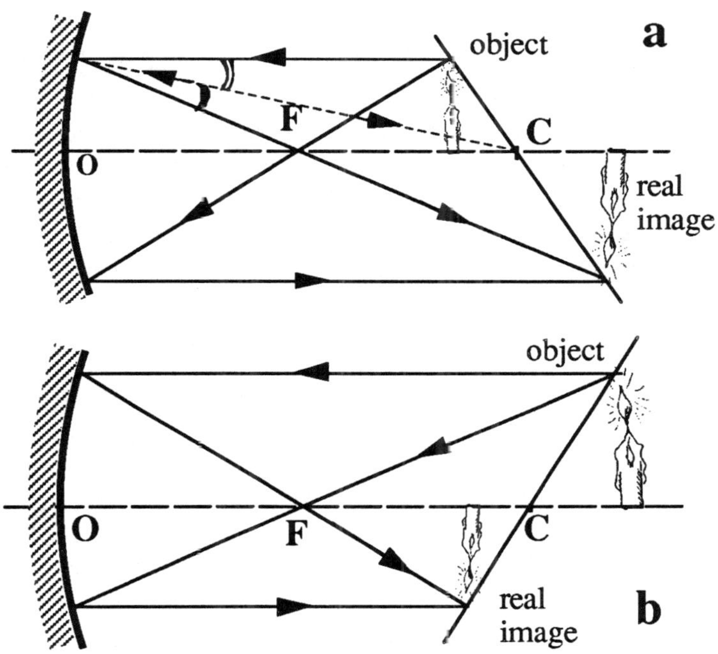

Fig. 3.7. Real image in a concave mirror.

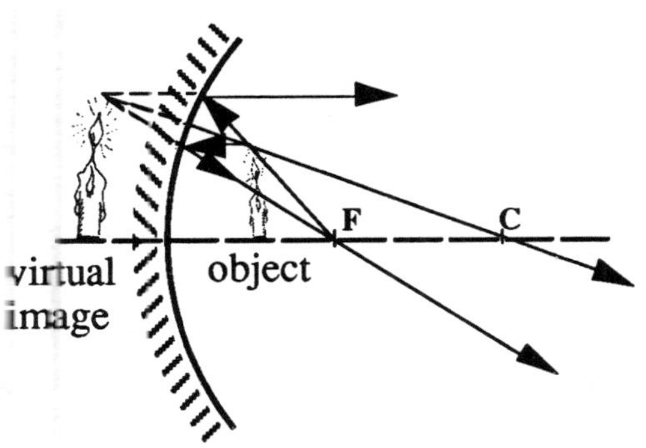

Fig. 3.8. Virtual image in a concave mirror.

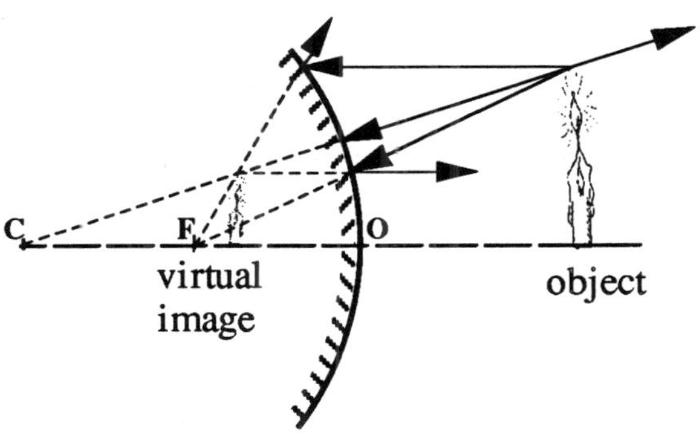

Fig. 3.9. Virtual image in a convex mirror.

4. How do you know your mirror is plane?

lab

Teacher. You have been given flexible mirrors. Bend them in different ways and watch the image of yourself. Make notes, for instance, "if the shape is concave, the image is enlarged," etc. Release the mirror and determine whether it became flat. How would you do it?

Mary. I will move it to and from my face. If it is still curved, the image will depend on the distance.

Michael. We did it and found it isn't easy to return a flexible mirror into its original flat state.

Ruth. I will try this idea at home with our dressing mirror. Mom is complaining about gaining weight, but perhaps the mirror is slightly concave. In this case, she would be happy to know that something in our house is of poor quality!

III. BURNING and MAGNIFYING with MIRRORS

1. Had Archimedes actually burned the Roman fleet?

history

Euclid knew how to build images in spherical mirrors. The name of Archimedes, another great ancient mathematician, is associated with the burning capability of concave mirrors. Unlike his colleagues, he did not shun solving practical problems. When his native Syracuse, a Greek colony in Sicily, was attacked by the Romans, Archimedes applied all of his great talent for defending his city. Various military machines invented by him horrified Romans. His greatest exploit, according to some ancient historians, was to set fire to the Roman fleet by means of polished copper shields acting as mirrors. One of the stories of his death says that when after a long siege Romans took the city, Archimedes was immersed in his thoughts. When he saw a soldier raising his sword to kill him, the only thing the old man asked him, was to hold it a moment until he finished his demonstration.

The legend on Archimedes' mirrors was questioned in the 17th century when scientists realized that the rays reflected far from the center of a mirror intersect at points other than the focus ("spherical aberration"). Due to spherical aberration, the mirror produces not a bright point but a luminous circle, the radius of which increases with the radius of the sphere. Since the amount of incoming energy is the same, the greater this circle the lesser the energy concentration and so is the temperature. To burn a ship at a large distance the mirror must have a large focal distance which means a large radius of curvature, but such a mirror would produce a large focal spot and will not be able to burn anything.

On the basis of these considerations ***René Descartes*** (1596-1650) and others denied the truthfulness of the story of Archimedes. However, ***Athanasius Kircher*** (1601-1680) together with his pupil Caspar Shott decided to check Archimedes' story experimentally. Kircher's idea was to use many smaller mirrors instead of one giant mirror. He erected a frame and positioned on it five identical plane mirrors so as to reflect light into the same spot. They produced a great amount of heat at a distance of more than a hundred feet. After Kircher traveled to Syracuse and found that the Roman galleys could not have been more than thirty paces from Archimedes, he decided that the event could have occurred. Unaware of Kircher's experiments, in 1747, ***Georges Buffon*** (1707-1788) constructed a system of 400 mirrors, each half a foot square, which melted lead and tin at a distance of hundred and forty feet from it and set wood on fire at a much greater distance.

In the second half of the 17th century instrument makers were busy with constructing single mirrors that produced an incredible amount of heat. Septala of Milan made a mirror 3.5 feet in diameter with a focal distance of 15 feet. Villete of Lyons manufactured a 2.5 feet mirror which burned at 3 feet. It melted any metals and vitrified stones. This mirror was purchased by Louis XIV. In 1687, mathematician ***Ehrenfried Walther Tschirnhausen*** (1651-1708) constructed a mirror 4.5 feet in diameter (made of thin copper), which burned at 12 feet. Its action was extraordinary: there was no substance which it could not vitrify, even asbestos.

2. Reflecting telescopes

The magnifying properties of a concave mirror were utilized in telescopes. In 1668, ***Isaac Newton*** (1642-1727) built the first mirror telescope, all by himself. It consisted of a metal spherical concave mirror placed in a metal tube and a small plane mirror set at the focus of the concave one. The secondary mirror is necessary to avoid having the light intercepted by the head of an observer. Newton's first telescope was only 6 inches long but it could compete with a 4 foot refractor. In 1671, he presented an improved version of his telescope to the Royal Society of London (Fig. 3.10). **history**

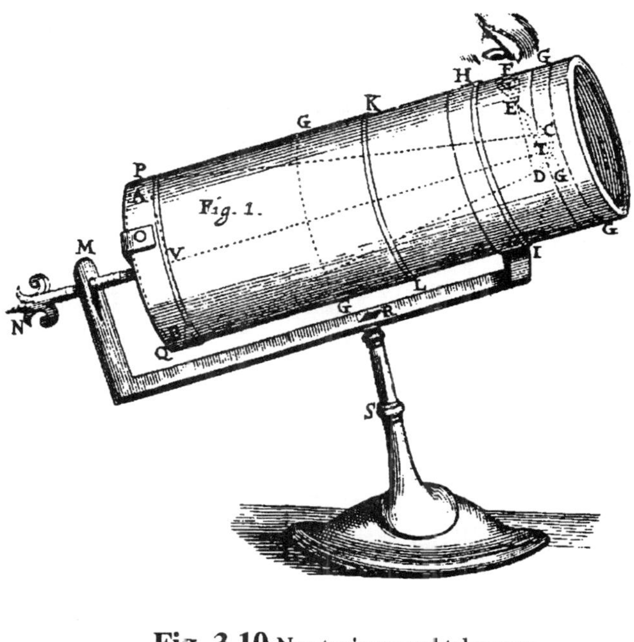

Fig. 3.10 Newton's second telescope. From *Phil. Trans.* 1672, 4004.

Newton was not the first, though, to conceive a reflecting telescope. A number of authors, including Descartes, suggested a mirror objective for a telescope, and in 1663, ***James Gregory*** (1638-1675) described a two-mirror telescope (Fig. 3.11b). However, he was unable to find an optician to make parabolic and ellipsoidal mirrors. In 1672, Frenchman ***Cassegrain*** offered one more version of a reflecting telescope (Fig. 3.11c) which had a parabolic reflector and a hyperbolic secondary mirror. All three kinds of mirror telescopes came into use since the 18th century.

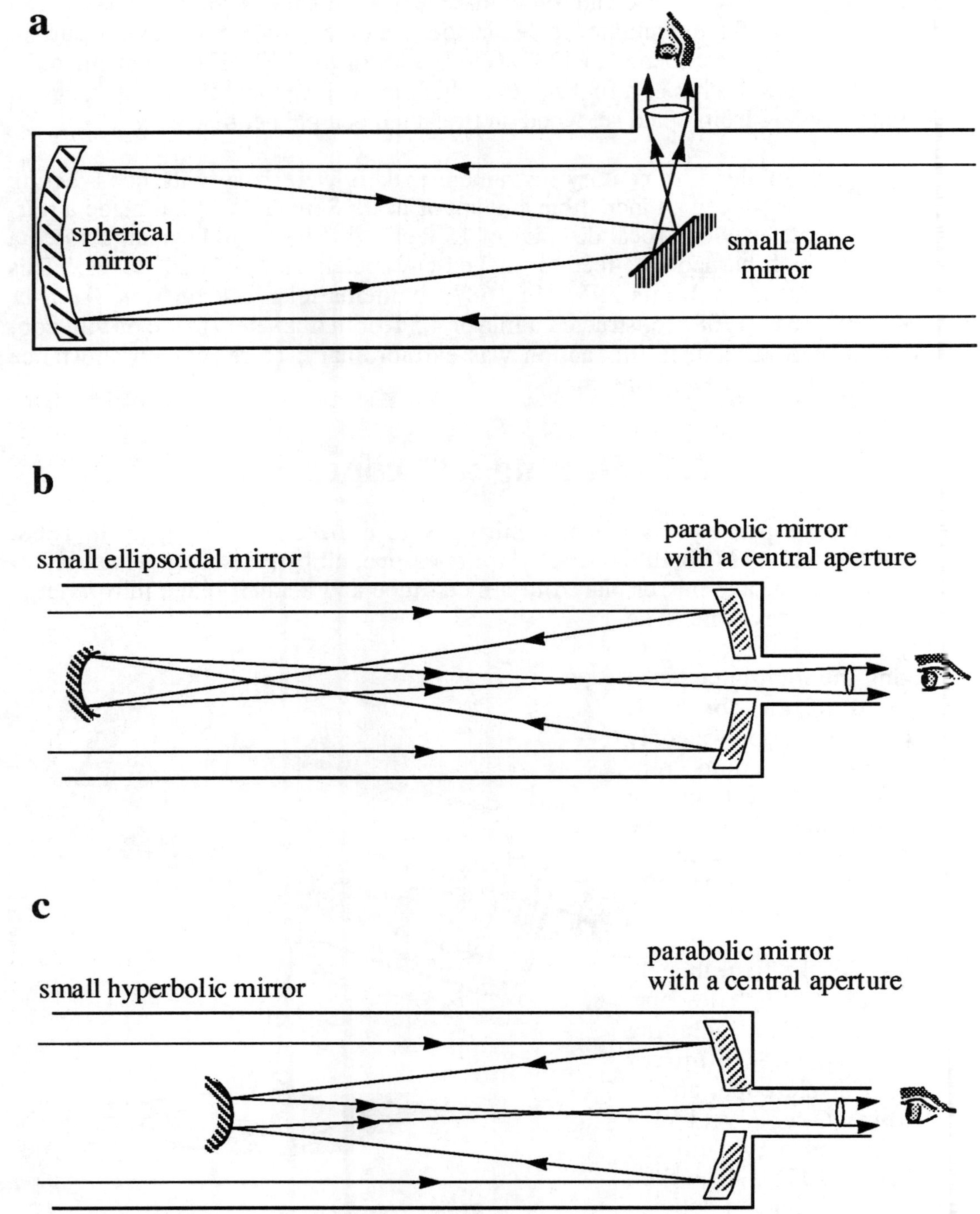

Fig.3.11 Reflecting telescopes: a) Newton's; b) Gregory's; c) Cassegrain's.

IV. PLAYING WITH A COSMETIC MIRROR

1. Which side is which?

Look into your cosmetic mirror and guess which side is plane and which is concave. Check your hypothesis by observing images in the mirror and projecting a bright distant object on a white screen (3.12). If you obtain an image, measure its distance from the mirror: this will be approximately the mirror's focal distance. Produce both an enlarged and a reduced images of a candle flame and check whether the object's position relative a focus corresponds the diagrams shown above. ***home***

2. Can you write?

You write without thinking, right? How about writing while looking into a mirror? Take a newspaper into you other hand and screen your writing hand from your eyes (Fig. 3.13). How long did it take you to write your name? ***home***

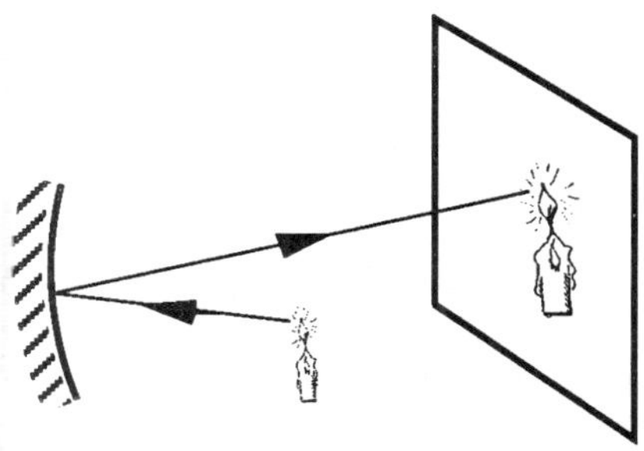

Fig. 3.12.

Fig. 3.13.

3. Newtonian telescope

Place a concave mirror against a window, at a distance about 2-3 m from it. Position yourself between the window and the mirror, somewhat to a side. Adjust the mirror so that you are able to see the objects outside the room. This is the simplest reflecting telescope. If you place a plane mirror in the focus of the concave mirror at 45°, you will be able to see the image without intercepting the light entering the telescope. This experiment can also be performed outside. While looking into a plane mirror held steady in one hand, turn slowly the concave mirror held in another hand so as to project different distant objects onto the plane mirror. Adjust the distance between the mirrors for a better view (Fig. 3.14). ***home***

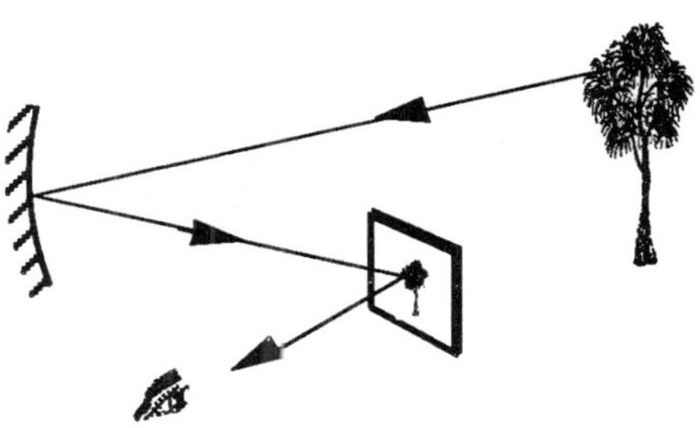

Fig. 3.14.

4. Burning mirror

lab
home

Burn a hole in a newspaper with a concave cosmetic mirror. Try large and small mirrors and explain why a large mirror works better. If you don't have a small mirror, cover up a part of a large one with paper. (Don't forget to examine whether such a cover affects the image produced by this mirror.) Let several people focus their concave mirrors on the same piece of newspaper. Does it help? Can you ignite the same object using several plane mirrors?

Warning: adults must be present.

BIBLIOGRAPHY

Plutarch's Lives V in Loeb Classical Library (Cambridge, MA: Harvard Univ. Press, 1990), pp. 471-491.
T. Needham, "Of a new mirror, which burns at 66 feet distance, invented by M. de Buffon...," *Phil. Trans. 44* (1746), 492-93.
G. Buffon, "Abstract from a letter...concerning his re-invention of Archimedes' burning specula," *Phil. Trans. 45* (1748), 504-505.
J. Parsons, "Observations upon Father Kircher's Opinion concerning the burning of the fleet of Marcellus by Archimedes," *Phil. Trans.. 48* (1753), 621-23.
F. Peyrard, "Description of a Burning Mirror," *Phil. Mag. 37* (1811), 133-146, 176-185.
R. Descartes, *Discourse on Method, Optics, Geometry, and Meteorology* [1637] tr. by P. Olscamp (Indianapolis: Bobbs-Merrill, 1965), pp. 127-49.
J. Priestley, *History of Light*, pp. 13-14, 120-22, 227-230.
Newton, "An Account of a New Catodioptrical Telescope," *Phil. Trans.*, 6 (1671), 4004ff; reprinted in I. B. Cohen (ed.) *Newton's Papers and Letters on Natural Philosophy* (Cambridge, MA, 1978), pp. 61ff.
H. King, *The History of the Telescope* (Cambridge, MA: Sky Publishing, 1955), pp.48-49, 71-81.

Chapter 4

REFRACTION of LIGHT

I. DISCOVERING REFRACTION

1. Single refraction

Teacher. We've learned that when light falls on a body some of it is reflected back. Are you aware that another part of incident light comes through the body?

Mary. I can see my face in a cup of water, which means that water reflects light. At the same time, I see the bottom of the cup.

Teacher. Right. Today, we are going to study what happens to light transmitted by a transparent body. We will begin with experiments. Like we did before, we will use two methods for finding the trajectory of transmitted light: displaying it directly with a bright beam of light and determining it indirectly from observations of a body and its image.

a. now you see it, now you don't

Teacher. Take a Styrofoam cup and make a horizontal mark on its inside wall, at about one half of its height. Place the cup on the table so as to see the mark; while holding your head straight move away from the table until you see the mark no longer. At this moment ask your partner to pour water into the cup: the mark reappears. The effect can be explained by the following diagram. Let **A** be the mark (Fig.4.1), **A'** its image, and **E**, the eye. We believe that light comes into the eye from the image along a straight line **A'E**. This ray emerges from water at **B**, but its true origin is **A**, and its true path is **ABE**. This means that light coming from an underwater object changes its direction when entering the air. Such a change of the direction of light crossing a boundary of two different media is called *refraction*. We've just observed that an underwater object appears to be located closer to the surface than it actually is. Thus, refraction moves the image away from the object. Let us now display the path of the refracted ray. ***lab***

John. Isn't it true that there are many other refracted rays coming from the mark **A**? If so, why don't we see **A** in all positions of the eye?

Ruth. Actually we do see it in many points, but all of these points lie above the line **BE**. If the eye is below this line, the refracted rays are intercepted by the cup.

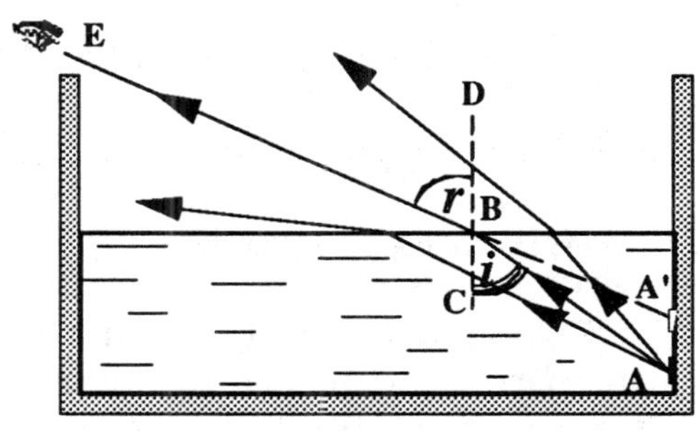

Fig.4.1

b. adventures of a laser beam

Teacher. What will happen to a laser beam in an aquarium? ***demo***

John. It will fry the fish.

Teacher. I doubt it, but you can investigate this later on. At the moment, however, we simply want to see the beam in water (no fish!) and air. As you can see, the beam is barely visible in water and practically invisible in air. To increase its brightness in water I put in a few drops of milk (the water must remain clear enough to read through it). I do this because light in itself is invisible, and we can only see bodies (particles of milk, for instance) illuminated by it. The closer the beam is to the front wall of the aquarium, the brighter it will appear to you. To make the beam visible in the air, we also need some particles. If we don't wont to fill the whole room with dust, we can fill the space above water with smoke and cover the aquarium with a cardboard sheet. Let us shine the beam through the wall of the aquarium so as to direct it from water into air. Can you see it in both air and water? You see that light changes its direction at the boundary of water and air. Imagine the refracted ray entering your eye (through a dark filter, of course). Where would you see the image of the light spot on the aquarium's wall, above the object or below it?

Mary. If we extend the refracted ray backwards it will cross the wall above the object, exactly as we observed in the previous experiment.

Teacher. Bear in mind that the actual path of light is similar to the diagram Fig.4.1, which we derived from indirect observations. Since in this experiment light goes from water into air, **AB** will be the incident ray and **BE** will be the refracted ray. Let us draw **DC** perpendicularly to the surface of water at the point of refraction **B**. The angles **i** and **r** that this perpendicular makes with the incident and refracted rays, respectively, are called the *angle of incidence* and the *angle of refraction*. Let us observe these angles in an experiment.

c. which angle is larger?

demo

Teacher. I am sending a beam from air into water (from a rarer to a denser substance) at a small angle of incidence. We see the incident and refracted rays and another ray in the air (Fig.4.2a). What is it?

John. This is the ray reflected at the surface of water.

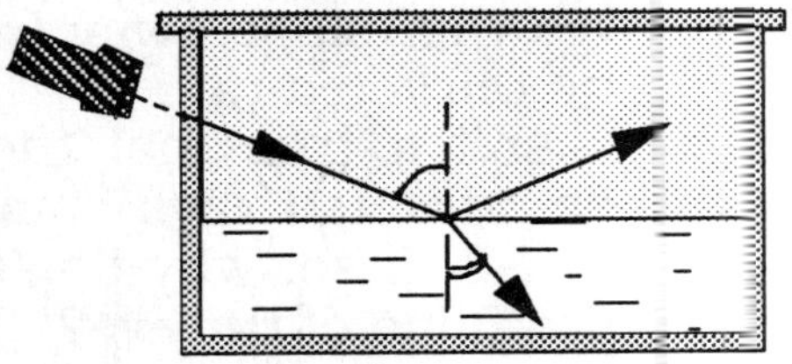

Fig. 4.2a

Teacher. Right. Your task will be to watch the incident ray and the rays refracted and reflected by water. Look at Fig. 4.1 and try to visualize the perpendiculars to the refracting surface and the angles of incidence and refraction. How do the angles of reflection and refraction change when I increase the angle of incidence?

Mary. When the angle of incidence increases both the angle of reflection and the angle of refraction increase as well.

Teacher. True. And now I want you to compare the angle of refraction with that of incidence.

Dorothy. The angle of refraction is smaller than the angle of incidence.

Teacher. I will vary the angle of incidence, and you answer the same question.

Michael. Apparently, the angle of refraction is always smaller than that of incidence.

Teacher. O.K. Let us now see what happens when light comes from water to air, which means from a denser substance to a rarer one (Fig.4.2b). You have to answer the same questions as before: 1) how do the angles of reflection and refraction change with an increase in the angle of incidence? and 2) how does the angle of refraction compare to the angle of incidence?

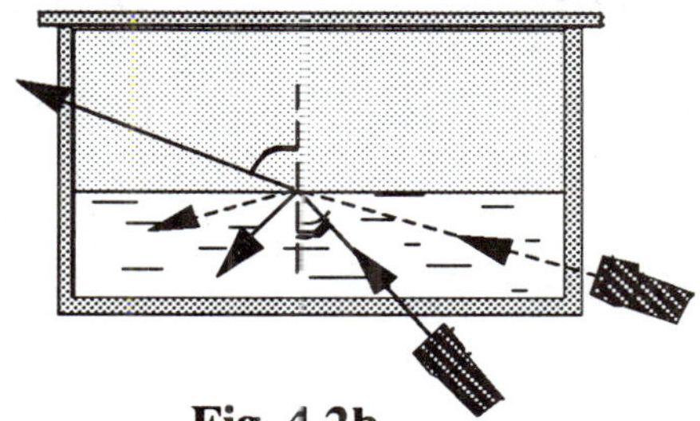

Fig. 4.2b

John. As in the previous experiment, both the angle of reflection and the angle of refraction increase with the angle of incidence. However, now the angle of refraction is always larger than the angle of incidence.

Mary. I've noticed something strange: at a large angle of incidence the refracted ray disappeared. ?

Teacher. Let us study this effect closely. I will slowly change the angle of incidence when shining the beam through the side wall, and you watch carefully the refracted ray shortly before it vanishes. This may give you the clue.

David. The larger the angle of incidence, the closer the refracted ray comes to the surface of the water. At some moment, they will coincide, which means the angle of refraction r=90°. If we continue increasing the incidence, then the angle of refraction must exceed 90°. How can this be?

Ruth. It looks like the refracted ray returns to water.

Teacher. Very good! You've discovered the so called *total internal reflection.* You have also found the following important results:

1. **when the angle of incidence increases the angle of refraction increases too;**
2. **when light is coming from a rarer medium to a denser one, the angle of refraction is smaller than the angle of incidence; if light is coming from a denser to a rarer medium, the angle of refraction is greater than that of incidence; or, in other words, in a denser medium light always forms a smaller angle with the perpendicular than in the rarer medium.**

So far, we've discovered that refraction affects the image of a point, let us now see what it does to a body.

d. broken pencil

Teacher. Fill a styrofoam cup with water and put a pencil in (Fig.4.3). Look at it from above and aside. Do you see anything unusual? *lab*

Mary. The pencil appears broken where it enters water but each of the two parts looks unchanged. This is probably the result of refraction, for each point of the immersed part of the pencil appears higher than it is supposed to be.

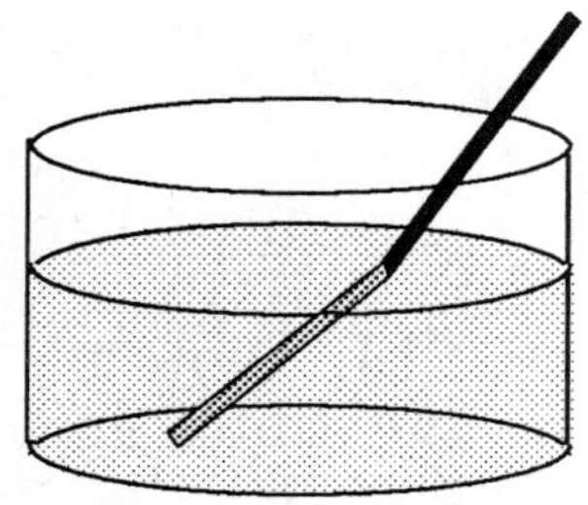

Fig. 4.3

Dorothy. Well, when I look at the pencil from above it doesn't appear broken.

good point!

Ruth. That doesn't disprove anything because the rays coming from under water perpendicularly to its surface are not refracted.

discussion

2. Two refractions

Teacher. In many practical cases we have more than one refraction. For instance, when we look through a transparent body light crosses two surfaces, which means two refractions. We have discovered that a single refraction moves the image away from the body. What will two refractions do to a body?

Michael. Nothing, as proven by a window glass.

Ruth. A change of its shape, as you can see when looking through a bottle.

Teacher. To check who is right, let us explore the phenomenon systematically, beginning with a prism.

a. seeing through a prism

lab

Teacher. Take a glass prism, hold it right before your eye so that its longer edge (*refracting edge*) is vertical, and look at a narrow vertical object. Can you see it?

Dorothy. I cannot, at least, not through the prism.

David. I've found how to do it. If the object is to my right, I look straight ahead and turn the prism around its vertical axis until I see the image.

Teacher. Good! Be careful, however, not to confuse the refracted image with the reflected one.

Mary. I also see the image, only I put the object to the left of me and look ahead. I guess I know how to tell the refracted image from the reflected one: the refracted image is **colored** and **distorted** (the straight edge became curved). Is this true?

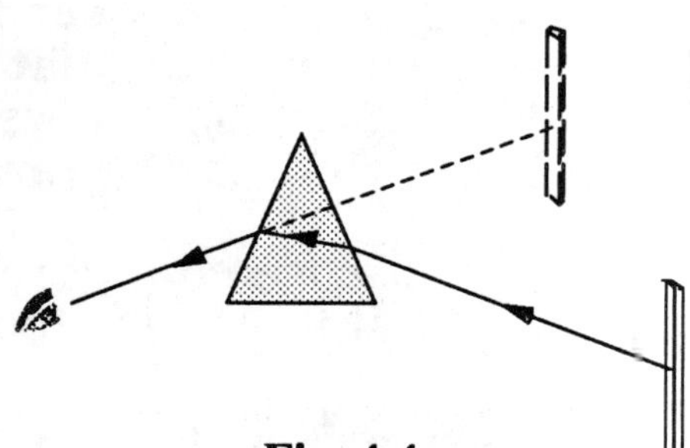

Fig. 4.4.

Teacher. Yes. Now, on the basis of all your observations, we can draw the following diagram of refraction in a prism (Fig. 4.4).

John. Can we prove this diagram with a laser?

Teacher. Let's try.

b. beaming through a prism

Teacher. I place a prism inside an empty aquarium, fill the aquarium with smoke, and send a laser beam through the prism. The directions of the incident and refracted rays are shown in Fig.4.5a, which resembles Fig.4.4. Now I am going to fill the aquarium with water and place inside it a hollow empty prism (to prevent it from floating I put in some lead shot). Can you predict how the direction of the refracted ray will change?

David. In the previous case, the prism was denser than the surrounding medium and the refracted ray deviated towards its base. Now, the prism is rarer than the medium around it and the refracted ray should recede away from the base, as in Fig.4.5b.

Teacher. Let 's run an experiment. What do you see?

Dorothy. David was right.

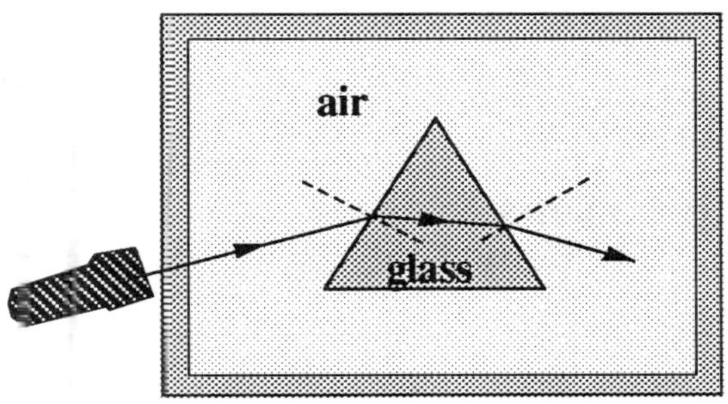

Fig. 4.5a. View from above.

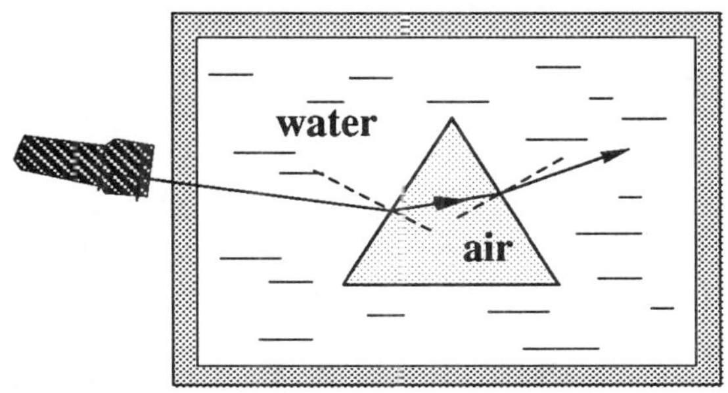

Fig. 4.5b. View from above.

c. can two refractions cancel one another?

(advanced level)

investigation

Preliminary part

Teacher. Do two refractions in a prism produce a stronger effect than one?

Mary. Yes. A separation of the image from the object is much greater, and the image is distorted.

Teacher. Do other clear bodies produce the same effect as a prism?

John. A window pane does not move the images of things outside or change their shape.

Teacher. How can you prove this?

Ruth. In the summer, if you open a window and look at a tree through the glass and through an opening, you will not notice any difference.

Teacher. Very good! So what does make a difference?

David. Perhaps, it is the shape? A prism is triangular in the cross-section, while the window pane is rectangular. Besides, a soda bottle, which is round, also distorts everything you see through it.

Ruth. How about thickness? The window glass is thin and a prism is thick.

Mary. How about the material? Water in an aquarium does not distort things although it is much thicker than the prism.

John. I am totally confused. Glass is good (no distortions) when in the window but bad when in a prism; a thick material is bad in a prism but good in an aquarium; water does not distort when in an aquarium but does so when in a bottle. It seems that the only thing we agree upon is that a rectangular shape (window pane and aquarium) does not change images while triangular and circular ones do.

?

Teacher. To sort things out, let's summarize how many variables are involved in your prior observations.

Mary. We have **shape**, **material**, and **thickness**.

Teacher. Let's begin with material.

Main part

1. material

Equipment. 60° solid prisms made of glass and plastic, and hollow one made of plexiglass; water, sugar syrup.

Preliminary experiments

Ruth. Let's compare a glass prism with the plastic one.

Michael. Both distorted the images to about the same degree.

Hypothesis

Dorothy. The material is not important.

Test

David. We filled a hollow prism with tap water and observed the same sort of distortions as before. The result was the same when we replaced tap water with sugar syrup.

Conclusion ➡ | The choice of material is not important for removing the distortion.

2. *shape*

Equipment. Jars and glasses of various shapes, including rectangular ones, water; a glass prism, microscopic slides.

Preliminary observations

Mary. We can compare a prism, a bottle, and a rectangular aquarium (or a small tank), all filled with water.

Michael. The prism and the bottle distort images but the aquarium doesn't.

Hypothesis

Ruth. There are shapes that make the refracted image different from the body.

Test 1

John. Let's try wine glasses and chemical flasks filled with water.
Dorothy. They all change the image.

Test 2

David. Perhaps solid substances differ from liquids. Let's compare a glass prism and a microscopic slide.
Michael. The slide does not distort the image while the prism does.

conclusion ➡ A rectangular body does not distort refracted images.

Discussion

David. I disagree because our group discovered that a rectangular tank can change the shape. When I held a nail near its narrow side and looked through its wide side at a very large angle, I saw the lower part of the nail wider than the upper part.
Ruth. It makes sense. You looked through a corner of the tank, which is triangular in shape and thus works as a prism.
Teacher. Very good! How can we change our conclusion to incorporate this new finding?
John. The image does not change only if we look through parallel planes. Bottles and cups have round walls, thus they always distort. Thus, we have to modify the conclusion as follows.

general conclusion ➡ **Two consecutive refractions do not distort the image only if they occur at two parallel planes.**

Mary. Now I know why the window pane is made in the form of plane-parallel sheet: any other shape would distort everything we see outside.

Discussion

Teacher. In the beginning, we've discovered that refraction moves the image of each point away from the object. If images of different points shift unequally, the body will appear deformed. Does the absence of such deformation in a plane-parallel plate necessarily mean that two refractions together don't move the images of separate points at all?
David. There may be two possibilities: 1) two refractions move the image of each point in the opposite direction; and 2) they move all the images into the same direction but by the same amount. In both cases, there will be no shape change. ***good point!***
Teacher. Excellent! How can we find out experimentally which one is true?
Ruth If all points are shifted equally, the whole image is shifted without changing its form; if each point remains at rest, the image coincides with the body. Thus, our original problem led us to a new one.

Formulating a problem

Can refraction shift an undistorted image?

Teacher. Very good! Who knows how to discover the image shift?

Selecting a procedure

good point!

Mary. I would look at the object simultaneously through parallel planes and above them.

Preliminary experiments

Equipment. An aquarium, a small rectangular tank, a microscopic slide; a pencil, a nail, a book, and other objects.

John. Our group used a pencil and found it unchanged when seen through an aquarium.

Dorothy. We observed a nail and found its lower part to be slightly displaced to the left from the upper part.

Teacher. To resolve the discrepancy let different groups share their procedures with one another and repeat the experiment. What do you see now?

John. If we look at a pencil straight on, we see no change; but if we look at it at an angle, we see the lower part separating from the upper one.

Mary. This is our result too. Apparently, we have another variable, the angle at which we see the image, or the angle of refraction.

3. *angle*

Hypothesis

Ruth. When we look straight ahead, the angle of refraction is zero, and the shift is zero too. We may suppose thus that the shift increases with the angle of refraction.

Test

good point!

David. Hold a pencil vertically against the middle of a side of an aquarium, slowly move your head to the right and to the left, and watch the distance between the object and its image; repeat the experiment with other objects.

Dorothy. We experimented with a pencil and a ruler and confirmed the hypothesis.

Conclusion ➡ When the angle of refraction increases, the distance between the object and its image increases too.

Discussion

?

Mary. Does this mean that every time you look through two parallel planes at a fairly large angle, you should see a separation between a body and its image? This is certainly not true for an open window

Teacher. A microscopic slide is no thicker than a window pane. Why don't we experiment with the slide rather than rely on recollections of things seen through a window? Look through a microscopic slide at wide and narrow objects. What do you see?

David. We looked at a nail at a very large angle, and it appeared definitely split. A book shows no effect. Apparently, narrow bodies are more suitable for observing the split, and certainly, the thickness of a plane-parallel plate is a variable. ☜ *good point!*

4. thickness

Hypothesis

Ruth With an aquarium we didn't have to look almost parallel to its wall to see the effect. Perhaps the thicker the body the easier to notice the split.

Test

Mary. We observed a nail through a small rectangular tank both lengthwise and widthwise. To keep the same angle we marked the position of the tank's front wall and looked at it through two apertures which made a line inclined to the tank by 75°.

Michael. We experimented with an aquarium and found the hypothesis to be true.

conclusion ➡ The greater the distance between parallel planes, the greater the separation between the object and its image (at the same angle of refraction).

general conclusion ➡ **A plane-parallel plate shifts the image without changing its shape, and the magnitude of the shift increases with the angle of refraction and the plate's thickness.**

Discussion

Mary. I think there is another way of testing the hypothesis about the angle of refraction and the thickness. If the image doesn't change its shape, all of its points are equally shifted from their respective object points. This means that the emerging ray should be parallel to the incident ray but shifted relative to it. ☜ *good point!*

Teacher. Excellent! Who will suggest a procedure?

Test 2

John. Empty an aquarium, place inside it a rectangular tank filled with water, and fill the aquarium with smoke. If we shine the beam through the tank lengthwise and sidewise, the result must be different. Of course, we must keep the same angle.

Michael. We've observed that, indeed, the incident and emerging rays are parallel, and their split increases with the angle of incidence.

Dorothy. We found that the beam's split is greater when light passes through a greater thickness of water.

conclusion ➡ The beam's shift increases with the angle of incidence and with the medium's thickness.

Discussion

math is a friend!

David. It's a nice experiment, but one could obtain all these results without any experiments. Let **MNM'N'** be a plane-parallel plate of the thickness **t**. If we apply the law of refraction to points **B** and **C** (Fig.4.6) and use the property of parallel lines **MN** and **M'N'**, we can easily show that **AB** is parallel to **CD**. It is also clear from the diagram that reducing the plate's thickness to **t'** will reduce the shift too.

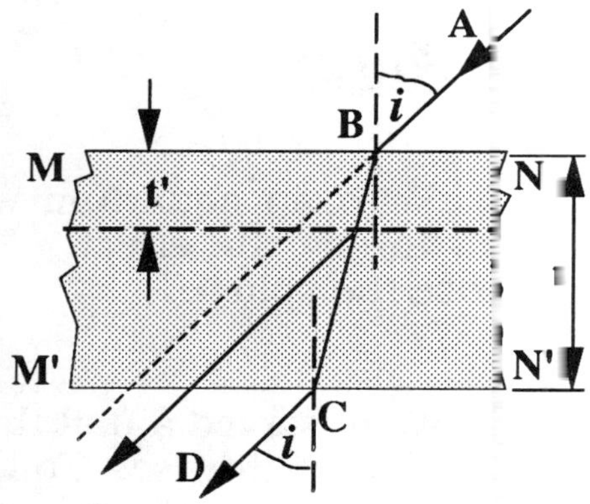

Fig. 4.6. Plane-parallel plate.

important

Teacher. Sometimes, as in this case, it's an excellent idea to start with mathematics because mathematics can suggest a good experiment. However, mathematics by itself does not prove anything in science: only experiment provides the ultimate proof.

II. THE LAW OF REFRACTION

1. The law

The law of refraction describes how the direction of refracted light changes along with a change in direction of incident light. The law consists of two parts.

1. For the given two media, whatever the direction of incident light

$$\sin i / \sin r = \mathbf{n} = \mathbf{const} \qquad (4.1)$$

(If the first medium is vacuum, the constant **n** is called the *index of refraction* of the second medium.)

2. The refracted ray lies in the plane formed by the incident ray and the perpendicular to the surface at the point of incidence.

Each substance has a specific index of refraction. Here are some examples.

Substance	**n**
Vacuum	1.000
Air	1.0003
Water	1.33
Glass	1.5
Vegetable oil	1.47
Gelatin	1.52
Sugar (41%)	1.4

If the indexes of refraction of two media are n_1 and n_2, respectively, and light passes from the first to the second medium, then

$$\sin i/\sin r = n_2/n_1 \tag{4.2}$$

When a ray of light is coming from a denser medium to a rarer medium beginning with a specific angle of incidence i_0, the refracted ray is reflected back. The angle of total internal reflection can be found from the equation

$$\sin i_0 = n_2/n_1 \tag{4.3}$$

where n_1 and n_2 are the indexes of refraction, $n_1 > n_2$ (Fig.4.7). Total internal reflection has various applications.

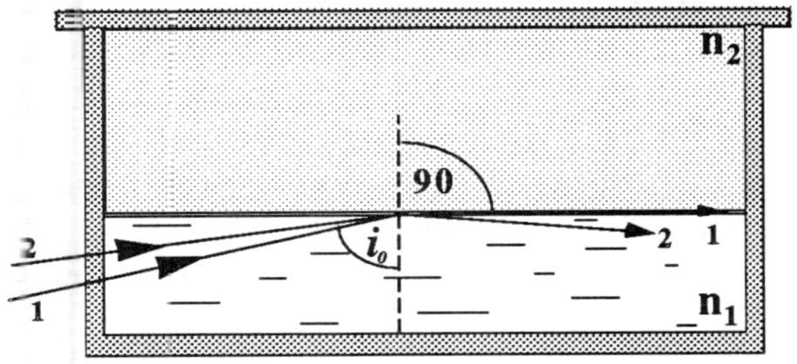

Fig. 4.7. Total internal reflection.

2. Applications

a. right-angle prism

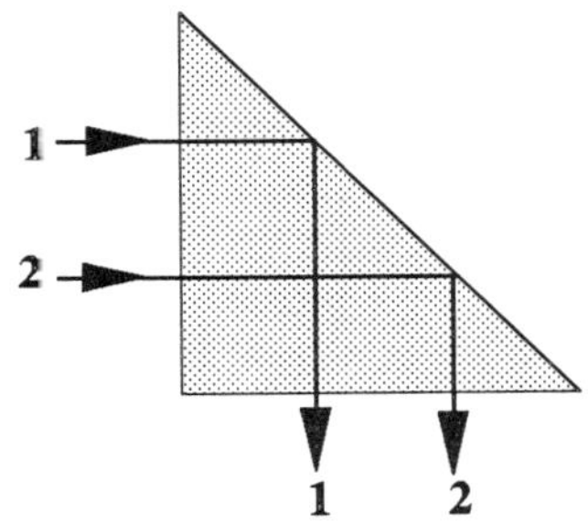

Fig. 4.8.

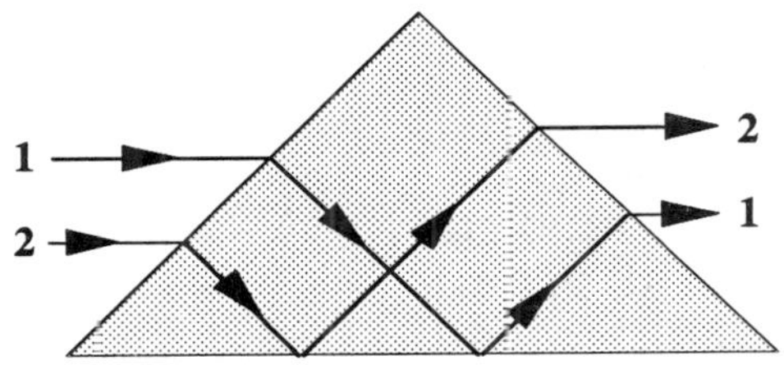

Fig. 4.9.

Due to the total internal reflection, a right-angle prism changes the direction of light in a photographic camera by 90° (Fig. 4.8) or inverts the image in a telescope (Fig. 4.9).

b. fiber optics

Modern optical fibers transmit light across long distances. A fiber is made of glass of the index of refraction n_1 and coated with a substance which has a lesser index of refraction n_2. All rays which enter such a fiber at an angle greater than the angle of total internal reflection will be reflected back and continue to travel along the fiber (Fig. 4.10). Such rays are called

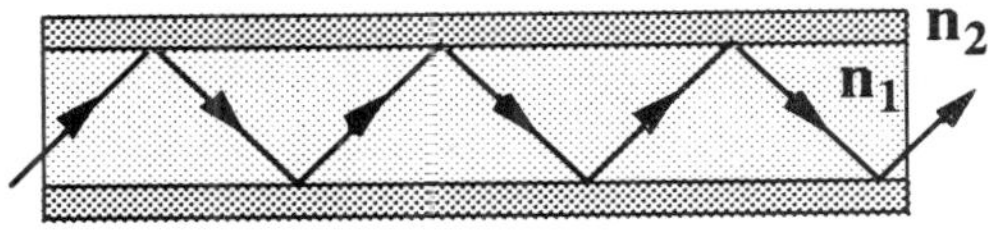

Fig. 4.10.

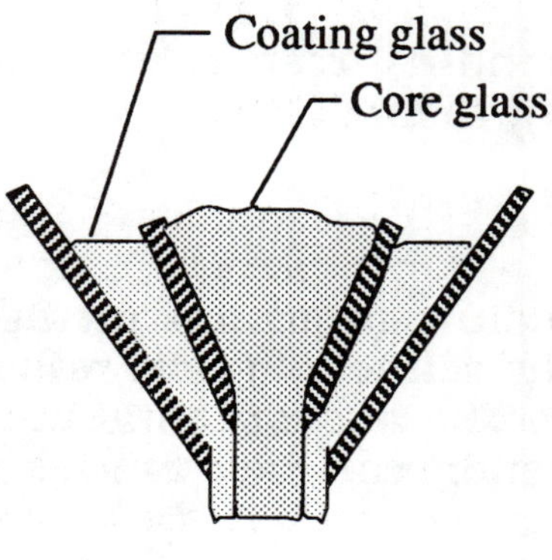

Fig. 4.11

"trapped." The fiber is produced by melting the glass core and coating glass in two crucibles (Fig. 4.11) and drawing them simultaneously. An optical fiber may be bent without a significant loss of light.

Single fibers are used in scanners where they are connected to photometric devices. They also transport radiation for heating purposes, for soldering delicate electronic systems, or for retinal surgery with a laser. To increase the amount of information or energy transported, fibers are combined in bundles or cables. Fiber cables are used for communications, fiberscopes allow to see inside human body or industrial objects.

c. mirage

When Napoleon's army occupied Egypt in 1798, soldiers frequently saw pools of water and lakes in a desert, but when the thirsty men approached them, the lakes disappeared. Napoleon had several prominent scientists with his expedition. The mathematician Gaspard Monge (1746-1818) explained the phenomenon (he called it *mirage*) by an abnormal refraction of light in the atmosphere. When the ground is very hot, the temperature of air near the ground gradually decreases with altitude, and its optical density increases. For this reason, light coming from the object **AB** is refracted as shown in Fig.4.12, and the eye **E** imagines light coming from the direction **A'** as if reflected by water at the ground level.

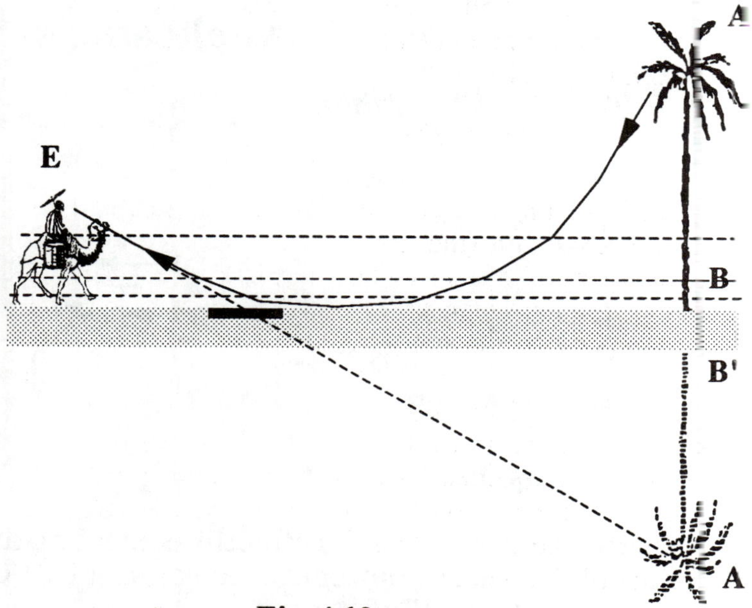

Fig. 4.12. Mirage.

III. REFRACTION and the NATURE of LIGHT

1. The discovery of the law of refraction

Refraction has been known since antiquity; Aristotle, for instance, mentioned that a straight stick partly immersed in water appears to be broken. Long interested in atmospheric refraction, Ptolemy conducted a series of measurements of the angles of refraction corresponding to the angles of incidence 10°, 20°, 30°,...80° when light passed from air to water, from air to glass, and from water to glass. He expected the ratio of the angle of incidence (**i**) to the angle of refraction (**r**) to be constant for the given two media. However, the only thing he was able to deduce from his measurements was that when the angle of incidence increased, the ratio of this angle to the angle of refraction also increased. Subsequently, scientists adopted this ratio to be constant for angles below 30°, which is a good approximation. history

The true law of refraction was discovered experimentally by ***Willebrod Snellius*** (1591-1626). He did not publish it, but it became known to some people after his death and before Descartes published the law in 1637 in a different form. Some scientists even claimed that Descartes was familiar with Snellius' work but no evidence to that was presented. By Snell's law (Fig. 4.13),

$$CE/CF = \text{const},$$

or, the ratio of the cosecant lines of the angles of incidence and refraction (the ancient trigonometry used the length of specific lines instead of trigonometric functions) is constant for the given two media, which means that it does not depend on the angle of incidence.

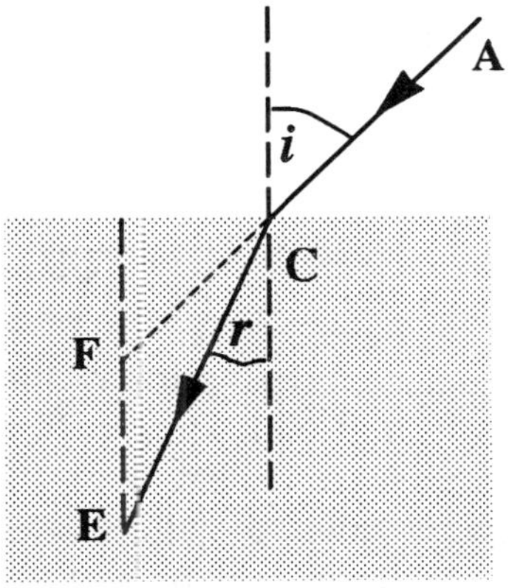

Fig. 4.13. Snell's demonstration of the law of refraction.

René Descartes (1596-1650) was born at La Haye in France in a family of minor nobility. He received a modern secondary education at the Jesuit school of La Fléche in Anjou, where he studied languages, mathematics, physics, and philosophy. In 1616, he graduated from the University of Poitiers with a degree in law. After that he enrolled as a volunteer in the army. Soon, however, he became more interested in mathematics and science than in pursuing a military career. On November 10, 1619, after a day of concentrated thinking he decided that he had to discover true knowledge by himself and to achieve this he had to question everything known in philosophy and science. That night Descartes had three consecutive dreams which reinforced him in his decision. That made such an impression on him that he vowed a pilgrimage to Our Lady of Loretto. However, he did not start his program immediately. First, he spent nine years studying, primarily mathematics, as the result of which he discovered analytical geometry. To execute his program he went to Netherlands where he could enjoy much more freedom of thought than in France and where he lived almost until his death.

Descartes' greatest contributions were in philosophy, mathematics, mechanics, optics, and physiology. His approach to the problem of refraction exhibits such typical features of his method as a tendency to mathematize physical phenomena and to proceed from the general concepts to particular ones. Although Descartes' formulation of his law

$$\sin i / \sin r = \text{const} = \mathbf{n}$$

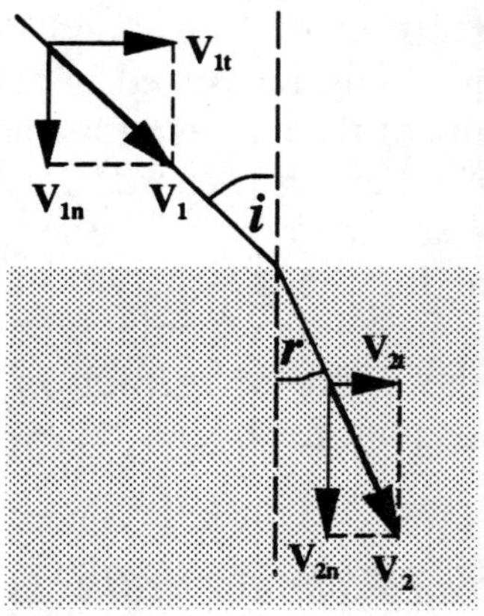

Fig. 4.14. Descartes' demonstration of the law of refraction.

important

looks similar to Snell's law, there is a profound difference between them. Descartes gave the first physical explanation of refraction: **light changes its direction when it changes its velocity**. He presumed that when light crosses the boundary of two media only the normal component of its velocity $\mathbf{v_n}$ (Fig.4.14) changes but its tangential component $\mathbf{v_t}$ remains constant. This means that

$$v_{1t} = v_{2t}, \quad v_{1t} = v_1 \sin i;$$

$$v_{2t} = v_2 \sin r; \quad v_1 \sin i = v_2 \sin r;$$

$$\sin i / \sin r = v_2 / v_1; \qquad (4.4)$$

Thus, the relative index of refraction of two media equals the ratio of velocities of light in these media.

2. Where does light travel faster?

history

Descartes' derivation of the law of refraction raised a number of objections. The French mathematician ***Pierre Fermat*** (1601-1665) found the preservation of the tangential velocity in different media utterly unconvincing. He adopted Hero's law that light always takes the shortest path and tried to apply it to refraction. In 1662, he arrived at the hypothesis that light always selects such a path between any two given points which takes the least time. One will understand how to pass from this hypothesis to the law of refraction by considering the following analogy by Fermat. Imagine a messenger who proceeds from point **A** (Fig.4.15) in a meadow, where his velocity is v_1, to point **B** on a sand beach, where his velocity is $\mathbf{v_2}$. At which point must he cross the boundary of the two media to accomplish his mission in the shortest time? For instance, if AD=DB, it is obvious that the straight line **AB** is not the solution, for it is preferable to extend the route in grass and reduce it in sand. Fermat proved that the necessary solution is provided by such a point **C** at which eq.(4.4) is true.

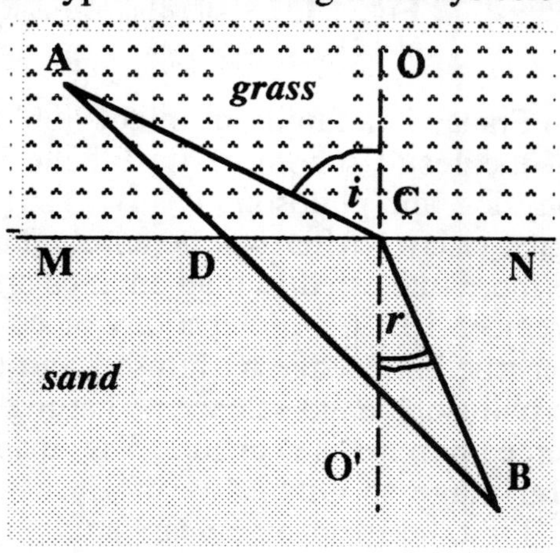

Fig. 4.15. Illustration of Fermat's principle.

It is interesting that one of Fermat's objections to Descartes was that resolving the velocity of light into two components is meaningless if the velocity is infinite, as Descartes stated. It is possible, however, that Descartes' term "infinite" meant simply extremely high. It was not until 1676 that the Danish astronomer ***Olaf Roemer*** (1644 -1710) proved that the velocity of light is finite and measured it. Cartesians objected to Ferma that light has no free will to choose one way or another, but soon his principle of the least time was accepted as a useful mathematical (not physical!) concept. Isaac Newton rescued Descartes' derivation by explaining the conservation of the tangential velocity. According to him, light changes its velocity due to an attraction from a refracting body. Since this force is perpendicular to the surface of the body, it does not affect the tangential component of the velocity.

important

Christiaan Huygens (1629-1695) offered another explanation, based on his wave theory. Huygens was born in The Hague in a family of politicians and diplomats. While studying law at Leyden University he became very interested in mathematics. The mathematician Marin Mersenne (1588-1648), who corresponded with the 17-year old Christiaan, predicted that he will become a new Archimedes. Huygens began with mathematics and mechanics and then switched to optics. In 1652, he discovered how to reduce spherical aberration of spherical lenses and decided to build an improved telescope. He learned how to grind high quality lenses and in three years he built his 12-foot telescope. With this telescope, one of the best in Europe, he discovered a satellite of Saturn. Soon he built a 24-foot telescope which allowed him to discover the Saturn's ring. From astronomy his interests returned to mechanics: he made important contributions to pendulum clocks and spring watches, discovered the laws of centripetal force and of the center of vibrations. By 1666, Huygens became so famous that he was invited to Paris to lead the newly founded Academy of Sciences. It was there that he presented in 1678 the wave theory of light (he published it in 1690), based on what became known as *Huygens' Principle.*

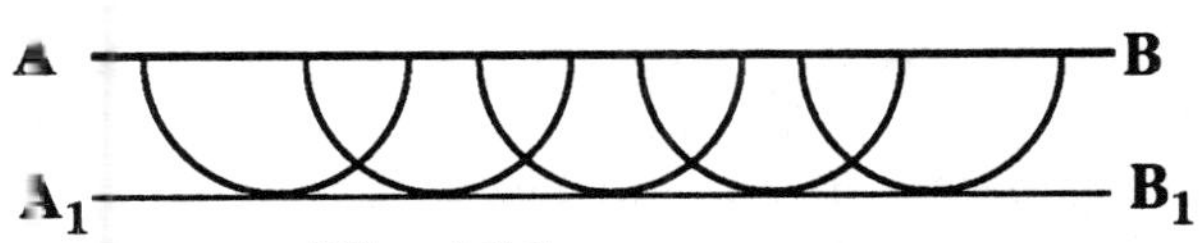

Fig. 4.16. Huygens' principle.

According to this principle, every point of a wave front can be treated as a source of secondary waves, moving only forward, and the common envelope of all secondary waves represents a new wave front. For instance, let **AB** be the initial wave front (Fig. 4.16) which travels with the velocity **v**. If we want to know its position after time **t**, we draw secondary waves of the radius R=vt with centers on **AB**: their envelope **A´B´** will be the new wave front. Let us apply this principle to refraction of light which passes from a rarer medium where its velocity is v_1 to a denser medium where its velocity is v_2 ($v_1 > v_2$). If **AB** and **A´B´** (Fig. 4.17) are two positions of the incident wave front separated by time **t**, the distance between them is $R_1=v_1t$. The corresponding refracted wave fronts will be **AC** and **A´C,´** the distance between which is $R_2=v_2t$. Since $R_1 = AA'\sin i$ and $R_2 = AA'\sin r$, $R_1/R_2 = \sin i / \sin r$, and $R_1 / R_2 = v_1 / v_2$. Thus,

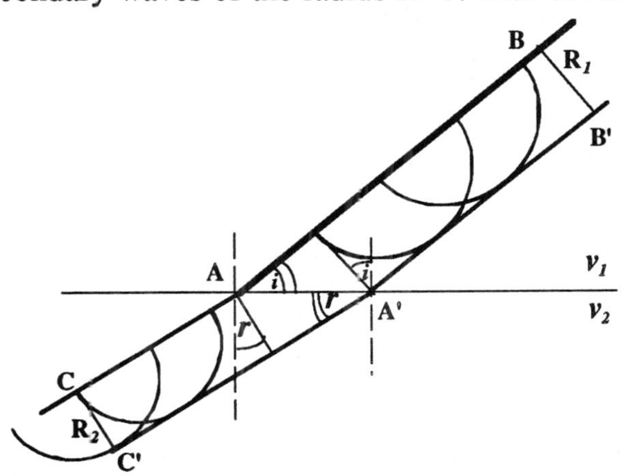

Fig. 4.17. Refraction according to Huygens.

$$\sin i / \sin r = v_1 / v_2 \; ; \qquad (4.5)$$

math is a friend!

(Bear in mind that the direction of propagation of light in each medium is perpendicular to the wave front.). Huygens' result looks similar to that of Descartes, however,his ratio of velocities is inverted. This means that to Huygens (and Fermat) light travels slower in a denser medium while, according to Descartes and Newton, light travels faster in a denser medium. For the next century and a half this distinction became an important issue in the debate between the emission (or corpuscular) theory of light and the wave theory.

3. Waves or particles of light?

history

From the 17th century on, among the hypotheses on the nature of light the chief rivals were the *corpuscular* (or *emission*) hypothesis and the *wave* hypothesis. According to the emission hypothesis, light was a stream of very small particles emitted by a luminous body and traveling to the eye. Newton was the best known supporter of this hypothesis in the 17th century. In the wave hypothesis, what was transferred from a luminous body to the eye was a motion rather than matter. This motion spreads like sound or waves on the surface of water. It was Descartes who originated this hypothesis, although he applied it to pressure rather than motion; Hooke and Huygens developed the concept further. The light waves were supposed to travel in the *ether*, an extremely elastic and rare medium which penetrated all bodies as well as vacuum. There was a certain parity between the two hypotheses in the second half of the 17th century because both explained equally well reflection and refraction of light. In the 18th century, however, the emission hypothesis dominated the scene, and only a few scientists (Leonard Euler included) continued to support the concept of light waves. Here are some of the arguments used in the debate between the two hypotheses. The *emissionists* stated that, no matter how small the resistance of the ether to the motion of the planets, after thousands of years the change in their orbits could become noticeable. The *undulationists* countered this by assuming that the ether had an extremely low density, too low to have any impact on planetary motions. In their turn, they pointed out that since the particles of light travel with such a high velocity they must communicate a considerable momentum to illuminated bodies. Experiments, however, did not show any trace of the pressure of light. The *emissionists* answered that if the mass of light particles is extremely small, their momentum should be insignificant. Euler indicated that since light particles took away some of the sun's mass, its brightness must decrease with time. The *emissionists* replied that if the mass of these particles were extremely small the diminution of the sun's mass would be unnoticeable even after a thousand years. One of the major objections to the emission hypothesis was that if two beams of light particles cross, their collision would cause scattering of light and produce distorted images. The *emissionists* pointed out that the distance between particles in a ray of light may be so large that particles of one beam could pass through the interstices between the particles of another beam without any collisions. Since most of the arguments on both sides were then unverifiable by experiment, the controversy about the nature of light had a rather abstract character. The emission hypothesis enjoyed more popularity since it explained some optical phenomena unexplained by its rival, including stellar aberration, dispersion, colors of thin films, diffraction, and others (see Chs. 13-15). The explanations were qualitative, of course, but that was the accepted standard for the eighteenth century.

interesting!

IV. VARIOUS EXPERIMENTS

1. Playing with prisms

a. right-angle prisms

1. How to set up a right-angle prism so as to see the objects rotated by 90°? by 180°? *lab*
2. Can you set up two 90° prisms so as to make a periscope?

b. equilateral prisms

Place two 60° prism together and look through the parallel sides. Move the double prism parallel to the eye so that you can see the object through it and above it Can you see the shift of the image? Does this shift depend on the angle you look at? *lab*

2. Measuring the index of refraction

lab

Measure the index of refraction of a liquid. Take a small rectangular container that you used before (you can use an aquarium as well), place it on a sheet of paper, and mark its sides **MN** and **PQ** (Fig. 4.18). Place the pins **1** and **2** before the container about 10 cm apart and looking through it move the container left or right until you see their images. Hold your eye close to the table and sight the lower parts of the pins.

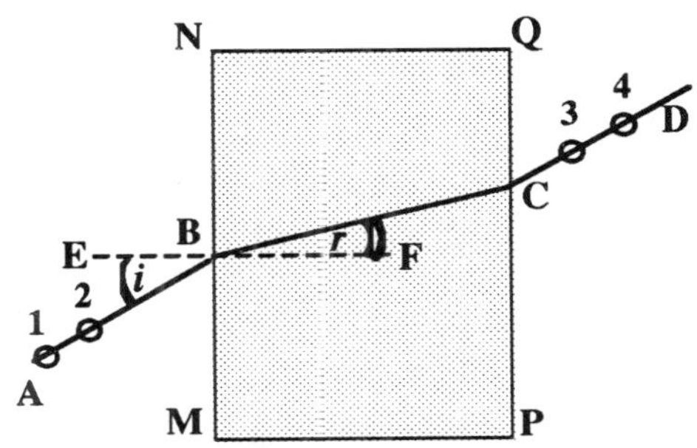

Fig. 4.18.

Place the pins **3** and **4** about 10 cm apart between your eye and the container so as to cover the first two pins. Remove the container and draw the lines **AB** and **CD** representing the incident and emerging rays to their intersection with the prism. Connect **B** and **C**: **BC** will be the ray refracted in the liquid. Draw the perpendicular **EF** to the surface **MN** and measure the angles **i** and **r**. Calculate the index of refraction **n** from eq.(4.1). Repeat the experiment at different angles of incidence, calculate the index of refraction for each angle and take the average. Measure the index of refraction of different liquids: water, alcohol, vegetable oil, corn syrup. Do you find any relation between the density and the index of refraction? Check whether the emerging ray **CD** is parallel to **AB**. Does the shift of these two rays depend on the angle of incidence?

3. Why do wine glasses distort images?

We have already observed how a glass filled with water distorts images. How about an empty glass? Examine wine and water glasses you have at home and establish which of them distort the images and why. *home*

4. Mirage

home On a hot day bring your eye close to the surface of an asphalt road and look far ahead. Can you see pools of water? You may see a reflection in those pools of the cars passing by. When your car is hot, look along its horizontal surfaces. Why do all the surrounding bodies appear to be moving? Suggest another experiment to check on your hypothesis.

BIBLIOGRAPHY

R. Descartes, *Discourse on Method,* pp. 65-82.
"Descartes, René du Perron," in C. Gillispie (ed.), *Dictionary of Scientific Biography* v. 4 (New York: Ch. Scribner's Sons, 1971), pp. 51-65.
Ch. Huygens, *Treatise on Light* [1690] tr. by S. Thompson (Chicago, n.d.), ch.I.
M. Cohen & I. Drabkin, *Source Book*, pp. 271-82.
E. Grant, *Source Book*, pp. 420-35.
J. Priestley, *History of Light*, pp. 100-106.
S. Omar, *Ibn al-Haytam's Optics*, pp. 124-46.
Leo Levi, *Applied Optics* v.2 (New York: J. Wiley & Sons, 1980), pp. 176-208.

Chapter 5

LENSES

I. THEORY

1. Images made by lenses

a. seeing through a lens

Teacher. You see in a box before you a few round-shaped clear bodies called *lenses*. Before looking through them, guess whether they would show objects correctly or distorted?

Ruth. Since the surfaces limiting these lenses are not parallel, we should expect a distortion.

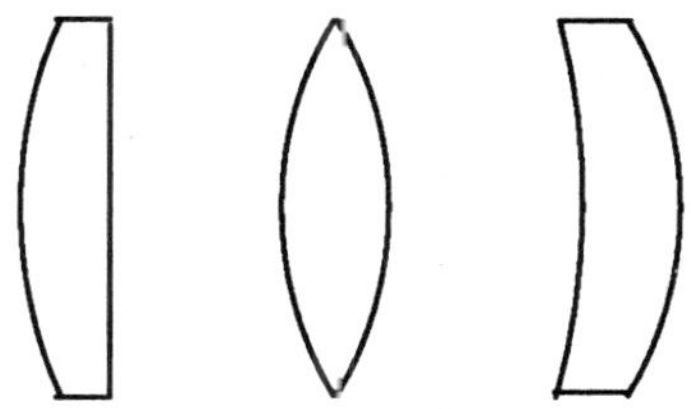

double-convex
plano-convex convex-concave

Fig. 5.1. Converging lenses.

Teacher. Before you start verifying this prediction, I'll show you how to distinguish two groups of lenses. At this overhead projection you see that some lenses are thicker in the middle than around the rim; they are called *convex* (Fig. 5.1). Other lenses are thinner in the middle than around the rim; these are called *concave* (Fig. 5.2). Now, look through various lenses you have and compare the action of these two groups.

John. Ruth was right, because we've found that all lenses produce distorted images. It appears that convex lenses magnify the objects while concave lenses reduce them. ***lab***

Mary. I disagree. Our group found that some convex lenses don't change the object (a tree outside the school), while most of them show the same object smaller and upside down. We also found that sometimes the images are blurred.

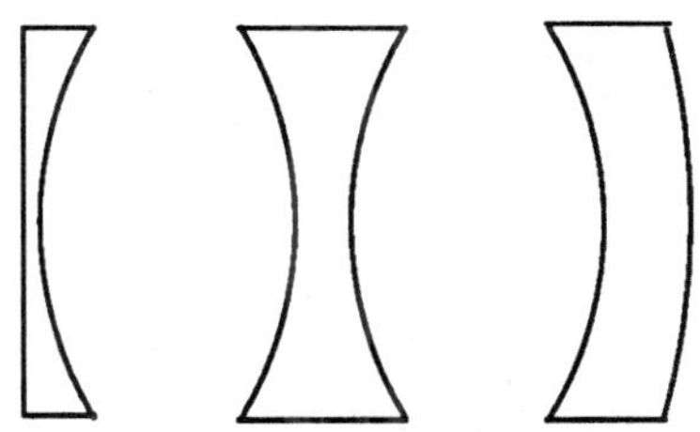

double-concave
plano-concave convex-concave

Fig. 5.2. Diverging lenses.

David. When we moved a lens to and from the eye the same object appeared blurred at one distance and sharp, at another. We have also observed that the same convex lens made a direct, although blurred, image when close to the eye and an inverted sharp image when far from the eye.

Ruth. Our convex lenses produce only direct images, and so do the concave ones. The only difference between them is that convex lenses magnify while concave lenses reduce the objects, in which we agree with John. It looks like convex lenses differ no more from the concave ones than they differ among themselves.

Teacher. Since some findings appear to contradict others, we need additional tests. Let us use the projection method which was very helpful in studying reflection and refraction. First, project distant objects (a window or a light) and close objects (a candle flame) when holding a lens in your hand. Use a sheet of white paper as a screen. The first question is: can you obtain both direct and inverted images with the same lens? Is the result true for every convex lens?

John. We've found that Mary was right about convex lenses producing only inverted images. In addition to this, we discovered that images of distant objects formed by convex lenses are always much smaller than the objects, while close objects appear magnified.

Ruth. However, our group discovered that sometimes a close object doesn't project any image!

David. We agree that any convex lens can produce images both larger and smaller than life. It all depends on the distance from the object to the lens.

Teacher. The next question is: what sort of an image can you obtain with a concave lens?

Mary. We haven't got any!

David. Neither have we. This is strange! Is it not true that since we see through a concave lens, we can compare the eye to a screen? Now, if the lens casts an image on the eye, why does it not do the same on a screen?

b. converging lens

Teacher. To answer your question we have to look into the theory. Let me show you a familiar overhead of a prism (Fig. 4.2) and ask you to recall which way the refracted ray deviates from the original direction.

John. A prism, which is denser than the surrounding medium, refracts light towards its base.

Teacher. Right. Now, let us first introduce two new terms. We will call the *lens' axis* its axis of symmetry **SS'**, and the *main optical axis*, the line **MN** drawn through the lens' center perpendicularly to the lens' axis. Imagine a convex lens cut by lines parallel to the main optical axis, as shown in Fig. 5.3. The central block is a parallelepiped, while the other parts are prisms. How will these parts refract a slightly diverging (or parallel) light beam?

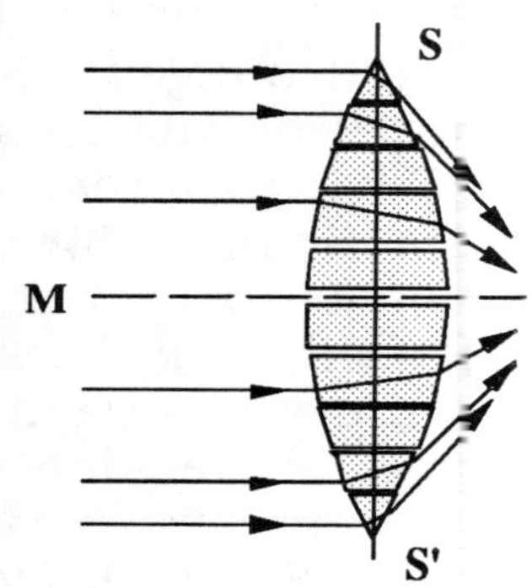

Fig. 5.3.

Ruth. We know that a parallelepiped does not change the direction of light, and the prisms will refract these rays towards their wider bases, meaning towards the main optical axis.

Teacher. Very good! Thus, a convex lens, which is denser than the surrounding medium, converges a parallel or slightly diverging beam of light, and for this reason it is called *converging*. The point where these refracted rays intersect the main optical axis is called *focus* (**F** in Fig. 5.4). The distance between a lens'

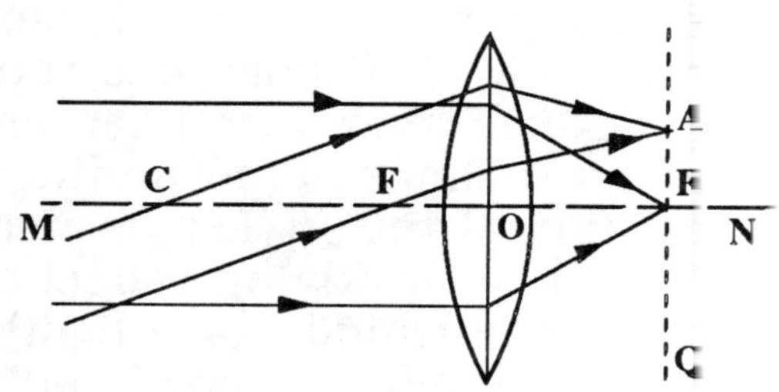

Fig. 5.4. Focus and focal plane.

focus and its center **O** is called *focal distance*. A plane drawn through a focus **F** perpendicularly to the main optical axis is called *focal plane* (**AQ**). Parallel rays that form an angle with the main optical axis intersect at some point **A** on the focal plane. A converging lens transforms two rays originating from the same point **A** into two intersecting refracted rays (Fig. 5.5). Their point of intersection **A'** is the *image* of the original point. Intersecting refracted rays create an image on a screen called *real.*.

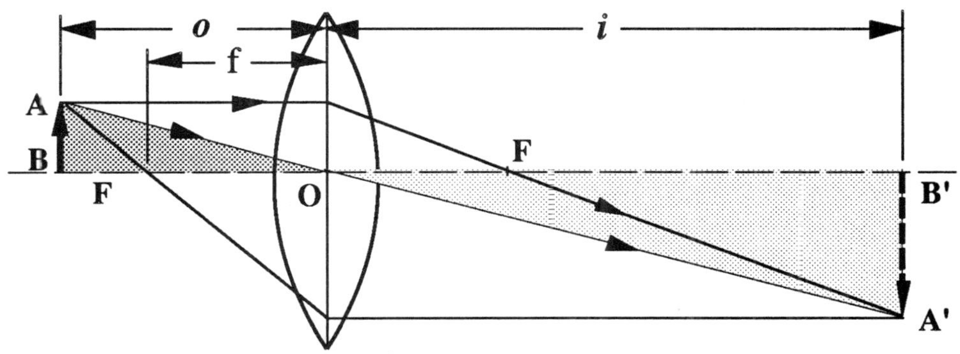

Fig. 5.5. Building a real image

Thus, to build a lens image of a point, we draw two incident rays through this point and find the intersection of the corresponding refracted rays. This procedure is greatly simplified when we use **special** rays proceeding through the given point as follows (these **rules** are based on experiment and similar to those for spherical mirrors):

1. **if an incident ray is parallel to the lens' main optical axis, the refracted ray goes through a focus;**
2. **if an incident ray passes through a focus, the refracted ray will be parallel to the main optical axis;** and
3. **a ray passing through the lens' center does not change its direction.**

Teacher. Let us now apply these rules to a converging lens (Fig. 5.6 and Fig. 5.7). What can you tell me about the image?

John. The image is real, it is on the other side of the lens relative the object, it could be magnified or reduced, and it is inverted.

Teacher. How do these diagrams correlate

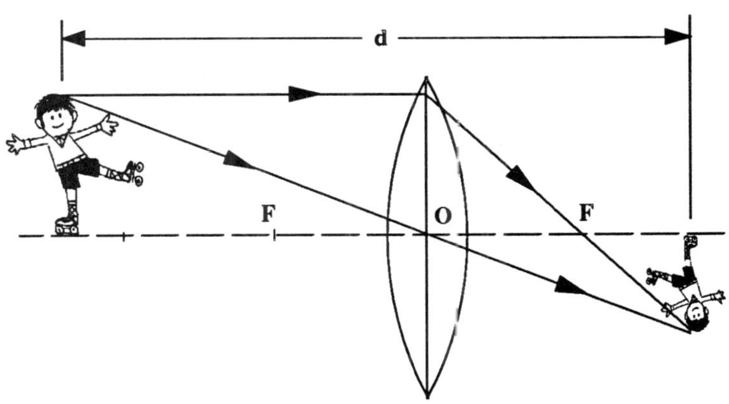

Fig. 5.6. Real image (reduced).

discussion

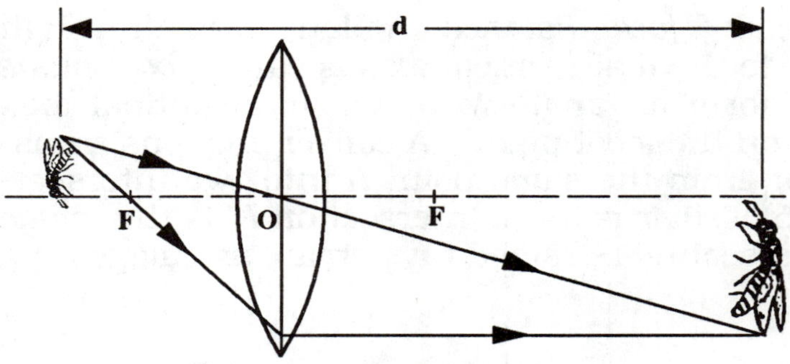

Fig. 5.7. Real image (magnified).

with our experiments?

Michael. They fully agree with our observations.

David. Can we use other rays than these three types?

Teacher. You can use **any** rays. However, to find the direction of a refracted ray corresponding to an **arbitrary** incident ray, you have to measure the angle of incidence and calculate the angle of refraction.

important

Dorothy. We'd better stick to the special rays: it's much faster!

Ruth. I have a problem: in my diagram the refracted rays do not intersect at all, perhaps because the object is too close to the lens. This is strange, though, because I can see through this lens my notes and other close objects.

Teacher. Good point! Let us reproduce your case on the board. The solution of your problem is simple: if two rays don't intersect, their continuations do, and their intersection will be the image point (Fig. 5.8). What can you say about this image?

Michael. It is magnified, and it lies on the same side of the lens as the object. What does that mean? Can we see it on a screen?

John. I placed a screen where the image was supposed to be and found none.

David. It would be fun if you did. Don't you see your screen blocks light coming from the object to the lens?

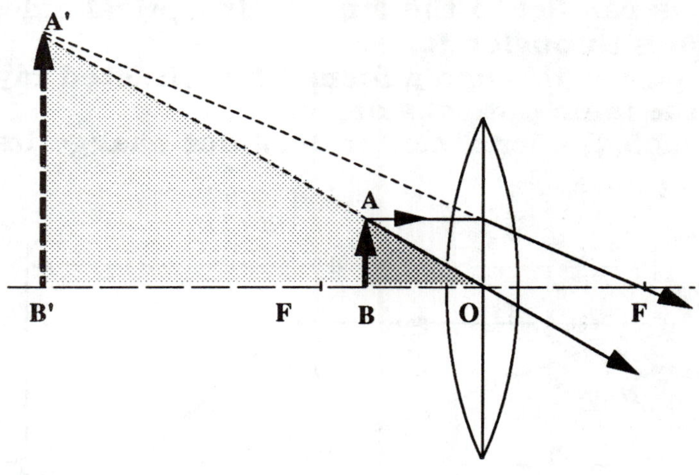

Fig. 5.8. Virtual imags in a converging lens.

Teacher. The image **A'B'** shown in Fig. 5.8 is called *virtual*, which means "unreal." A point is unreal when it is an intersection of *continuations* of refracted rays rather than the rays themselves. A virtual image cannot be projected onto a screen. Can you say how close the object should be brought to a convex lens to make a virtual image?

math is a friend!

Mary. It must be closer than a focus. It is clear from geometrical considerations that when an object crosses a focus the real image changes to a virtual one.

Teacher. Excellent! This is a good example of the power of mathematics. Let us check Mary's prediction in an experiment. First of all, you have to measure the focal distance of your convex lens. To collect refracted rays into a focus, we need a light beam parallel to the main optical axis, which is achieved when the source of light is very far from the lens. You can use sunlight, for instance, or light from a window.

John. We've found that a candle flame placed between a focus and a lens produced a uniformly lit spot instead of the image. Thus, Mary was right.

c. diverging lens

discussion

Teacher. Let us now "dissect" a concave lens in the same way we did with a convex one. What will be the result?

Mary. In this case, the refraction towards the bases of prisms means refracted rays directed away from the main optical axis (Fig. 5.9).

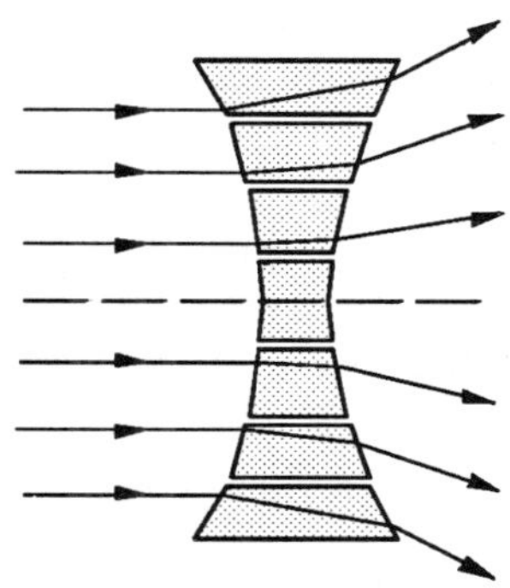

Fig. 5.9.

Teacher. That is why a concave lens is called *diverging*. Since diverging refracted rays do **not** intersect, they cannot draw any picture on a screen, and, as a result, a concave lens does not project any image on the screen.

Dorothy. Does it produce a *virtual image* ?

Teacher. Let us see it in a diagram. Draw two rays from the point **A**, one through the center, and the other parallel to the main optical axis (Fig. 5.10). How will they go after refraction?

Mary. The former ray will proceed without a refraction, while the latter ray should go through a focus, but which one will it be?

Teacher. Remember that the focus in question must lie either on the refracted ray, or on its continuation. Thus, we have to draw a refracted ray (approximately), according to the type of the lens (converging or diverging). In the Fig. 5.10, the refracted ray must go through the left focus. What sort of an image is this?

John. The image is on the same side of the lens as the object, besides, it is obtained with the aid of continuations of rays. Thus, the image is virtual.

David. It appears from Fig. 5.10 that whenever we place an object, its image is virtual. Why do we need lenses that can't project images?

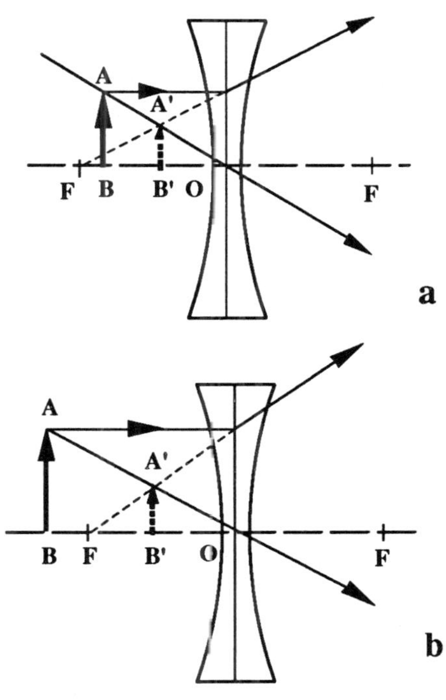

Fig. 5.10. Virtual images in a diverging lens.

Teacher. With a diverging lens, one can **see** a virtual image in space. Actually, one can even project a virtual image. If you place a convex lens behind a concave one (Fig. 5.11), the latter can converge the rays and produce an image on the screen (A_1B_1 is a virtual image produced by the diverging lens, and it serves as an object for the converging

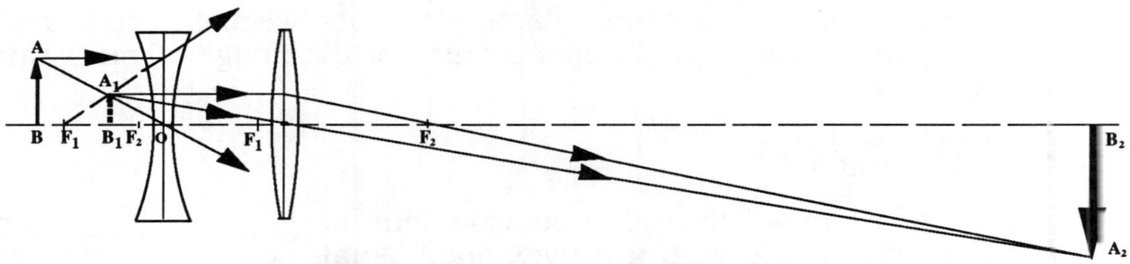

Fig. 5.11. A combination of converging and diverging lenses.

lens, which transforms it into the real image A_2B_2). The eye also plays the role of a converging lens, which draws the image on the retina.

?

Dorothy. Now I understand why we can see through a concave lens.

Ruth. Does it always work? I've just tried to combine a converging and a diverging lenses and couldn't project any image.

2. Basic equations

theory

Teacher. You can investigate this puzzle experimentally, but mathematics can lead you to the correct solution faster than experiment. Let **o** and **i** be the distances (Fig. 5.5) from the lens to an object and its image, respectively (**i** is negative if the image is virtual), **r** and **r´** the radii of curvature of the two lens surfaces (negative if the surface is concave), and **n** and **n´** the indexes of refraction of the lens and the surrounding medium. It can be shown that

$$1/o + 1/i = 1/f \qquad (5.1)$$

which is called "the lens equation," and that

$$(n - n')\,(1/r + 1/r') = 1/f \qquad (5.2)$$

known as the "lens-maker's equation."

Optometrists call the magnitude 1/f the **optical power (P)**. Thus,

$$P=1/f \qquad (5.3)$$

Optical power is measured in **dioptres** when the focal distance is given in meters. For instance, the power +2 d corresponds to focal distance 1/2 m, or 50 cm. The optical powers of several consecutive refractions add up (the sum is algebraical, with "+" corresponding to converging refraction and "-" corresponding to diverging refraction). Thus, for a combination of **thin** lenses, that is, the lenses the thickness of which is small as compared to their radii of curvature, the total power will be

$$P = P_1 + P_2 \tag{5.4}$$

This is useful for measuring the focal distance of a diverging lens, which cannot be done directly. A diverging lens is combined with a converging lens so as to produce a real image. If the focal distance of this system is **f** and the focal distance of the converging lens is $\mathbf{f}_1$, then, according to eqs.(5.3) and (5.4), the focal distance of the diverging lens $\mathbf{f}_2$ will be

$$1/f_2 = 1/f - 1/f_1, \tag{5.5}$$

Because the power of a converging lens is positive and that of a diverging lens is negative, these lenses are called sometimes **positive** and **negative,** respectively.

Ruth. Probably in my experiment the total optical power was negative, and the combination worked simply as a diverging lens which cannot project any image.

good point.

Teacher. You are absolutely right! When measuring focal distance of a diverging lens pick a converging lens of a small focal distance, then you have a better chance that its high positive power will exceed the negative power of the diverging lens.

Another example of the power of mathematics can be seen in predicting whether an image will be magnified or reduced. The linear magnification is the ratio of the length of similar lines **A´B´** in the image and **AB** in the object. One can see (Fig. 5.5) that the magnification depends on the positions of the object and the image, for A'B'/AB = i/o, thus

math is a friend !

$$M = i/o \tag{5.6}$$

Michael. Does it mean that the image of a distant object is always reduced rather than magnified?

Teacher. Correct.

3. Lens aberrations

Teacher. The equations given above are based on geometrical considerations. They show us where an image is located and how big it is, but they tell us nothing about its visibility. In fact, sometimes an image is so blurred that neither its shape nor size can be determined with certainty. This results from lens aberrations, and we will now discuss the causes of some of them.

important

a. spherical aberration

important

Teacher. I've already mentioned that focusing ability of spherical mirrors is limited by their *spherical aberration*. A similar phenomenon takes place for lenses too, for a beam of light, parallel to the main axis of a convex lens, produces a bright circle instead of a point. The reason for this is that not all incident rays which are parallel to the main optical axis intersect it after refraction in the same point: the ray AB, which is close to the main axis and thus called *paraxial*, is refracted towards **P**, while the ray **CD**, falling at the edge of the lens, crosses the axis after refraction at **Q** (Fig. 5.12). The distance **PQ** is the measure of spherical aberration. It is clear that this distance increases with the diameter of a lens, and Kepler suggested to reduce it by applying before a lens a diaphragm with a small opening which let in only paraxial rays. If a beam of light is parallel to the main axis, the farther a ray is from it, the greater the angle of incidence (and refraction) it forms. Thus, one can give another interpretation of the cause of spherical aberration: it is produced by rays which form large angles of refraction. This implies that spherical aberration increases with the index of refraction and with the curvature of the surface. If the two surfaces of a lens have different curvatures, this aberration depends on which side of the lens is directed to the luminous source. One of the ways of reducing refraction is by combining a converging and diverging lenses: while a single converging lens would refract the extreme rays **AB** and **CD** towards **P** and **Q** (Fig.5.13), the two lenses refract **CD** towards **R** which is closer to **P** (the latter point changes its position very little).

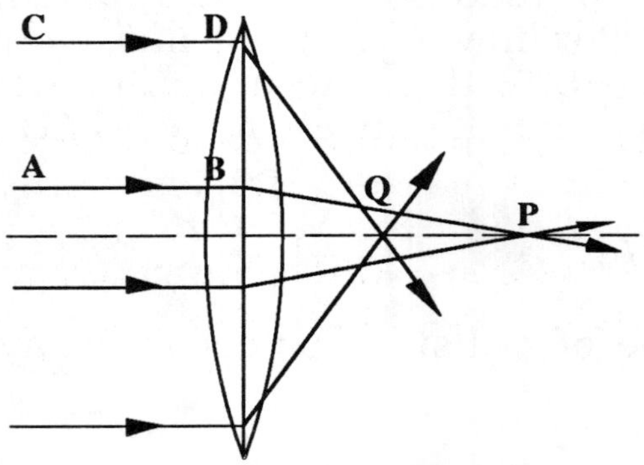

Fig. 5.12. Spherical aberration.

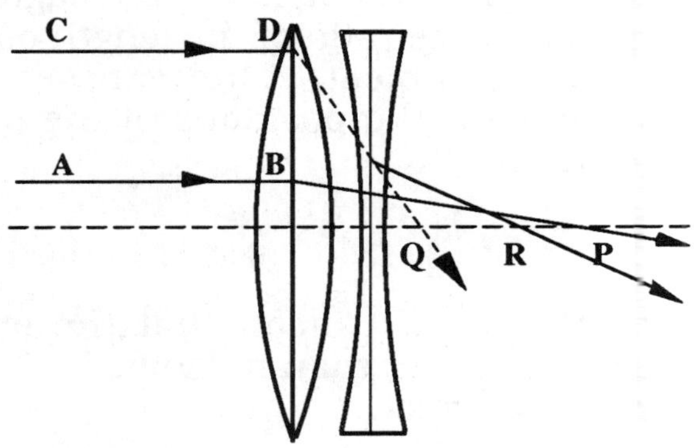

Fig. 5.13. Correcting spherical aberration

John. Is it possible to make a lens without any spherical aberration?
Teacher. Yes, according to Descartes, if the lens surfaces are paraboloids, hyperboloids, or ellipsoids.
Mary. If so, why all the lenses we see around are spherical?
Teacher. Because spherical lenses are much cheaper to make.

b. astigmatism

Dorothy. My eyeglasses are supposed to correct an *astigmatism*. What does it mean?

Teacher. This aberration takes place if the lens is asymmetrical about the ray of light. A cylindrical lens, for instance, can act as either a spherical lens or a plane-parallel plate: it leaves unchanged the lines parallel to the cylinder's axis, and magnify those which are perpendicular to this axis. For instance, a square **ABCD** produces rectangular

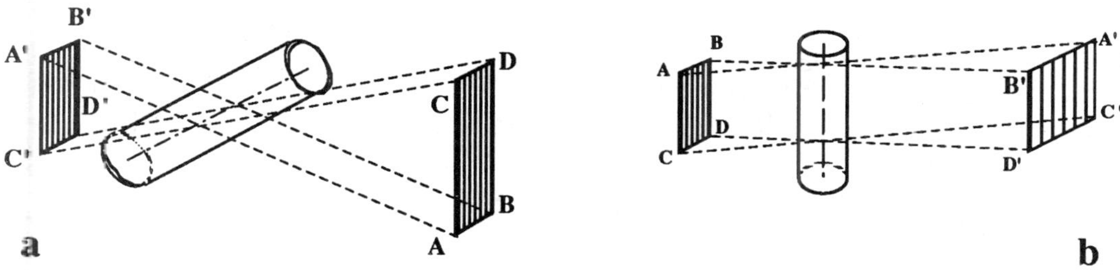

Fig. 5.14. Astigmatism.

images extended vertically (Fig. 5.14a) or horizontally (Fig. 5.14b), depending on whether a cylindrical lens is placed, respectively, horizontally or vertically. Thus, Dorothy, if your eye acts as a cylindrical lens, to correct it you need a cylindrical lens too. And here is a home assignment for all: to determine how two cylindrical lenses must be arranged about one another to produce no distortion at all.

c. chromatic aberration

John. My older brother made a simple telescope, and the image in it is colored. Why is it so?

Teacher. This is *chromatic aberration*. When using colored light, you can see that the same lens focuses rays of different colors at different distances from the lens (Fig. 5.15). Since white light consists of rays of various colors (see Ch. 7), the image produced by white light is enlarged, blurred, and its edges are colored. Different materials refract light of the same color differently, thus it is possible to combine lenses made of different materials (Fig. 5.16) so as to reduce chromatic aberration.

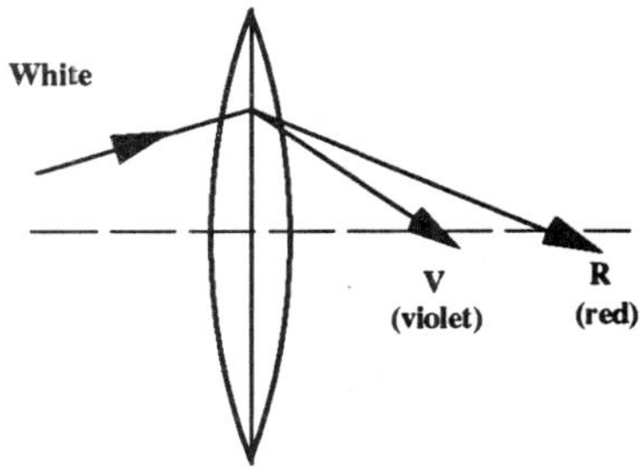

Fig. 5.15. Chromatic aberration.

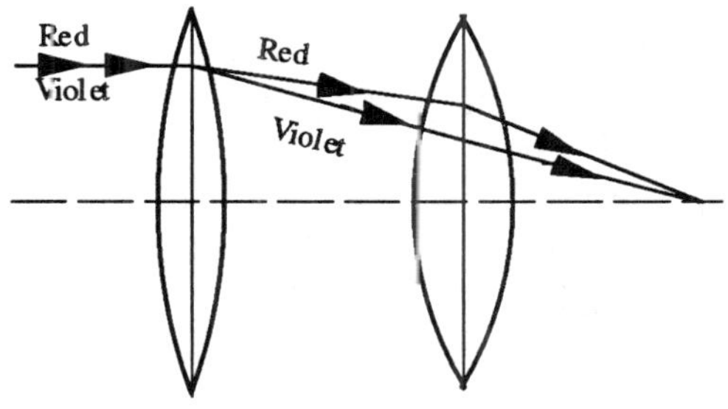

Fig. 5.16. Correcting chromatic aberration.

II. HISTORY

1. Ancient magnifiers

history and discussion

Teacher. Since both reflection and refraction were known in antiquity, can we expect the ancient philosophers to be familiar with lenses as much as with mirrors?

David. Indeed, if they burned things with mirrors, they could do it with lenses too. I remember seeing in a museum lens-shaped glass objects: they could be lenses.

Mary. And I saw in a museum engravings of the 4th century B.C., so fine that one simply couldn't make them without a magnifier.

Teacher. As to the burning ability of lenses, there is documentary evidence that Greeks were aware of it. In ***Aristophanes***' (c.380 BC) comedy *Clouds* one of the characters intended to melt a legal document (they were written on clay tablets) by collecting sunlight with the help of a globe sold in pharmacies). ***Gaius Plinius Secundus*** (ca. 23-79 AD), also known as ***Pliny the Elder,*** said: "**glass globes containing water become so hot when they face the sun that they can set clothes on fire.**" (Book XXXVI, 199); and "**I find that among doctors there is considered to be no more effective method for cauterizing parts that need such treatment than by means of a crystal ball so placed as to intercept the sun's rays.**" (Book XXXVII, 27-29)

Unlike the heating ability of lenses, the subject of their magnifying is far from being clear. ***Lucius Seneca*** (3 B.C.- 65 A.D.)noted that: "**apples appear much larger when seen through a glass**" [Liv. I, III-8]; "**every object seen through water is considerably enlarged. The letters, however minute and indistinct, appear larger and more readable when one sees them through a glass globe full of water. Apples appear more beautiful if they swim in a glass. The stars seen through a cloud appear to be larger. It is that our visual rays when passing through a humid medium lose some of their firmness and cannot grasp reliably what one wants to see.**" [ibid., VI-3]. It appears that Seneca attributed the magnifying property to a "humid medium" itself rather than to a peculiar shape of the vessel. There is no evidence that solid lenses had been ever used as magnifiers. On the contrary, there is a suggestion, based on a passage from Pliny the Elder, that the emerald used by Emperor Nero was to correct his shortsightedness, which means a reduced rather than magnified image.(Ford, 16) However, another translation of this passage emphasizes the **reflecting** properties of Nero's emerald: it could be a concave mirror rather than a lens! (Pliny, Book XXXVII, 215). Some historians argued that the fine engravings could have been done by very short-sighted artisans, and that the ancient technology did not allow a proper polish of glass, Besides, if you remember, David, the museum lens-shaped glass objects are not very transparent, thus they cannot work as lenses. On the other hand, it is difficult to say whether the lack of transparency of the artifacts was original or resulted from the aging of glass. Whatever the case with glass, it is clear from the passages cited above that ancient artisans were able to make burning lenses of rock crystal. It appears that no ancient philosopher was able to combine the knowledge of refraction and the existence of clear lens-shaped bodies so as to discover the magnifying properties of lenses.

interesting!

2. Reading glasses

history

Ruth. If this is true, there could be no eyeglasses in antiquity.

Teacher. Right, the eyeglasses appeared comparatively late. Alhazen was the first who explicitly described a magnified effect of a glass sphere, or, more exactly, of a larger segment of a sphere. ***Roger Bacon*** (1214-1292) suggested to people with poor vision to use a small segment of glass globe for reading. In his view, a smaller segment produces an image closer than the object, while a larger segment makes an image (also magnified) farther than the object. He asserted that the magnification was due to refraction but could not explain how. A glass segment was laid down on the object or held in hand close to it. However, people who needed a magnifier all the time - the old monks who copied books - had to free their hands, thus they started attaching the glass segments to the head. That is how the first convex eyeglasses appeared around 1280. The concave lenses were developed much later. Paradoxically, the invention of reading glasses did not immediately lead to a theory of lenses or an improvement of the theory of vision. Since lenses were of poor quality and distorted the images of bodies, it was difficult to suppose an analogy between the eye and a lens. Besides, the distortions made a lens a poor tool for scientific investigations. Thus, for about three centuries scientists did not study lenses, and their improvement proceeded purely empirically by craftsmen.

interesting!

Francesco Maurolico (1474-1575) supposed that the need in correcting the eye results from an abnormal shape of the crystalline lens. According to him, in the case of nearsightedness the crystalline is more spherical than normal, thus it converges the rays to a focus before they reach the retina. Farsighted people have the lens flatter than normal, which collects refracted rays beyond the retina. In a nearsighted eye the refraction in the crystalline is too strong, thus it must be reduced by concave glasses; the refraction in a farsighted eye is too weak, and it must be increased by convex glasses. Maurolico did not give any ray diagrams in the modern meaning of this term and spoke only in terms of converging and diverging light rays, or increasing and diminishing refraction. He also explained why the nearsightedness sometimes improves with age: a relaxation of the crystalline diminishes its curvature, and this increases the distance at which the rays converge. When Maurolico completed his book in 1555, he possessed several important ideas which if known could have accelerated the development of optics. Unfortunately his book was published only after his death in 1611 (he himself did not have the money to do it and had to rely on benefactors), and his results were superseded by those of Kepler.

3. The eye as a lens

history

An analogy between the eye and a camera obscura attracted a number of followers in the 15th century, including Leonardo da Vinci (1452-1519) and Porta. Porta compared the pupil to the hole in the camera and the crystalline, to the wall receiving the image. The analogy was limited, however, because a camera obscura created an inverted image. Following Ibn al-Haytham, Porta and others placed the seat of vision in the crystalline. The main reason for not choosing the retina was that if the image was created behind the crystalline, it had to be inverted, which appeared to contradict the daily experience. To avoid this inversion Ibn al-Haytham used only the rays entering perpendicularly to the cornea and made the seat of vision at the anterior surface of the lens. However, in the

16th century an attempt to explain the propagation of light inside the eye without refraction appeared artificial, and some scientists, the anatomist ***Felix Platter*** (1536-1614) for one, suggested that the retina rather than the crystalline is the true seat of vision. The problem was to conceive a satisfactory theory of the refraction in the eye which would explain. the formation of an inverse image on the retina. That was left for ***Johann Kepler*** (1571-1630).

Kepler is best known as the founder of modern astronomy: he discovered three laws of planetary motion. He grew up in a poor and unhappy family. He did not see much of his father, a mercenary soldier who eventually abandoned his family. After graduating from secondary school, Kepler studied at the University of Tubingen. In his course of astronomy he learned about the Copernican system and in a few years he became its most ardent and influential champion. Meanwhile, he did not think of science: he intended to become a clergyman. Chance terminated his study of theology and sent him to teach astronomy in Graz. With a few students to teach, young Kepler took on astrology and predicted for the next year a bitter cold, peasant revolt, and Turkish invasion: it all turned out to be true. Kepler believed in the harmony of the universe and a link between the universe and an individual. However, he did not take astrology seriously: rather, it became for him a source of income to support his large family, especially when he was not paid his salary for several years in a row. Besides, it was his success with horoscopes rather than his astronomical discoveries that secured him an employment with rulers of various German states. Being a Protestant was a difficult lot early in the 17th century and Kepler had to change his residence several times. On top of all his other troubles, he had to spend a great deal of time and money to save his mother who was accused of witchcraft. When the famous astronomer died in 1630, he left his family penniless.

Fig. 5.17. The formation of the image in the eye. From Kepler, *Dioptrice* (1611).

Kepler presented his new ideas on vision in his book *Paralipomena a Vitellion* ("Additions to Witelo") published in 1604. Witelo was a Polish scientist of the late 13th century who wrote a book on optics, which familiarized European scientists with Alhazen's ideas. Kepler suggested that every point of an object has a single image on the retina. He drew a cone of rays with an apex at a given point and the base on the pupil and considered their refraction by the cornea and the crystalline. He showed that the rays originating from the same point focus at a single point of the retina if the posterior surface of the crystalline is a hyperbolic cone (Fig. 5.17). Kepler imitated the refraction in the eye by refraction in a water filled flask and found that it produces sharp images if the light beam is limited by a stop (the "pupil"). He asserted that the image created on the retina is inverted, and if we see it direct it is due to a correction by the brain. Kepler also introduced in 1611 the modern method of constructing images in a lens, which he needed to understand how Galileo's telescope worked. He knew that a convex lens gathers a beam parallel to its main axis in a single point and assumed the existence of one-to-one correspondence between a point object and its image. He suggested that the location of the image point could be determined as an intersection of two refracted rays coming from the same object point (Fig. 5.18).

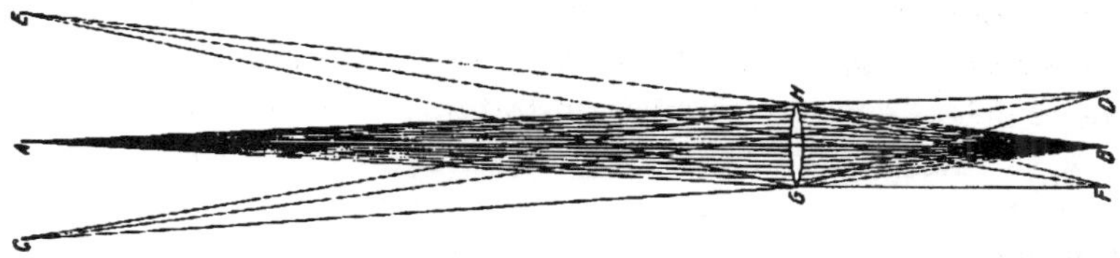

Fig. 5.18. The formation of the image in a lens
From Kepler, *Dioptrice (1611)*

4. Various applications

In the 17th century, lenses were primarily employed for eyeglasses as well as for telescopes and microscopes (see Ch. 6). In addition to these, they found other, sometimes quite curious applications. For instance, the camera obscura was improved by adding a lens in front of the pinhole which made the image brighter and more detailed (Cardano, Porta, and others). Such an apparatus was used, for instance, by artists to copy a landscape. The heating capability of mirrors and lenses became of a great interest to scientists, especially chemists. Tschirnhausen made a double-convex glass lens 3 feet in diameter and 12 feet in focal distance. Its focal circle was 1.5 inch in diameter, and with an aid of another lens he reduced it three times. The combination produced the same results as Tschirnhausen's great mirror described in Ch. 4. To melt larger quantities of metals, the French chemist ***Antoine-Laurent Lavoisier*** (1743-1794) decided to build a lens with a greater focal spot. Since it was difficult to make a large mass of glass sufficiently homogeneous, the lens was built out of two glass sheets 1.6 cm thick bent around a sphere of 5 m in diameter. The two parts were connected into a lens of about 1.2 m in diameter, with the space between the surfaces filled with alcohol (Fig.5.19). This lens, combined with a smaller one, could fuse an iron bar of 10 cm long and 0.8 cm wide in 5 minutes, and it could even destroy diamonds.

history

wow!

Fig.5.19. Lavoisier's giant lens. From A.Lavoisier, *Oeuvres*, v.3

5. Fresnel lens

history and discussion

Mary. Speaking of burning lenses, I've seen a device, quite flat, which people called *fresnel lens*. It is being used, I was told, to make solar furnaces. What I don't understand is how a flat body can work as a lens.

Teacher. Nowadays, these devices are also employed in overhead projectors: it is the one you lay a transparency on. Let us check whether it works as a lens. How can we do it?

Michael. First, let us try to burn something.

Teacher. O.K. I will remove the projecting lens from the top of the projector to prevent any confusion between its work and that of the fresnel lens. Here I have a piece of a newspaper which I will hold over the fresnel lens tilting it to the class and move it up and down. What do you see?

Dorothy. We see a big bright light spot which becomes smaller and brighter when the paper moves higher. Wow! Smoke!

Ruth. Since the material is clear, this device cannot be a mirror, thus it must be a lens.

John. Can it make images?

Teacher. Find out for yourselves: I am giving you several fresnel lenses.

David. When we looked through this device holding it far from the face, we saw an inverted image. We were also able to get an image of a window, which was also inverted. Thus, it is indeed a lens, a converging lens. But how does it work?

Teacher. You can see and feel many parallel grooves cut in a clear plastic. Their shape is prismatic, thus we have a multitude of concentric prisms. If a convex lens is cut as shown in Fig. 5.20, a considerable part of the material of the lens can be removed without changing its refracting power. This may be very useful when it is necessary to build a very large and light lens because a large lens could be compounded of a number of spherical segments.

Such a lens (Fig.5.21) was first built in 1823 by ***Augustin-Jean Fresnel*** (1788-1827), known as one of the founders of the wave theory of light. Fresnel was not only a brilliant theoretician but also a very good experimenter and engineer. When the French Government decided in 1819 to improve French lighthouses, it gave Fresnel, an engineer of the state Corps of Roads and Bridges, the commission to cre-

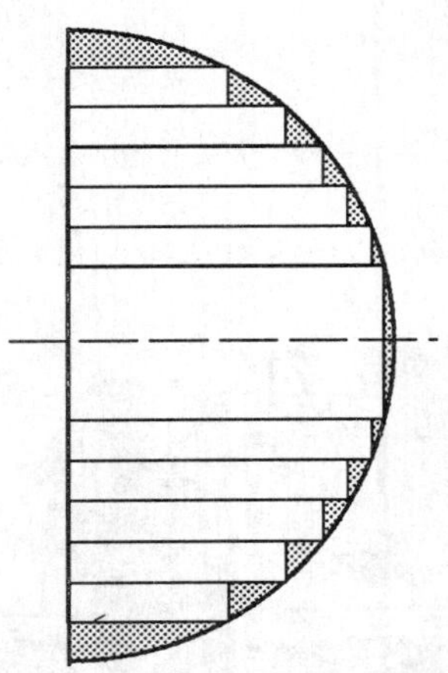

Fig. 5.20. A compound lens.

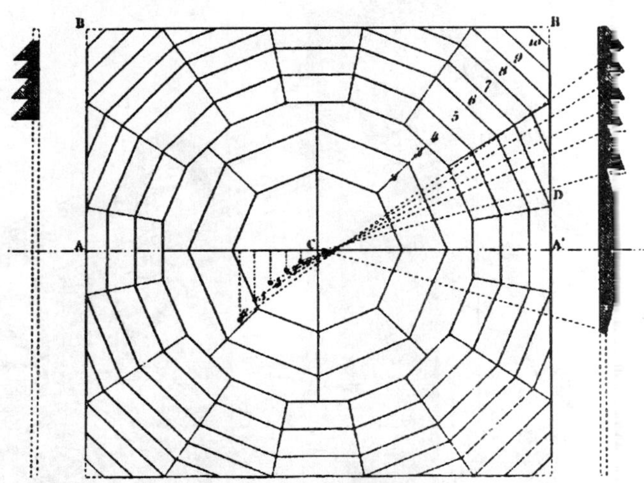

Fig. 5.21. Fresnel's compound lens. From A. Fresnel, *Oeuvres complètes*, v.3 (Paris, 1872).

are a better source of light for a lighthouse. Fresnel decided to replace parabolic metal reflectors with a ring of lenses surrounding the source of light. Since lenses had to be large (about 1 m) and thick, they could absorb much of the light. To overcome this difficulty Fresnel offered to remove extra glass and to compound lenses of small pieces. He was unaware that the idea was not a new one; however, he was the first to realize it in practice. The new source of light proved to be very successful, and later it was used in lighthouses throughout the world even in this century.

III. EXPERIMENTS

1. Measuring focal distance and magnification

(student instruction sheet)

Equipment.
Lenses, lens holder, a ruler or tape measure, white screen, candle..

a. the easy method

Procedure. When holding a lens in your hand project the sun or another bright distant object on a white screen (Fig. 5.22a) and move the lens until the image becomes the sharpest (for the sun. it will be the smallest). Ask your partner to measure the distance between the lens and the screen. This distance will be the lens' focal distance. *lab home*

conclusion ➡ **The simplest way to measure the focal distance is to obtain an image of a *remote* bright object on a white screen and to measure the distance between the lens and the screen**

b. the precise method

Take a lens of a considerable curvature and place it at about 10 cm from an object (a candle flame). Move a screen behind the lens until you obtain a sharp image (Fig. 5.22b). Measure the distances **o** and **i** from the lens to the object and the image, respectively, and calculate the focal distance **f** using eq. (5.1). Now, move the object 50 cm away from the *lab*

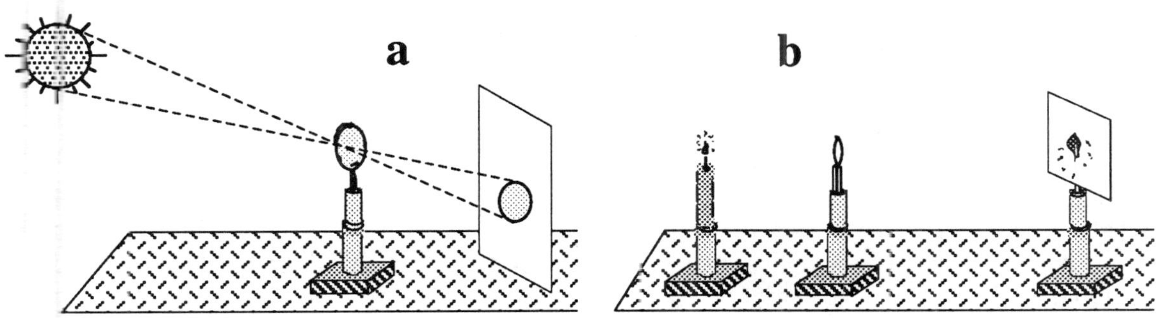

Fig. 5.22. Measuring focal distance of a converging lens.

lens, adjust the screen to see a sharp image and compare the new distance **i** from the lens to the screen to focal distance **f**. Repeat the same when the object is 2 m away from the lens. You can see that the farther the object is from a lens, the closer the distance between the lens and the screen comes to the focal distance. This is the ground for the rule given above.

Note. A fine object can be made out of an arrow drawn on a ground (frosted) glass with a marker. Make a housing for a desktop light of a cardboard box, cut a small (2 cm) window in it, and attach the glass to this window. Instead of a ground glass, you can use a waxed paper mounted in a slide frame.

c. making Gullivers and Lilliputians

lab

Everything can be made gigantic or tiny... with a single movement of the same converging lens! Take the lens you used in the previous experiment. Set a screen and an object (a candle flame) at four times (or more) the focal distance of the lens from one another, and **do not move them** afterwards. Move the lens between the object and the screen until you obtain an enlarged image; then move the lens until you obtain a reduced image. In each position of the lens measure its distances from the object (**o**) and the screen (**i**), as well as the sizes of the object and the image. Calculate the magnification in each case as found from measurements and from eq. (5.6). Compare the distances **o** and **i** in both cases. Do you see anything interesting about the results (compare Figs. 5.8 & 5.9)?

d. diverging lens

lab

Since a diverging lens does not produce a real image, to determine its focal distance one has to combine it with a converging lens. First, place the selected diverging lens very close to a short-focus converging lens and by moving both lenses together obtain a magnified image (Fig. 5.23). Measure the distances **o** and **i** from the midpoint of the lens system to the candle and to the screen, respectively, and calculate the focal distance **f** and power **P** of the system. If you cannot obtain such an image, replace the converging lens with the one with a shorter focal distance. Then obtain a magnified image by means of a converging lens alone and calculate its focal distance $\mathbf{f}_1$. Use equation (5.5) to calculate the focal distance of the diverging lens $\mathbf{f}_2$.

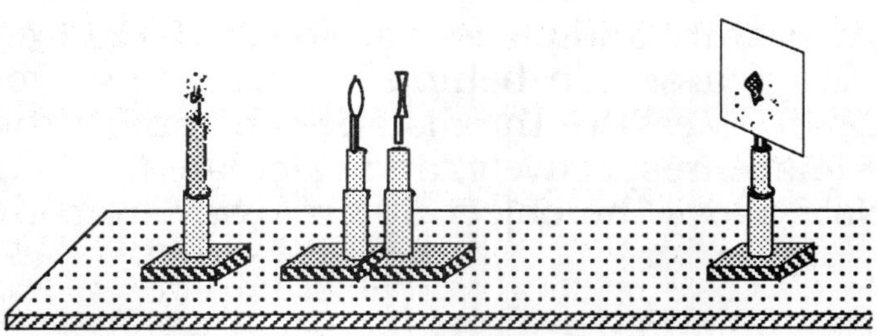

Fig. 5.23. Measuring focal distance of a diverging lens.

e. comparing focal distances

If it is necessary to estimate which of the two lenses has a smaller focal distance, place them side by side and looking at a nearby object (print, for instance) and select the one which has a greater magnification. Remember, however, that you look for a larger image only if the lenses are converging; among diverging lenses, a stronger magnifier is the one which produces a **smaller** image.

interesting!

2. Burning power of a lens

Preliminary part

Background

investigation

Teacher. Some of you have probably tried to burn something with a lens. Would you share with us when it was easy and when it was difficult?

Michael. It is easy to burn wood but not a white paper.

John. Small lenses don't work.

Teacher. We are not going to really start a fire. It is sufficient to obtain smoke. Thus, we can formulate the problem as follows.

Formulating a problem

How to obtain the first smoke by means of a lens and sunlight in the shortest time possible.

Selecting variables

Teacher. It appears from your experience that the solution depends on both the body to be burned and the lens. Let us begin with the body. Which properties of materials can affect the result?

Dorothy. Color. Dark bodies absorb light better.

Ruth. Perhaps, the amount of moisture in the body. Dry materials are easier to burn.

David. Possibly, the smoothness of its surface. If it reflects much of the light, it will take longer to burn it.

Teacher. Thus, we would have to investigate several variables related only to the body. Since our primary concern in this unit is properties of lenses rather than materials, it is preferable to exclude the part on materials. How can we do it without misleading ourselves?

Mary. If we keep the same body in all experiments, we'll eliminate its material as a variable.

Teacher. Very good! Which properties of lenses do you believe to be important?

Michael. Size.

Teacher. O.K. We can start with a single variable, the size of a lens. However, before we begin experimenting we have to think about the best procedure to follow.

Selecting a procedure

John. Take a piece of a newspaper (it burns well), lay it on the ground, focus a lens on it and measure the time necessary to get the first smoke.

Main part

a. size of the lens

Preliminary experiments

Teacher. Try two lenses of different size.

Dorothy. We had lenses of 4 cm and 1 cm in diameter. The bigger lens worked fine, but the small one didn't work at all.

Hypothesis

Larger lenses ignite bodies quicker than the smaller ones.

Test

Michael. Where can we obtain lenses which have the same material and shape but different diameters?

David. Perhaps we can do with a single lens by changing its open area. For instance, we can make 3 light stops from pieces of an index card about 6 cm x 8 cm each and cut square openings in their middle with the side of 10 mm, 20 mm, and 30 mm. The stop can be attached to the lens holder with a piece of scotch tape so that the opening were against the lens' center (Fig. 5.24).

good point!

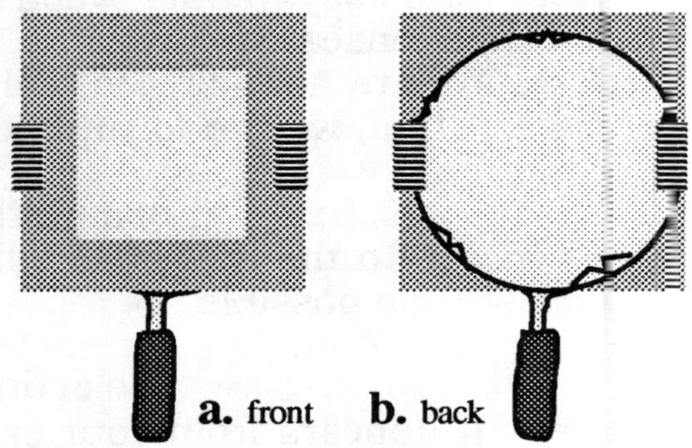

Fig. 5.24. A lens with a stop.

Mary. We've tried this, and our experiments confirmed the hypothesis.

conclusion ➡ The larger the open surface of a lens, the greater its igniting power.

b. shape of the lens

Preliminary experiments

Teacher. Try a couple of larger lenses of the same material and about the same size but of different curvature. Compare their focal distances.

John. It appears that those lenses which are thicker in the middle (have a greater curvature) work faster than others. We found that the more curved lenses have shorter focal distances.

Hypothesis

Dorothy. The greater the curvature of a lens the better it burns.

Test

Mary. To be sure that the lenses have the same open area, test them with the same aperture in the stop attached to them (it must not exceed the size of the smallest lens).

Michael. We've observed that more curved lenses work better.

conclusion ➡ The burning capability of a lens increases if its curvature is greater, or its focal distance smaller.

c. lens' material

Discussion

David.. If focal distance is involved, according to the lens-maker equation, the index of refraction must be of importance. Doesn't this mean that we have to study the lens' material?

Teacher. You are right, we should. To do this we would probably need a more precise way of comparing the action of different lenses than timing the first smoke. Any ideas?

Selecting a procedure

Ruth.. Since burning is the result of heating, why can't we measure the degree of heating? For instance, we could focus the lens on the bulb of a thermometer (Fig. 5.25) and measure the temperature rise for a given time. Actually, this method is applicable to the previous parts too, and it should work even for small lenses which don't burn.

good point!

Teacher. Great idea!

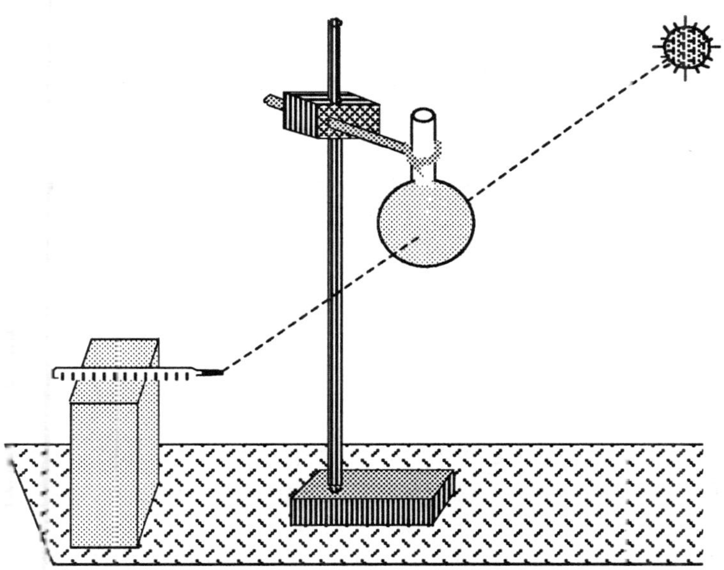

Fig. 5.25. Measuring the heating power of a lens

John. But where can we get several lenses of the same size and curvature but made of different materials?

Mary. I know! We can make liquid lenses by filling the same flask with different liquids.

good point!

Teacher. Excellent!

Dorothy. If we have several identical flasks filled to the same level with different liquids, it would be much faster than pouring liquids in and out of a single flask and wash it afterwards.

Preliminary observations

Teacher. Very good! Take two round-bottom flasks, fill one with water (less than half-full) and the other with equal amount of sugar syrup and compare their heating properties.

Michael. The sugar lens heats faster.

Hypothesis

David. The index of refraction of sugar syrup is greater than that of water (see Ch. 4), which means that, according to eq. (6.2), the sugar lens has a shorter focal distance than the water lens. On the basis of part **c**, a sugar lens must heat faster. Thus we can suppose that **the greater the index of refraction the greater the heating power of a lens.**

Test

Dorothy. We found that a vegetable oil lens works even better than a sugar lens.

conclusion ➡ The heating power of a lens increases when its index of refraction increases, or when its focal distance decreases.

general conclusion ➡ **A good burning lens must have a large diameter and a small focal distance**

?

Teacher. If oil is somewhat colored, the **color** may be another variable to investigate. For instance, does coloration produce a sufficient absorption to affect the heating action?

3. Why do apples appear bigger in water?

advanced

Preliminary part

Background

Teacher. As I've mentioned before, ancient philosophers were not certain of why apples appear bigger when placed in a glass jar filled with water. Seneca apparently attributed the effect to some property of water. Let us check whether he was right.

investigation

Experiments

Michael We can start with the original experiment by placing an apple in a jar with water.

Dorothy. Indeed, the apple appears bigger in water. Since it floats, we compared the upper and lower parts of it.

John. I put an orange in water: the result is the same as with apples.

Mary. To be sure that this is not an exclusive property of organic matter, we tested a Coke can. We discovered that its submerged part was obviously larger than the one above water.

Selecting a procedure

Ruth. Since the effect is the same for different objects, we could use the one which is more convenient. Apples and oranges are of different size and rounded, which makes them difficult to compare. We need something identical for all groups and it should be of the same width throughout its height, something like a pencil or a can.

Michael. We've placed a pencil vertically in a jar with water and found almost no magnification.

Dorothy. Where did you put it inside your jar?

Michael. Near the front wall.

Dorothy. We had it near the back wall and found a considerable difference in the width.

Ruth. Apparently, we have here another variable, the **position** of the body inside the jar. This means that thick objects, such as a Coke can, are inappropriate, and we need a thinner object, such as a ruler or a pencil.

good point!

John. We've discovered that moving a pencil away from the eye, even a little, increased the visible width of its underwater part. Thus, a ruler may be better because it is thinner, but we must make sure to place it always in the same position, for instance, in the center of the jar.

David. If we want to compare the change of the visible width under different circumstances, we need a device to measure it. For instance, we can take a short ruler with millimeter divisions and tape it perpendicularly to a pencil or another ruler standing vertically in water. When we place this "cross" in a jar with water the eye will project the submerged part on the scale which is above water level, and we will take readings for the width of the vertical stand both above and under water (Fig. 5.26).

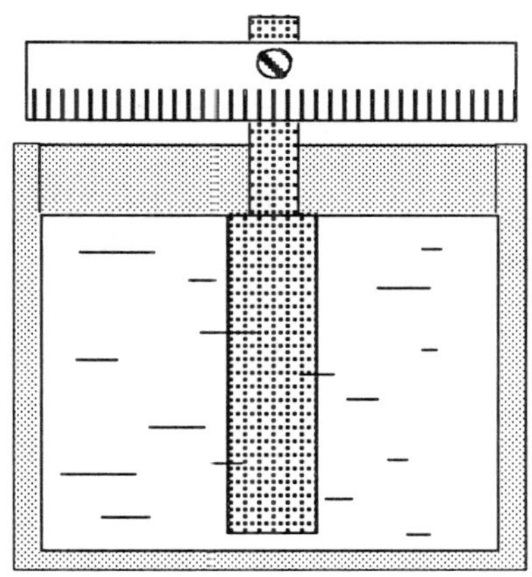

Fig 5.26. A 'cross' to measure refraction.

Selecting variables

Ruth We have found so far two variables: 1) water (according to Seneca), which we will generalize to a **liquid**, and 2) the **location of the object.**

David. The object's position refers to an optical effect, while the ancient philosopher apparently meant a material effect: an impact of water on apples. For this reason, I would delay the study of the position until we exhausted all possible material influences.

Mary. This can include not only a liquid in a container but the container itself.

Main part

a. liquid

Preliminary experiments

David. The original suggestion was about properties of water. However, why don't we check other liquids as well.

Michael. We tried a Coke. The effect was the same.

Hypothesis

Water is not the only liquid which magnifies the objects immersed in it

Test

Dorothy. We experimented with sugar syrup and saline water, trying to maintain the same level of liquid. The result was the same.

conclusion ➡ All liquids magnify the bodies submerged, and the effect does not differ much for different liquids.

b. container

Preliminary experiments 1

good point!

Mary. If liquid is not the main cause of the effect, perhaps the vessel is. Since we have no idea of what can be of importance in a container, let us try different jars, wine glasses, and cups made of glass and ceramic. We fill them with water, submerge the "cross" and look for a change.

Michael. We observed the effect in a glass jar but not in a ceramic pitcher.

Hypothesis 1

David. The effect is present only in clear materials.

Test 1

John. We used metal containers and different plastics cups. Clear plastics work fine, metals and opaque plastics don't work at all.

conclusion ➡ The effect is observed only in transparent containers.

Preliminary experiments 2

good point!

Ruth. We don't see the expansion effect in opaque containers. Although we see the underwater part of an object placed in an opaque container, it doesn't appear to be larger than the rest of the object. Unlike clear vessels, in opaque containers we see the submerged part of the object from above rather than through its walls. Perhaps this is our clue.

Hypothesis 2

If you look from above, the effect disappears.

Test 2

Dorothy. We repeated the experiment with clear containers when looking from above rather than through the wall. The submerged part was the same as the top.

conclusion ➡ | A liquid produces no magnification by itself.

Preliminary observations 3

Mary. If we look at the submerged part only through water, we don't see any effect. On the other hand, if we look through both water and the container's wall, we see the expansion.

good point!

Hypothesis 3

Dorothy. The effect is produced by the container's wall.

Test 3

Michael. We emptied a glass, placed the "cross" inside it and looked at the "cross" through the glass. No effect.

conclusion ➡ | The container produces no magnification by itself.

c. liquid & container

Preliminary experiments

Teacher. Since neither water nor the container taken alone produce the phenomenon, it should result from some property of their combination. Go back to the original experiment and try something new, like moving the "cross" inside the glass to and from the eye.

Dorothy. We've seen that the magnification is minimal, if any, when the "cross" is close to the front wall of the container and maximal when it is close to the back wall.

Ruth. When moving the "cross" we change the thickness of water between the eye and the cross. However, this factor did not play any role when we looked from above. What is different now is the shape of the surface of water through which we look: from above, the surface of water is plane, and from a side, it is cylindrical.

good point!

Hypothesis

The magnification is produced by a liquid plano-convex lens limited by the wall and the cross.

Test 1.

David. I moved the "cross" beyond the outer wall and found that its lower part continued to expand when the "cross" receded from the container. Thus, there is no difference between the action of water inside the container and the container with water as a single unit.

conclusion ➡ The visible expansion of the underwater part of a submerged body is caused by a liquid lens.

general conclusion ➡ **Ancient philosophers were wrong. A body placed in a clear liquid appears larger then its true size only if the container has convex walls and we look through these walls.**

Theoretical explanation

math is a friend !

Teacher. In Fig.5.27, **Q** is the eye, **AB** is the object placed near the back wall of a cylindrical container filled with water, and **CD** is the same object close to the front wall. Let us draw the ray **AC** parallel to the main optical axis: after refraction it will follow **CQ**. The rays **AO** and **CO** will not change their direction after refraction. The image of **AB** will be **A'B'**, and and the image of **CD** will coincide with **CD**. We see that the image increases when the object is farther from the front surface.

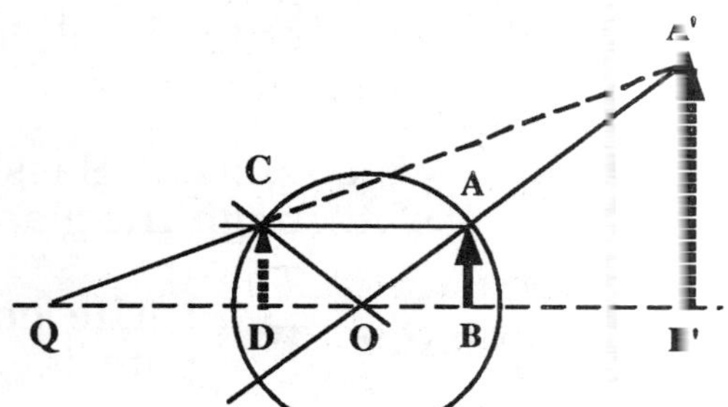

Fig. 5.27.

4. Home experiments

a Take a few uncolored wine glasses or clear bottles which have a smooth surface (when removing a label from a bottle, remove the glue too). Look through each of them at different objects. If you see any distortions of the objects, investigate their causes. **Tip**: look through different parts of a glass (bottle), move it to and from the eye, left to right, and also turn it around the line of vision.

b. Find a wine glass which has a semi-spherical part and fill this part with water. Bring it near the eye and look at a distant object which has a distinction between its top and bottom. While slowly moving the glass away watch how the image of this object changes. If this reminds you of a lens, what sort of lens is it? What other experiments can you offer to test your hypothesis? Measure the focal distance of the glass. Does it depend on the amount of water in the glass (use a measuring cup)? If you screen the left half of your glass, which part of the image you see through it disappears, left or right? Answer the same question when projecting a candle flame on a white screen.

c. Take two clear cylindrical water bottles of the same size, fill up each bottle with water and plug it with a stop. Prove that such a bottle works as a converging lens. Place a bottle vertically before your eye and look at a distant object (a tree, a building, etc.). How does its shape change when you place the bottle horizontally? What sort of aberration is this? Which dimensions of the image are affected by the bottle when it is held horizontally? Vertically? Can you reduce this aberration by applying two bottles with water? If "Yes," how?

Teacher's note. Being a cylindrical lens, a water-filled bottle produces astigmatism. To reduce it, hold two bottles perpendicularly to one another and find a proper distance between them.

d. Find a clear glass jar with a wide neck and fill it by about one half with water. Use it to project an image of an electric light (or the sun) on a piece of paper. Measure the focal distance of this lens. Take an empty 10-ounce water bottle and place it inside the jar. The image disappears. Investigate possible causes: 1) did the focal distance of the system change? 2) did the system become a diverging lens? 3) any other ideas? How would you test each of these hypotheses?

Teacher's note. The empty bottle placed in water acts as a diverging lens. Thus we have a combination of a converging and a diverging lenses which may be either a converging lens or a diverging one. In the former case, the focal distance increases and the image can be found further away from the bottle. In the latter case, however, there is no real image.

e. Watch in a camera-obscura the images of the objects at different distance, for instance 1 m and 10 m, and compare their sharpness. Replace the window cover with another one which has a larger hole (about 5mm). Answer the same question as before. How did the images change otherwise? Now, tape a converging lens to the hole. What is the effect on the image? Compare the sharpness of the images made by the objects close to the camera with those far away from it. Imagine that you attached a photographic film to the back of the camera obscura. What will be the advantages and disadvantages of such a photo-camera when you have a small opening and no lens, or a large opening with a lens?

Teacher's note. Increasing the aperture makes images much brighter but more fuzzy. Adding a lens allows to produce bright images of the objects at specific distance from the camera. Objects at distances very different from this one will appear fuzzy. In a photo-camera, a larger aperture with a lens reduces the exposure time but it also reduces the field of view; while a pin-hole without a lens provides a very large field of view (all objects appear sharp) but requires a very long exposure.

BIBLIOGRAPHY

Pliny, *Natural History* v. 10 (books XXXVI-XXXVII) in Loeb Classical Library (Cambridge, MA: Harvard Univ. Press, 1962).

L. Seneque, *Questions Naturelles* t.1, (Paris: Société d'édition "Les Belles-Lettres,"1929), III-9, VI-5, pp. 23, 32.

Aristophanes, "The Clouds," Loeb Classical Library: Aristophanes, v. 1 (Cambridge, MA: Harvard Univ. Press, 1960), pp. 266-401.

J. Kepler, *Dioptrice* [1611] (facsimile: Cambridge, Heffer, 1962), pp. 393-402.

A. Lavoisier, "Réflexions sur les expériences qu'on peut tenter l'aide du miroir ardent [1772]," *Oeuvres*, v. 3 (facsimile; New York: Johnson Reprint, 1965), pp. 261-348.

V. Ronchi, *Nature of Light*, pp. 53-93.

R. S. Clay & T. H. Court, *The History of Microscope* (London: Holland Press, 1975), pp. 1-5.

E. Rosen, "Did Roger Bacon invent eyeglasses?" *Archives Internationales d'Histoire des Sciences* 7 (1954), 3-15.

E. Rosen, "The invention of eyeglasses," *Journal of the History of Medicine & Allied Sciences* 11 (1956), 13-46, 183-218.

Albert van Helden, "The invention of the telescope," *Trans. of the Amer. Philos. Soc.* 67, pt.4 (1977), 5-64.

B. Ford, *Single Lens: The Story of the Simple Microscope* (New York: Harper & Row, 1985), 12-16.

R. Longhurst, *Geometrical and Physical Optics* (London: Longman, 1973), chs. I, II, XV.

Chapter 6

TELESCOPES and MICROSCOPES

MAGNIFYING an OBJECT

1. Simple magnifier

discussion

Teacher. What do you do to see more clearly the small print or other fine details?

Dorothy. We can look at it through a magnifier.

Teacher. What does it mean *to magnify*?

John. To make an image larger than the object.

Teacher. This is one way of expressing it, which is true in the case of a print. There is also another way of presenting this concept, which requires some understanding of how the eye together with the brain estimate the size of objects. When comparing the eye to a converging lens and the retina to a screen (Fig. 6.1) one sees that the linear sizes of the object **H** and its image **h** are related as follows:

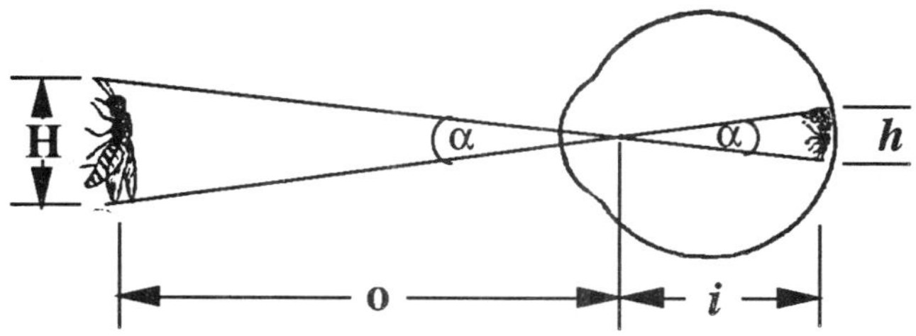

Fig. 6.1.

$$H/o = h/i = \tan \alpha \qquad (6.1)$$

where **o** and **i** are the distances from the eye to the object and its image, respectively, and α is the angular size of the object, or the angle at which the eye sees the object. At the same angle the image is seen from the pupil. Since the length of the eye is constant, the larger the object (or, the larger its angular size), the larger its image on the retina. Thus, we estimate the size of an object by the size of its image on the retina. However, the image increases when the same object approaches the eye. Thus, **what the eye is actually measuring is the angular size of an object rather than its linear size**, and the purpose of all optical instruments is to increase the angle at which the object is seen. If we look through a converging lens (Fig. 6.2), the angular size of the image α is greater than the angular size of the object α_1. The simplest optical instrument is a *simple mag-*

important

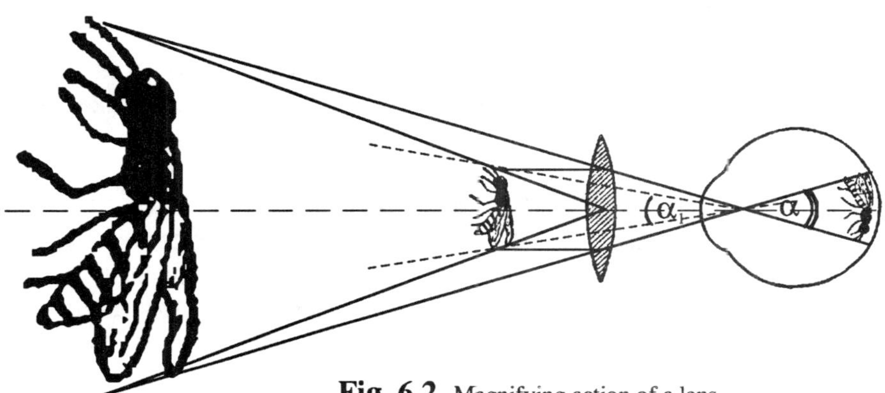

Fig. 6.2. Magnifying action of a lens.

lab

nifier. Take a converging lens and place it near the eye so as to see the small print magnified. Measure the distance between the lens and the object and compare it to the focal distance of the lens. What are the results?

Michael. Our lens had a focal distance of 4 cm, and it was about 3.5 cm from the object.

Ruth. We had a lens with a focal distance of 10 cm, placed 9 cm from our object.

Teacher. Thus, you set an object slightly closer to the lens than its focus. As we already know, in such a case the image will be virtual, direct, and magnified. According to eq.(5.6), its linear magnification is M = i/o. When the eye is most relaxed (focused on infinity), or well relaxed (focused at the "distance of the best vision" **D**, which is about 25 cm for most people), the angular magnification **N** of a simple magnifier is

$$N = D / f \qquad (6.2)$$

Thus, the shorter the focal distance the greater the magnification. A similar relation is also true for a diverging lens, although for such a lens we speak of **reduction** rather than magnification.

2. Telescopes and microscopes

Preliminary part

Discussion

investigation

Teacher. If you want to magnify a distant tree rather than the print, can you use a simple magnifier? Experiment and report the results.

Michael. It doesn't work. When a lens is close to the eye the image is blurred, and when the lens is far from the eye we saw an inverted and reduced image.

Teacher. How about using two lenses? The idea is to magnify the image in two steps.

David. Do you mean that the first lens produces an enlarged image and the second one uses this image as an object and makes a second image even larger than the first?

important

Teacher. Not exactly. You cannot produce an image which is larger than a building or a tree. What you can achieve is to increase its **angular** size: the first lens increases the angular size of the object, and the second lens does the same to the first image.

Formulating a problem

Teacher. An instrument for viewing distant objects is called *telescope.* Our purpose will be constructing a simple telescope.

Dorothy. A telescope? I don't see any tubes around to make a telescope.

Teacher. Tubes serve to hold lenses and eliminate stray light. The most important part is lenses. We are going to discover which lenses to use and how to place them.

Selecting the procedure

Teacher. Begin by holding both lenses near the eye. Keep one lens close to the eye and slowly move the other one away from the first until you obtain a sharp and magnified image. You can hold the two lenses in both hands. or put them in the lens-holders described in Ch. 5 (Fig. 6.3)

Experiments

John. We have tried a combination of a converging and diverging lenses and failed, regardless of whichever of the two we placed near the eye.

Mary. We were successful with two converging lenses. The magnification was a little more than 1, and the image was inverted.

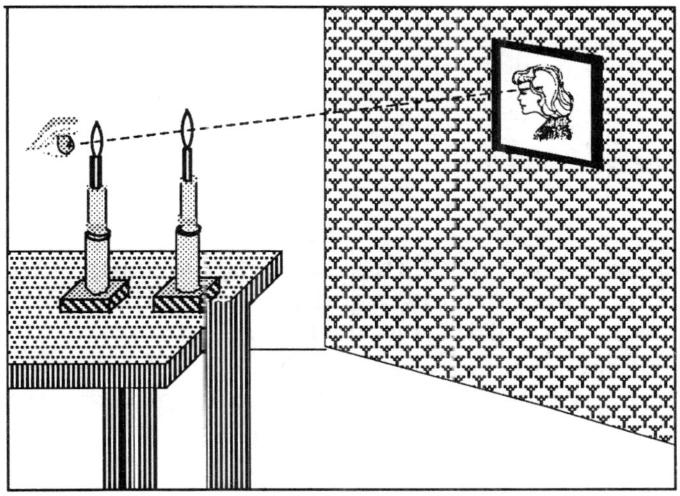

Fig. 6.3. A simple telescope.

Dorothy. We've also combined a converging and a diverging lenses. When the converging lens was near the eye, there was no magnification. When a diverging lens was near the eye, it worked with some converging lenses and it didn't with others.

David. We have discovered that a similar combination of lenses didn't work when both lenses were close to the eye and did when the second lens was very far from the eye.

Michael. We've tried two diverging lenses and found that they reduce the image even more than one of them.

Teacher. Thus, you have found that it is possible to make a telescope with two lenses, and that at least one of them must be converging. This lens should be on the object side, for this reason it is called *objective*, while the lens on the eye side is called *eye-piece*. Let us now investigate how to obtain a sharp image of the greatest size possible. You have found only two successful combinations, and we will begin with the one with the converging eye-piece, with which you had more success. Such a system is called *astronomical telescope*. How many variables do we have?

Selecting variables

John. Since replacing either lens affected the result, we can treat both the **eye-piece** and the **objective** as variables.

Teacher. Since your variables are lenses, what are you actually changing when replacing them?
Ruth. Focal distance?
Teacher. Yes.
Michael. Why don't we use the distance between the lenses as another variable?
Mary. Changing this distance is a part of the procedure. If we do it properly, which means changing the distance continuously from zero to a sufficiently large magnitude, we don't need an additional variable.
Teacher. This sounds reasonable. How are you going to proceed?

Main part

1. ASTRONOMICAL TELESCOPE.

a. objective

Preliminary experiments

Michael. We kept the same converging lens as an eye-piece and examined two converging lenses with different focal distances for an objective. An objective with focal distance 20 cm worked better than the one with focal distance of 10 cm.
Mary. We had objectives with focal distances of 50 cm and 30 cm. The lens with a longer focus created a larger image. Both images were inverted.

Hypothesis

The greater the objective's focal distance, the greater the magnification

Test

John. We obtained a larger image when the focal distance was 20 cm than when it was 10 cm. However, with a lens of 90 cm in focal distance we obtained no image at all.
David. We have had a similar case and discovered that a long-focus objective had to be moved farther from the eye than the arm's length. Thus we employed another person to hold this objective.

conclusion ➡ | The greater the focal distance of an objective the greater the magnification of an astronomical telescope.

b. eye-piece

Preliminary experiments

Dorothy. We will retain the same objective and alternate two eye-pieces of different focal length.

Michael. In our experiment, we had lenses with focal distances of 3 cm and 10 cm, and the former produced a larger image than the latter.

Hypothesis

Ruth. The shorter the focal distance of an eye-piece, the greater the magnification.

Test

David. We had eye-pieces with focal distances of 5 cm and 15 cm, and found the hypothesis to be true.

John. We tried to compare eye-pieces with focal distances of 4 cm and 20 cm, however, the latter one did not produce any image.

Mary. What was the focal distance of your objective?

John. About 10 cm.

Mary. Try an objective with a longer focus, we also had trouble with a short-focus objectives. Actually, we should have paid more attention to the conclusion of the previous part that objectives with a large focal distance work better.

John. Indeed, we have just tested a 100 cm objective, and it worked fine. Thus, we also confirm the hypothesis.

conclusion ➡ The shorter the focal distance of an eye-piece, the greater the magnification of an astronomical telescope.

Teacher. Now, we are going to study a system with a diverging eye-piece, called *terrestrial telescope*. We'll see whether its properties are similar to those of astronomical telescope.

2. Terrestrial telescope

a. objective

Preliminary experiments

Dorothy. We can probably do what we did before: keeping the same concave lens for the eye-piece and combining it with one of several different convex lenses for the objective. We've found that an objective with focal distance of 30 cm magnified more than the one with focal length of 15 cm.

John. We have discovered that a lens with a focal distance of 10 cm made a larger image than the one with the focal distance of 5 cm.

Hypothesis

Ruth. The greater the focal distance of an objective the greater the magnification of a terrestrial telescope.

Test

Michael. We experimented with lenses with focal distances of 20 cm and 50 cm and confirmed the hypothesis.

Mary. We obtained a similar result with lenses with focal distances of 10cm and 30cm but failed to produce a direct image when the focal distance was 3 cm.

Teacher. Mary, possibly your eye-piece is too diverging for your 3 cm objective. Try another eye-piece.

Mary. I will. Indeed, with a less concave eye-piece I can see a direct image.

conclusion ➡ The magnification of a terrestrial telescope increases with the focal distance of an objective lens.

b. eye-piece

Preliminary experiments

Teacher. How would you conduct this part?

John. I will keep the same objective and try two diverging eye-pieces with different focal distances. Since we know that increasing the focal distance of an objective improves the magnification, I would chose a lens with a focal distance of 30 cm or more.

Mary. Is there a simpler way of measuring the focal distance of a diverging lens than the one described in eq. (5.5).

Teacher. No. There is, however, a shortcut for **comparing** focal distances of two diverging lenses, which is sufficient for our purpose. Place two lenses side by side and compare the size of the two images of the same object. The smaller the image the greater the "magnification" (actually, reduction) of the lens, and, accordingly, the smaller its focal distance. In other words, **a diverging lens which produces a smaller image of the same object has a shorter focal distance.**

important

Ruth. We tried two eye-pieces and found that the one with a smaller focal distance made a greater image. The images were direct. Is that is why this telescope is called *terrestrial*?

Teacher. Exactly. One wants to see people and buildings erect, while one doesn't care about seeing the moon upside down.

Hypothesis

Dorothy. The shorter the focal distance of an eye-piece the greater the magnification of a terrestrial telescope.

Test

Michael. We tried two diverging lenses, and the results agree with the hypothesis.

Dorothy. Our experiments also support it; however, when we tried to see people in the room, we could not make their images sharp.

David. Aren't telescopes made for viewing **distant** objects? Perhaps you had a wrong eye-piece to see closer objects.

conclusion ➡ The magnification of a terrestrial telescope increases with the reduction power of the eye-piece. or with reducing its focal distance.

general conclusion ➡ Both types of telescopes have similar properties relating their magnification: **to increase the magnification of a telescope it is necessary to increase the focal distance of its objective and reduce that of the eye-piece.**

3. Compound microscope

Teacher. We've already established that a magnifier is employed to see small (not distant!) objects. What other device is being used for the same purpose?

Ruth. A microscope. It allows us to see much smaller objects than a magnifier.

Teacher. Can you guess why?

Michael. Probably, its magnification is greater because it has two lenses, like a telescope.

Teacher. Perhaps. If you decided to make a microscope, how would you select lenses to increase its magnification.

Mary. We could, probably, use the same procedure as for a telescope.

Teacher. Basically, yes, although you may not be able to use lens-holders with very short-focus lenses. Remember that your instrument will work as a microscope only if you **the object is close to its objective**. Bearing this in mind, I want you to do this investigation at home (I'll lend you the lenses) as an alternative to an investigation of a **simple microscope**.

David. Is a simple microscope not the same as a magnifier?

Teacher. No, despite the fact that either consists of a single lens. You will see the difference when you compare their work. Based on your investigation of a compound microscope, I want you to answer the following questions: 1) can you use a telescope made in prior experiments as a microscope? 2) can you use a microscope you have just made as a telescope? and 3) what are the key differences between a microscope and a telescope?

3. Ray diagrams

It was Kepler who gave the first ray diagrams for two-lens terrestrial and astronomical telescopes (Figs. 6.4 & 6.5). However, they are short of some important details, and we will use more modern drawings.

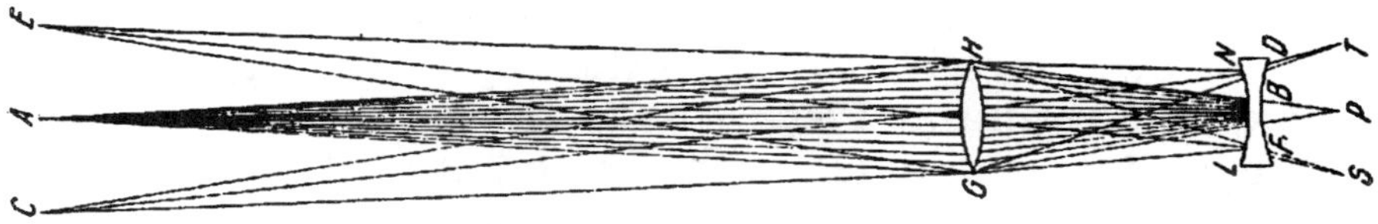

Fig.6.4 Kepler's diagram of a terrestrial telescope. From J. Kepler, *Dioptrice* (1611).

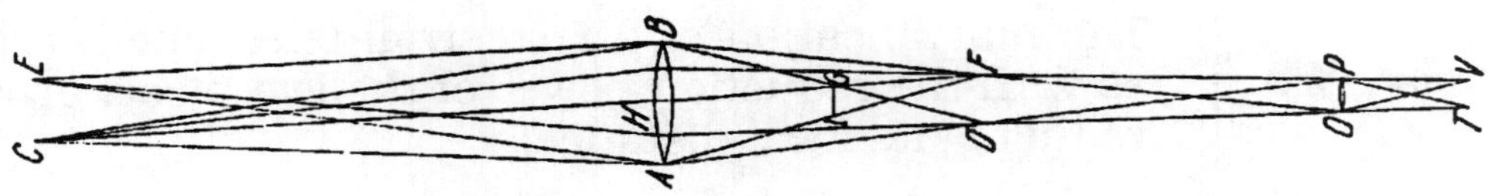

Fig.6.5 Kepler's diagram of an astronomical telescope.
From J. Kepler, *Dioptrice* (1611)

Let **ABC** be an object, **DEF** the objective lens, and **KLM** the eye-piece. Since the object is very far from a telescope (practically, in infinity) the real image **IG** that the objective produces will be in its focal plane. This image will serve as an object for the eye-piece which creates the virtual image **RST** (Fig.6.6a). The foci of the objective and of the eye-piece coincide, and thus the length of the Keplerian or astronomical telescope equals the sum of the focal lengths f_o and f_e of the objective and the eye-piece.

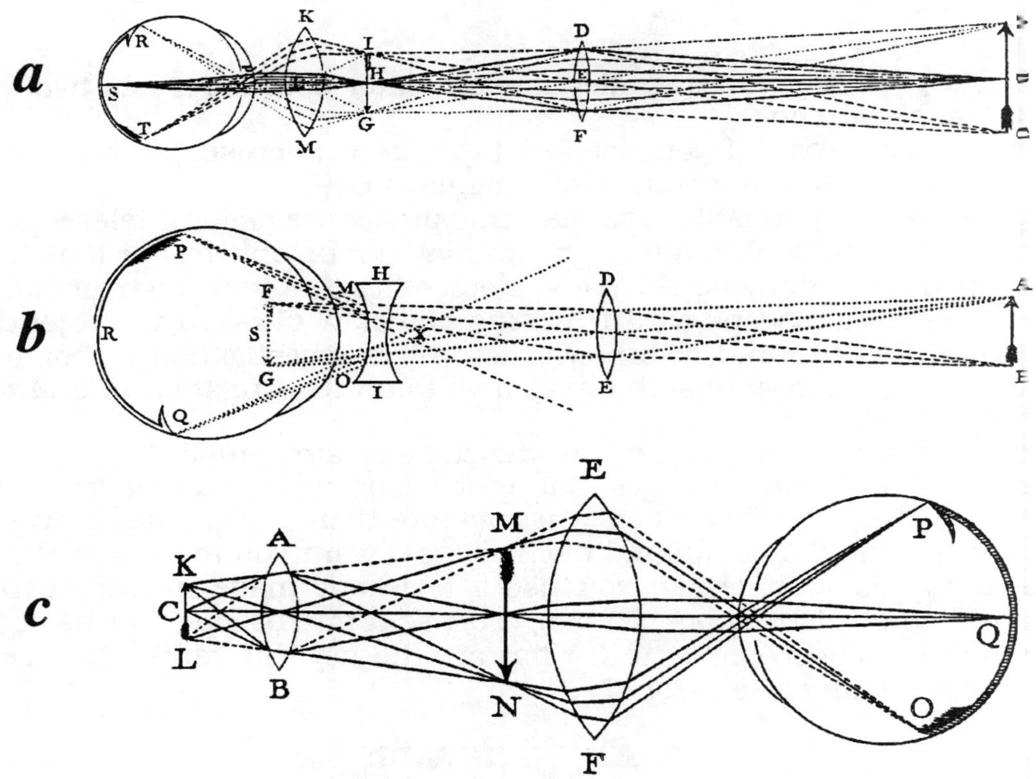

Fig. 6.6. a) Keplerian or astronomical telescope; b) Galilean or terrestrial telescope; c) compound microscope. From J.Priestley, *History of Light* (1772).

In a Galilean or terrestrial telescope (Fig. 6.6b), let **AB** be an object. **DE** the objective, **HI** a diverging eye-piece, **FG** the image, produced by the objective. The eye-piece **HI** diverges the rays forming the image **FG** so that they meet the retina at **PRQ**: this is the final image. Now the back focus of the eye-piece coincides with the focus of the objective; thus the length of the telescope equals the difference of the focal lengths of the objective and of the eye-piece.

In a compound microscope the distance of an object **KL** (Fig. 6.6c) from the objective **AB** exceeds a little its focal distance. As a result, we obtain a real magnified image **MN.** This image must be a little closer to the eye-piece **EF** than its focus. Thus, finally we obtain a larger virtual image **PQO** than it would be with the eye-piece alone.

The main difference between a telescope and a microscope is not so much in the focal distance of the objective lens but in the location of the object relative to it: in a telescope it is far away from the objective, while in a microscope it is very close. As a result, in a telescope the foci of the two lenses coincide, while in a microscope there is a certain distance between them. Thus, in principle, one can use a telescope for microscopical observations, although the magnification will be small. For a telescope, the linear magnification is of no importance (it is a reduction rather than magnification), and we use the angular magnification instead

$$N=\tan\alpha_2/\tan\alpha_1 \approx f_o/f_e \qquad (6.2)$$

where α_2 and α_1 are the angles at which the eye sees the image and the object, respectively.

II. FIRST TELESCOPES and MICROSCOPES

1. Galileo

It seems that some combinations of lenses had been in use by the end of the 16th century in Italy. The earliest known telescope is dated 1590, the maker unknown. By 1608, telescopes reached such a quality that their makers started applying for patents. The first such application by a spectacle-maker from Middelburg was mentioned on September 23, 1608 in the government correspondence of the Dutch Republic. Soon there were three contenders for the patent: Hans Lipperhey of Middleburg, Jacob Metius of Alkmaar, and Sacharias Jansen of Middleburg. The patent was refused on the ground that the device was already sold by other makers, in particular, in Germany. The new invention spread rapidly. By the spring of 1609, Paris instrument makers sold the new *spyglasses*, and by the summer the device had reached Italy. The first telescopes were marketed as an indispensable thing for the Army and the Navy. They were made of a combination of a convex and a concave lenses in a tube and magnified three to four times. The quality of the image was poor. history

The situation began to change after ***Galileo Galilei*** (1564-1642) took the matter in his hands. Galileo was born in Pisa in Tuscany and brought up in Florence. Since 1592, he was the Professor of Mathematics at the University of Padua, in the Venetian Republic. He heard the first rumor about the new invention in May 1609, and he started thinking about a possible design of the telescope and experimenting with lenses. Soon he realized that to increase the magnification he needed an objective lens with a longer focal distance and an eyepiece with a short focal distance. The spectacle makers did not produce such lenses, and Galileo had to learn himself how to grind and polish them. At the end of August 1609 he already possessed a spyglass which magnified nine times. He showed it to the Doge and the Senate of Venice, and they were greatly amazed to be able to see the sea vessels two hours before their arrival at the port. The military benefit of the spyglass appeared obvious, and when Galileo donated this instrument and the rights to manufacture it to the Republic, he received the reward he hoped for: tenure at the University and an increase of his salary. That was what he wanted because his financial situation was rather poor.

Galileo continued to work on improving the spyglass, perhaps he hoped to sell it to another Court. In November 1609, he completed an instrument which magnified 20 times, and he directed it at the Moon. He observed mountains and valleys on the moon and found that the Milky Way consists of great many stars. In January 1610, with a new instrument, which magnified 30 times, Galileo discovered four Jupiter satellites, a few months later he saw Saturn with two appendices (the ring), and discovered dark spots on the sun. In March 1610, Galileo published his first important results, including the discovery of the Jupiter satellites, in a book *Sidereus Nuncius* (Sidereal message). In the hope of winning favors from Cosimo Medici, the Grand Duke of his native Tuscany, he named the satellites "Medician stars" (there were four Medici brothers). Galileo also sent him his book together with the telescope to enable Medici to see for himself the wonders described in the book.

Galileo was not the first to direct a telescope towards the sky, there had been a few previous attempts to observe stars and the moon. Since those instruments had a low magnification, nothing essential was discovered that could not be done without a telescope. It was Galileo, who opened a new world to observers, and he deserves the title of the pioneer of telescopic astronomy. The publication of the *Sidereus Nuncius* made Galileo a celebrity overnight. While it took six months to get the first confirmations of his discoveries (there were no good telescopes readily available), the objections to them came immediately. Since new discoveries contradicted to the Aristotelian doctrine, philosophy professors denied them even before they had a chance to look into a telescope. And many simply refused to look into a telescope, claiming that it deceived the senses. That was understandable: for the first time an artificial device exhibited something inaccessible to the senses, and it was natural to question the reality of the phenomenon. Galileo's telescope (Fig. 6.7) consisted of a convex objective lens and a concave eyepiece. Such a telescope produced a direct image, which was convenient for terrestrial observations, but it was difficult to use because of its very narrow field of view. Two of Galileo's telescopes have been preserved. One magnifies 14 times and is 1.36 m long, its objective is 26 mm in diameter; its tube is made of wood and covered with paper. Another magnifies 20 times, it is of 92 cm long; the diameter

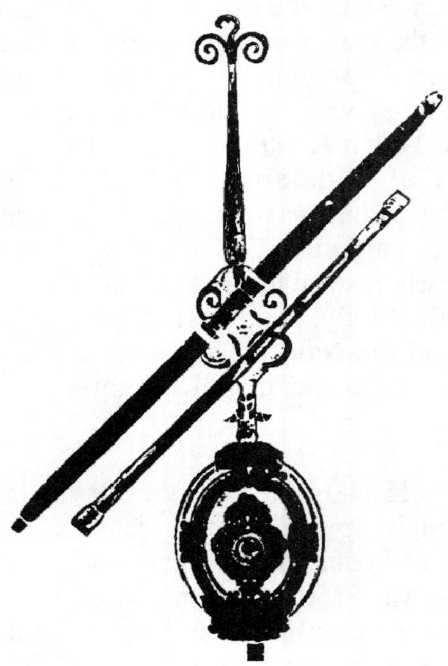

Fig. 6.7. Galileo's telescopes.

history

of its objective is 16 mm; the tube is made of wood and covered with leather.

The first theory of telescopes was created in 1611 by Kepler, and on its basis he suggested another telescope, with two convex lenses. To have the same magnification, it had to be longer than the Galileo telescope; on the other hand, it had a much larger field of view, which was convenient for astronomical purposes. The first Keplerian telescope was built in 1617 by Christoph Scheiner (1575-1650).

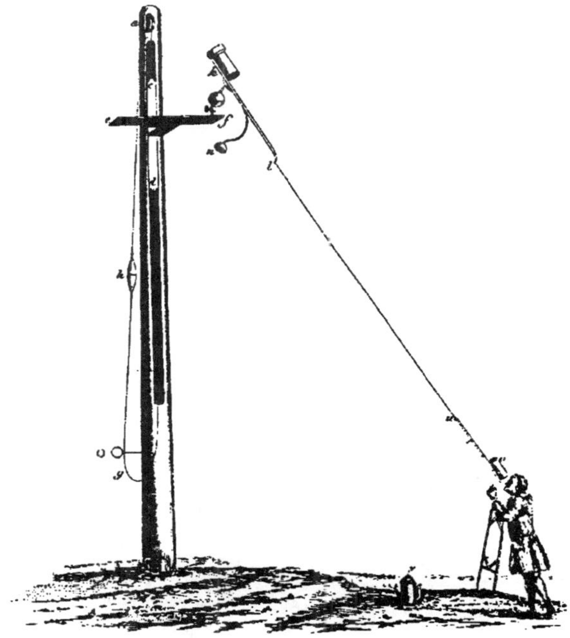

Fig. 6.8. Huygens' telescope.
From J. Priestley, *History of light* (1772)

To reduce spherical aberration the telescope designers tried to use objectives with a small curvature. This considerably increased their focal distance and the total length of a telescope. Johannes Hevelius of Danzig built a 150-foot telescope with a sectional wooden tube. Such long and heavy telescopes were difficult to set for observations and to control. Huygens came up with an ingenious idea of building a telescope without a tube. He placed an objective in a short tube and attached it to the top of a very long pole by means of a ball and a socket (Fig. 6.8). To turn the objective tube and align it with the eye-piece he pulled a silk string.

2. Simple microscope vs a compound one

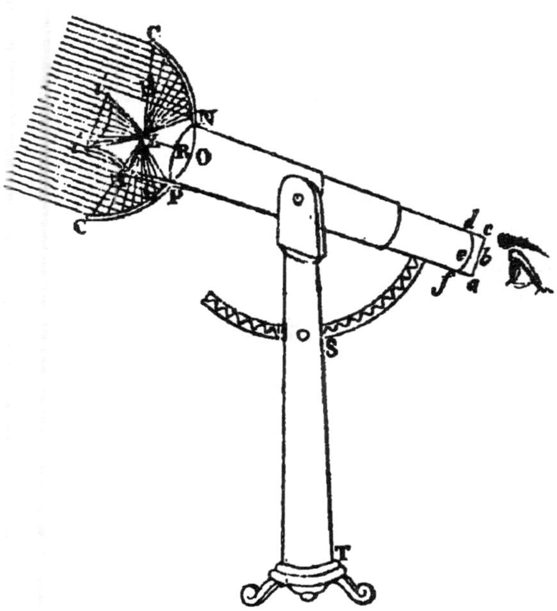

Fig. 6.9. Descartes' microscope.
From Descartes, *Discourse on Method* [1637].

Microscopes appeared at about the same time as telescopes. There is an opinion that the first compound microscope was built by Hans and Sacharias Jansens in 1590 or later, and it was a foot and a half long. In 1610, Galileo applied one of his telescopes (from 2 to 3 feet long) to microscopical observations. Other microscopes of that period were also quite long. For instance, one of the microscopes of Cornelius Drebbell was 6 feet long. Apparently, at least some of his microscopes (1619) had a convex eye-piece. Descartes (1637) described a microscope (he called it *telescope*, but it was designed for viewing very close objects), consisting of a hyperbolic concave eye-piece and a hyperbolic convex objective (Fig.6.9), the hyperbolic shape being selected to reduce spherical aberra-

history

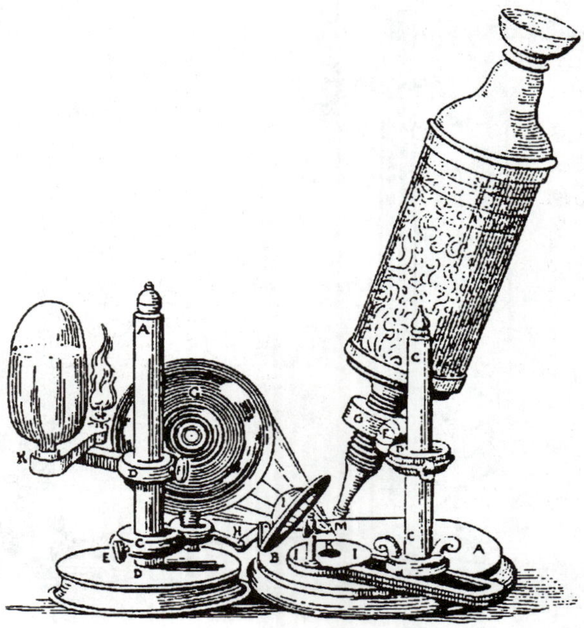

Fig. 6.10. Hooke's compound microscope. From Hooke, *Micrographia* (1665).

tion. A parabolic metal mirror and an additional convex lens directed sunlight towards an object placed in special holder. This project had never been realized. One of the most successful microscopes of the 17th century was the one built by ***Robert Hooke*** (Fig. 6.10), by means of which he collected the vast body of microscopical observations published in his *Micrographia*. His microscope had an objective of a very short focal length provided with a pin-hole diaphragm, which reduced spherical aberration; the eye-piece consisted of a large field-lens (to increase the field of view) and an eye-lens. To provide enough light through a very small diaphragm Hooke used condensors made of a round globe filled with water and a plano-convex lens, the light sources being a window or a lamp.

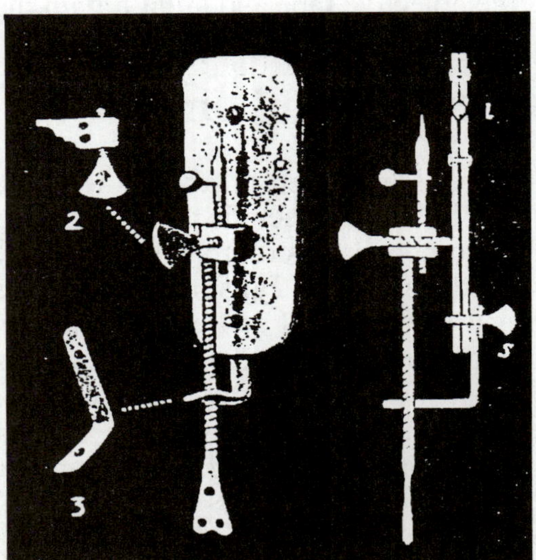

Fig.6.11 Leeuwenhoek's microscope. From C. Dobell, *Antony van Leeuwenhoek and his "little animals,"* (New York, 1932):328.

Although compound microscopes provided a considerable magnification, the images produced were of poor quality. The best results in the 17-century microscopy were achieved with one-lens microscopes. The first good simple microscopes were built in the 1670s by ***Antony van Leeuwenhoek*** (1632-1723). Leeuwenhoek was born in the Dutch town of Delft in the family of a basket maker. At the age of 16 he began to learn linen-draping. In the early 1670s he developed a technique (he never made it public) for grinding very small lenses that magnified up to 270 times. With such a lens he discovered microorganisms (1674) and made other valuable observations the accounts of which he regularly published in the *Philosophical Transactions of The Royal Society of London*. Leeuwenhoek' microscopes consisted of a metal plate with a small hole drilled through, where the lens was installed, and a specimen-holder with an adjusting screw (Fig. 6.11). Photographs obtained with some of his surviving microscopes and specimens by Brian Ford show details of about 1 mμ. ***Nicolas Hartsoecker*** (1656-1725), an astronomer and mathematician, replaced lenses with small globules of glass, made by melting them in the flame of a candle. A globule of about 2 mm in diameter magnified 100 times. His microscope had a condensor, a revolving diaphragm with holes of different size to control the amount of light, another disk with holes for fixing transparent specimens, and a lens. This type of an instrument, later known as the *screw-barrel microscope,* was made popular by an instrument maker James Wilson. In his 1702 microscope(Fig. 6.12) the *barrel* **AAAA** had an object-holder consisting of two thin brass plates **EE** supported by a spring **H.** A specimen was set in the hole in an ivory plate between two thin plates of glass or talc. The *screw* **D** carried at its end a lens set in an ivory frame. Since the lenses were screwed in they were easily interchangeable. To focus, the tube **D** was screwed in and out. At the bottom of the barrel Wilson set an illuminating lens.

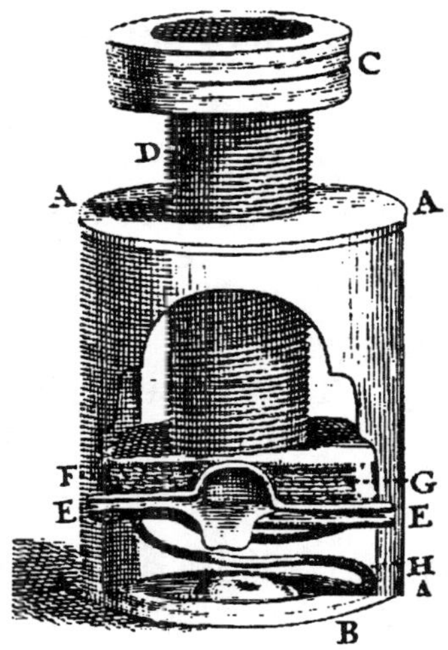

Fig.6.12. Wilson's screw-barrel microscope. From *Phil. Trans.*, 1702.

III. HOME EXPERIMENTS

1. Wine glass telescope

Take two wine glasses or clear bottles and fill them with water. Use one as an eye-piece and the other as an objective lens and try to obtain a distinct image of a removed object. Is the image magnified? Can you increase the magnification with the same pair of glasses?

2. Simple microscope

Teacher. It is much easier to make a very small liquid lens than a solid one. Apparently, the first liquid lens (of honey) was used by Augustin Fresnel in 1815. We will try several liquids.

investigation

Preliminary observations

Teacher. Place several drops of water on a microscopic slide. Look **through** the drops at small objects such as fine newspaper print (stock quotes, for instance), hair, thin wire, threads. Your eye must be close to the glass, and the glass must be close to the object (about focal distance of the liquid lens). To be sure you have not missed the proper position, place the glass on the object and slowly lift it while keeping your eye close to it. To compare the magnification look through different drops while keeping the glass at constant height over the table. What have you found?

Michael. Some drops magnify stronger than others. We've also looked at the print through an ordinary short-focus lens and found the image smaller than with drop-lenses.

Formulating a problem

Dorothy. Why do drop-lenses magnify more than the ordinary ones?

Selecting variables

Teacher. Probably the difference in magnification arises from the difference in properties of these two types of lenses. Which properties could be relevant for our purpose?

John. First of all, the size. We've noticed that smaller drops make larger images.

Mary. They also differ in material and shape. Thus we can start with such variables as 1) size; 2) material; and 3) shape.

Selecting a procedure

Teacher. Take a ruler with a millimeter scale and attach to it (along the divisions) with a scotch tape two wires or threads of different thickness within the range 0.1- 0.4 mm. Place the glass over the ruler so as to see the wire through the drop and the scale through the glass. In this way you will be able to project the image on the scale and find its width. Compare the magnification of different drops when holding the slide at the same distance from the object. Use a toothpick to make small drops.

a. size

Hypothesis

Dorothy. It appears from the general preliminary observations that the smaller the drop the greater the magnification.

Test

Ruth. Place 3 drops of water of different size (between 1 and 5 mm) in a row on the glass; likewise, make 3 drops of honey in another row. Compare the magnification **within each row** separately.

John. Our results confirm the hypothesis.

conclusion ➡ The smaller the drop the greater the magnification, and this is true for more than one substance.

b. material

Preliminary observations

Michael. We compared two small drops of the same size (1-2mm), one of honey and another of water. It was not easy to make equal drops, we checked them with a magnifier with a scale. Honey appears to work better than water.

Hypothesis

Mary. A stronger magnification is the result of greater refraction, thus we may suppose that substances with greater index of refraction produce greater magnification.

Test

Ruth. We've already proved the case for honey and water. Our group compared honey and glycerin. The hypothesis is correct.

John. We compared oil and sugar syrup and confirmed the hypothesis too.

conclusion ➡ | The magnification increases with the index of refraction.

c. shape

Preliminary observations

Teacher. While comparing drops of honey and water of the same size one can notice that the former is much more spherical than the latter, and this difference in shape could interfere with the difference in material. Thus, try to use materials of comparable viscosity, such as oil, water, and alcohol, which helps obtain the same shape. We can consider drops to be of the same shape if they have the same diameter and the same height (look at them from a side).

BIBLIOGRAPHY

Galileo Galilei, *Sidereus Nuncius, or the Sidereal Messenger* (Univ. of Chicago Press, 1989).

R. Descartes, *Discourse*, pp.157-160.

R. Hooke, *Micrographia*, pp. 3-4.

Albert van Helden, "The invention of the telescope," *Trans. Amer. Philos. Soc.* 67, part 4 (1977), 5ff.

R. Clay & T. Court, *History of the Microscope*, pp. 8-89.

B. Ford, *Single Lens*, pp.25-59.

V. Ronchi, *The Nature of Light*, pp. 61-109.

J. Priestley, *History of Light*, pp. 55-81, 119-23, 228-30.

A. Wolf, *A History of Science, Technology, and Philosophy in the 16th & 17th Centuries*, 2nd ed. (New York: Harper, 1959), pp.70-85.

H. King, *History of the Telescope*, pp. 34-65.

Chapter 7

COLORS

I. PRE-NEWTONIAN THEORIES OF COLORS

history and discussion

Teacher's note. Modern optics textbooks deal with colors in various sections, such as dispersion, interference, diffraction, and polarization. It was otherwise in the 17th century when color was a separate branch of optics, and every theory of color aimed at explaining all phenomena of colors.

Teacher. Colors have intrigued scientists since antiquity. The most popular theory was that of Aristotle, which survived until the late 17th century. He believed that colors result from mixing light with darkness (shadow) produced by reflection or absorption: "**when the sight is fairly strong the color changes to red, when it is less strong to green, and when it is weaker still to blue;**"(*Meteorologica*, 261) and "**when what is black and shady is mixed with light the result is red. Purple is gay and bright whenever the rays of the sun are a weak mixture of white and shady.**" [*Minor Works*, 9] Why are you smiling?

Ruth. It sounds weird.

Teacher. Really? How about a later interpretation of Aristotle's idea: colors appear at the boundary of a body and its shadow? Do you like this version better?

John. Not really.

Teacher. Well, let the experiment be the judge.

1. Colored borders

lab

Equipment. A 60° prism and a rectangular piece of white paper at least 1 cm wide.

Teacher. Hold the prism so that its refracting edge is directed towards the left edge of the paper and parallel to it. Look through the prism at the left edge of the paper held against a bright background (the sky or an electric light). Repeat the experiment when the paper is held against a dark background (a table, the floor, etc.). Describe what you see near the left edge.

Mary. Indeed, we saw colored fringes near the paper's edge: in the former case, the color was red and yellow, and in the latter case, it was violet and blue.

Teacher. So Aristotle was right after all?

Michael. Maybe he was. However, he did not mention any prism, did he? If so, why is this experiment relevant?

David. It is relevant to Aristotle's theory, for the colors we've just observed do occur at the boundary of light and shadow, the shadow being cast either by a body or by a background.

history and discussion

Teacher. For many centuries, scientists had debated whether the color of a colored body belongs to light or to the body. In Aristotle's view, in the act of reflection light copies [picks up] a body's color and brings it to the eye: "**when light falls on something, and being tinted by it, becomes reddish or greenish, and then the reflected light falls on another colour, being again mixed by it, it takes on still another mixture of colours.**"[*Minor Works*, 17] Do you agree with this?

Michael. I guess so. A body reflects green light to the eye because it **is** green.
Ruth. Do you mean the body is green even when no light falls on it?

?

Michael. Of course. Do you think a body can change its color after being exposed to light? If we see it green now, it was green five minutes ago, before we decided to look at it.
David. Can you prove that a body was colored before you first saw it?
Michael. I don't know.
Teacher. You have raised an interesting question which we are going to discuss further. And now, why don't we check on Aristotle's hypothesis. How can we do this?

2. Mixing colors according to Aristotle

lab

Equipment. A source of light in a box with a small opening, two sheets of colored paper, for instance, yellow and blue.
Mary. I would do it as follows. The room must be darkened. Set the yellow and blue paper sheets vertically against one another and about 10 cm apart. Set the light so that it shines at the yellow sheet but not at the blue one. Watch the blue sheet where it is reached by the light reflected from the yellow sheet. According to Aristotle, light will pick up first the yellow color and then the blue color. As a result, we must see a mixture of yellow and blue colors, which is green, as everyone knows.
John. We did as Mary suggested and found that a part of the blue sheet appeared to be colored yellowish green.

history *and discussion*

Dorothy. Thus, Aristotle was right, after all. However, I don't like the notion of light "picking up" a color from a body like a shoe picking up mud.
Teacher. Well, ancient scientists believed that we see a body's shape and color because its imprint enters the eye (imagine a moving color photograph!). A modified Aristotelian theory of the seventeenth century divided all colors into *true* colors (of natural bodies) and *apparent* colors (made by a prism, rainbow, soap bubble, and others). In this theory, light is needed only in creating apparent colors, while the true colors could be seen even in the dark. What is your opinion about this theory?
Ruth. Nothing can be seen in the dark. Perhaps they meant a low level of light rather than an absolute darkness. If so, we can check this hypothesis experimentally.

3. Colors in the dark

lab

Mary. Let us compare the colors of bodies seen in daylight and in a dark room.
Ruth. We've found that the colors hadn't changed, but the result is inconclusive because when it is really dark it is difficult to distinguish between say, yellow and white, or red and black.

good point!

David. We could eliminate much of stray light by placing colored objects inside a cardboard box with a small window, perhaps the one used for a camera obscura.

Teacher. Good idea, try it.

Dorothy. We've done it and discovered that, indeed, all objects lose their colors in the dark. As the saying goes, "in the dark, all cats look gray." That means that light is essential in the production of colors.

good point!

4. Mechanical models of colors

Teacher. Right. On this point, scientists came to an agreement. However, they disagreed on **how** an interaction of light and a body makes a color. Some scientists associated the variety of the true colors with the chemical content of bodies or the size and shape of their constituent particles. The so-called "mechanical philosophers," such as Descartes, Robert Boyle (1627-1691), and Robert Hooke (1635-1703), offered several **mechanical** models of the production of colors.

history and discussion

Descartes stated that light is a movement, transmitted from a luminous body to the eye by an extremely rare but highly elastic medium — the ether — much in the same way as water transmits hydrostatic pressure. The motion of particles of the ether is both translational and rotational, and the speed of rotation determines the color. The ether particles change their rotating speed according to the rotation of their neighbors. An obstacle stops one side of the particle and thus affects its rotation. That is why colors are often seen near the edge of a shadow. For instance, a prism **MNP** produces a spectrum **FGH** when light passes through an aperture **DE**, in the screen **NP**, the shadow been created by this screen (Fig. 7.1).

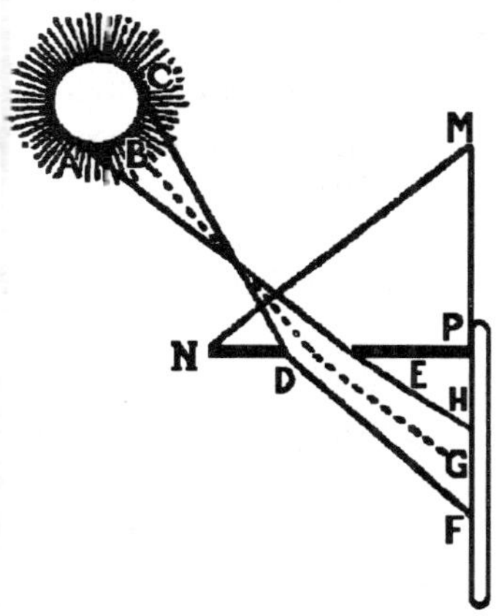

Fig. 7.1. Descartes' idea of dispersion. From R. Descartes, *Discourse on Method* (1637).

Ruth. How about colors of dew? The drops are not covered and produce no shadow.

Teacher. You are a keen observer! Hooke tried to account for the colors produced away from shadows. His explanations were based on the idea that light is an undulatory motion in the ether. In the case of refraction, he argued, the incident light is perceived as white because the *pulse* [wavefront] **CD** (Fig. 7.2) is perpendicular to the ray **CB**, which represents the direction of propagation of light. After refraction, however, the pulse **BA** makes an acute angle with the ray **BO**, and this is the cause of colors. The part **OS** of the *physical ray* [the space swept by the moving pulse] **PQ** which precedes the part **NK** experiences more resistance from the medium and becomes weaker, thereby creating the sensation of a blue color. The part **NK** makes a stronger impact

good point!

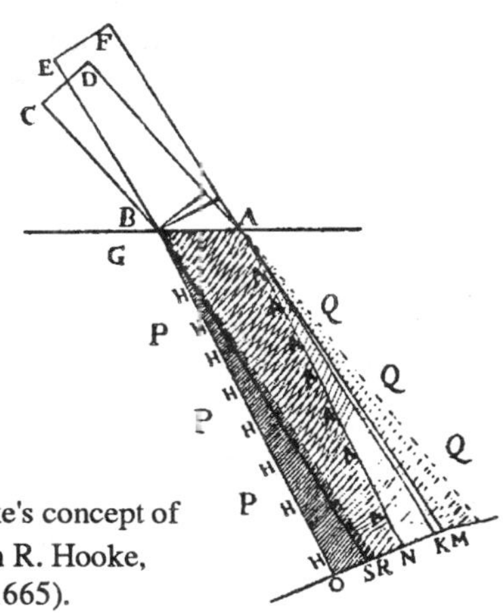

Fig. 7.2. Hooke's concept of dispersion. From R. Hooke, *Micrographia* (1665).

on the retina and produces a red color. Hooke also argued that Descartes was wrong about a plane-parallel plate not producing colors, since he found that a thin plate of mica does just that (*Micrographia*, 47).

Isaac Newton (1642-1727) turned the problem of color upside down: instead of explaining how colors are made of white light he assumed colored light to be the primary concept and explained how white can be made of colors.

II. NEWTON'S THEORY

1. Newton and his theory

history ***and discussion***

Teacher. Newton developed an interest in optics while a student at Cambridge. He received his baccalaureate in 1665 and then obtained an opportunity to concentrate on his new ideas in mathematics and physics: the plague began in England, and many people left cities for the countryside. The years 1665-66 spent in his home village were very fruitful for Newton: he conceived his theory of gravitation, discovered calculus, and developed a new theory of light and colors. In 1669, after becoming the Professor of Geometry at Cambridge, Newton included the new theory in his lecture course of optics. The attendance was poor, however, and the theory received little notice. To attract attention to it, Newton made and sent to the Royal Society of London a reflecting telescope, which, he claimed, was based on his theory. In 1672, the Royal Society approved the telescope and published Newton's paper on the new theory of light and colors. This publication provoked an animated debate which lasted three years and ended in favor of Newton's theory.

In this theory, color is an inherent property of light that does not depend on light being reflected, refracted, or partly intercepted by an obstacle. While all of his predecessors insisted that colored light is made from white light, Newton asserted that white light is composed of different colors.

Mary. How did he prove it?

Teacher. His proof consisted of three parts. First, he showed that white light can be decomposed into a number of rays of many colors, which he called *simple*.

Dorothy. Many colors? When looking through a prism at a body we saw only a few colored fringes at its edges.

Teacher. You did what scientists had done before Newton for centuries, without achieving much. Newton was able to make his discovery because he invented a better experimental procedure. He let sunlight into a darkened room through a hole in a shutter of about one quarter of an inch, placed a prism near the hole (Fig.7.3), and observed the image on a wall

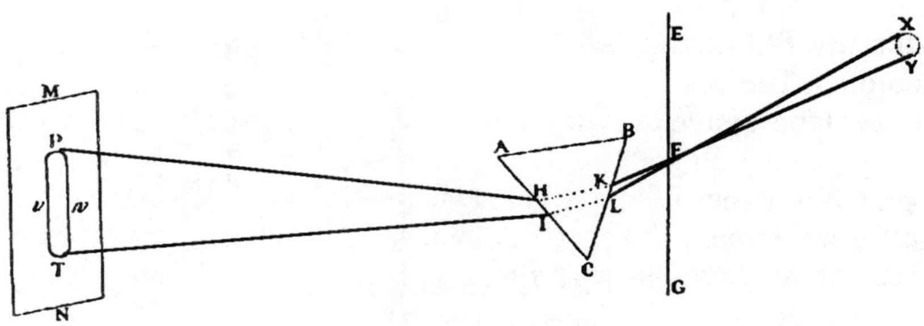

Fig. 7.3. Newton's idea of dispersion. From I. Newton, *Opticks* (1704).

twelve feet away from the prism. The image was rectangular with rounded ends, its length was about four times greater than its width, and it was colored. The colors changed continuously from red on one end (**T**) to violet on the other (**P**), and Newton usually distinguished seven colors: red, orange, yellow, green, blue, indigo, and violet. This image became known as *solar spectrum*.

Ruth. We have already proven that the image must resemble either the sun or the opening. Why does an interposition of a prism change the image?

Teacher. Because it creates colors. Let us see in an experiment whether or not the coloration and the image distortion are connected. We will work with sunlight, as Newton did. Unlike him, however, we will perform some experiments outdoors, for students like to do something outside the classroom.

2. Different refrangibility

a. using sunlight outdoors

Teacher. In all experiments with a prism we should eliminate reflections inside the prism which create false images. To achieve this, cover one side of the prism with a piece of black paper and attach it with a scotch tape to the triangular sides. Set the refracting edge of the prism (the one which is opposite to the blackened side) horizontally and perpendicularly to the direction of sunlight, with its blackened side downward. You can hold the prism in hand or use a floor stand **S** with a prism-holder described further. Cast the image on a white screen placed in a shadow; if necessary, create a shadow yourself. How many colors do you see? ***lab***

Dorothy. I can see red, yellow, green, and violet.

Teacher. Can you improve the image to see more colors?

Mary. We achieved this by turning the prism around a horizontal axis and making the image longer. In addition to the colors mentioned by Dorothy, we were able to see orange and blue.

Teacher. What is the shape of the image? Can you change it by screening a part of the prism?

Michael. We covered the front side of the prism with a piece of an index card carrying rectangular, triangular, and circular holes. They all produced rectangular spectra, with their length being the same as before but of a much smaller width.

Teacher. Do the images change if you hold a white screen very close to the prism?

Ruth. Yes, now they look exactly as the holes, which agrees with the rule we've derived earlier.

Teacher. Move the screen slowly away from the prism. Do the images change?

John. Yes, they do. Closer to the prism we saw a white image with red and violet margins, but at larger distances, the white part disappeared, and the whole image became multicolored.

history and discussion

Teacher. Moving a screen far from the prism was another improvement of Newton. Before him, when shining light through a prism on a screen scientists held the screen close to the prism and obtained the same image as if seen through it. To explain the change of the image at a large distance between the screen and the prism, Newton suggested that solar spectrum consists of many overlapping colored images of the hole. When they coincide completely, a mixture of colors creates the impression of white, if they overlap partially, colors become noticeable at the margins.

Mary. Why do we see more colors when the screen is farther from the prism?

John. What is the reason for multiple images?

Teacher. Newton presumed that light consists of rays of many sorts, which given the same angle of incidence refract at different angles (he called this property of light *different refrangibility*). Each image is made by light of specific refrangibility, and for this reason, the images are shifted about one another. The farther they are from the prism the greater the shift, which is why colors are less mixed at a screen farther removed from the prism. According to Newton, light of each sort has only one degree of refrangibility and always produces in the eye the sensation of the same color. Thus, the refrangibility of light is its **objective** characteristic, it does not depend on an observer being present. Since the nineteenth century, physicists have described this objective feature of light in terms of *wavelength*, but in Newton's time, the wave theory was not sufficiently developed, and he used the *refrangibility* instead. Unlike refrangibility, color is a **subjective** feature of light, for it does not exist outside the eye. In the visible spectrum, violet rays (**P**) have the greatest refrangibility, and red rays (**T**), the least (Fig. 7.3).

Teacher's note. The same discussion can take place when experimenting indoors, the only change is in the equipment used (as shown further, some of the apparatus employed outdoors is usable indoors too).

b. using sunlight indoors

lab

Teacher. Darken the room, at least partly, by screening the windows with sheets of cardboard. Make a hole in the cardboard of about 6 mm. Place the prism near the hole with its blackened side up and its refracting edge directed horizontally and perpendicularly to the beam of sunlight. Rotate the prism until the spectrum appears on a white screen several meters away from the prism. To hold a prism, use a universal stand **S** (Fig. 7.4) which consists of a base made of particle board and a PVC pipe. The floor version is about 1.5 m tall, and the desktop version is about 60 cm. At the top, the pipe is drilled through in a horizontal direction, and a

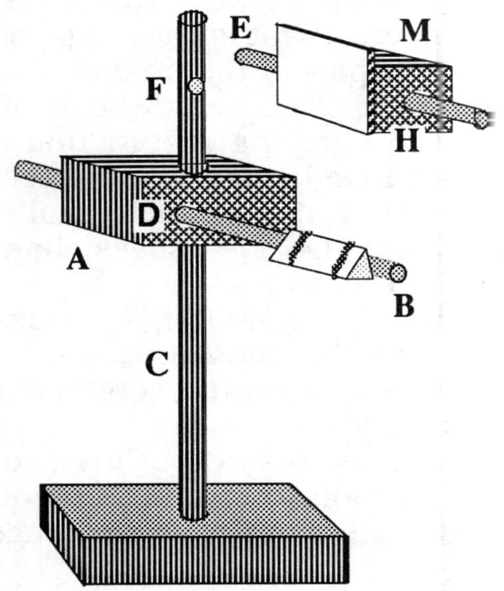

Fig. 7.4. Universal stand S with appliances.

dowel **E** 10 cm long is inserted into the hole **F** so as to be able to turn with some effort. The stand is also provided with a wooden slider **A**, which has two holes **C** and **D** drilled through in mutually perpendicular directions. The hole **D** serves to hold the dowel **B**. A prism is attached by means of two rubber bands or a reusable adhesive putty (for instance, *Plasti-tak*, *Fun-tak*, or "Tak", for short) either to the dowel **B** or to the dowel **E**, depending on whether or not you have to move the prism up and down. The friction in the holes must be such as to retain the slider and the dowel in the given position but to allow a rotation or a translational movement when necessary.

To experiment indoors you may need two floor stands: one to hold a prism and the other to hold a mirror (Fig. 7.5). The latter is necessary for those classrooms which never see direct sunlight. To bring sunlight in and retain its direction despite the sun's motion, you need a facility to turn the mirror around both horizontal and vertical axes. One possibility is to use a photographic tripod, to which the mirror is attached with a masking tape, or an reusable adhesive. Or, you can use the stand **S** with a dowel **E** inserted in the hole **F** at its top to support a mirror-holder **M** (Fig. 7.4). The mirror-holder consists of a rectangular wooden block with a mirror attached to it (the one used to study the law of reflection), which has a hole **H** drilled through along its longer side. The diameter of the hole **H** should be such as to allow the dowel to retain any given position and to turn around its axis when necessary. Darken the room, at least partly, by screening the windows with large cardboard sheets. Make a few holes in the cardboard of 5-10 mm in diameter at the level of about 85 cm from the floor (125 cm if the room is on the second floor). Send one student from each group outside to adjust the mirror following the sun's motion and keep the sunbeam on the designated hole. Tell students to rotate every 5 min between shining light in and performing the experiment. Inside the room, place near the hole the floor stand **S** with a prism-holder. Attach a prism to the prism-holder horizontally using rubber bands or a reusable adhesive ("Tak") so as to direct the beam towards the table. The subjects for investigations will be similar to those used outdoors, to which one may add the size and shape of the hole.

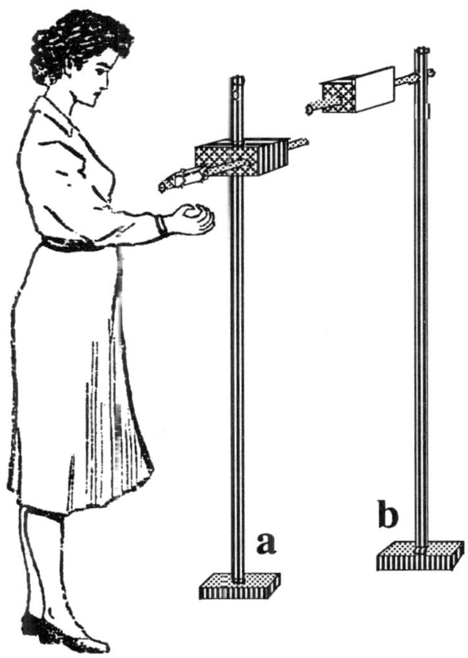

Fig. 7.5. Stands: a) with a sliding prism-holder; b) with a mirror-holder.

c. using an artificial light

Teacher. If sunlight is not available, you can do these experiments with an artificial light. As compared to sunlight, the light beam will be steady, however the colors in the spectrum will be less bright and less separated. You can use for a light source any ceiling/wall lamp-socket with a 40 w bulb surrounded with a cardboard housing, which has a small window (1 cm) at the filament level (the "light house"). Place a converging lens before this window so as to send a more or less parallel beam towards the prism (Fig. 7.6). Place the prism vertically on the platform set in a lens-holder, attaching it to the platform with a reusable adhesive ("Tak'). Adjust the height of the platform so as to bring the prism into the beam. Set a white screen (for instance, an index card 5x8 in.) as far from the prism as the table allows, attaching it to a lens-holder with the "Tak."

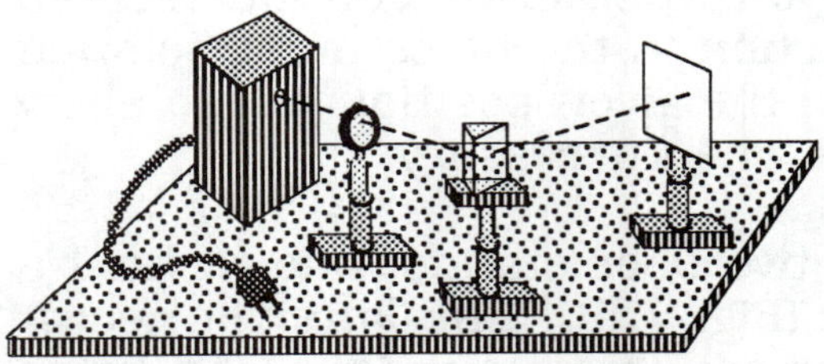

Fig. 7.6.

Michael. Why do we set a prism horizontally when working with sunlight, and vertically, while experimenting with the "light house"?

Teacher. With the "light house," you can make the beam of light horizontal and adjusted to the height of the prism and the lens. However, the beam of sunlight is coming at an angle, and only a horizontal prism can project it on a screen located on the table.

David. However, if we use an outside mirror to direct sunlight inside the room, we can make the beam horizontal and keep the prism vertically, can't we?

Teacher. Yes, if the room is on the first floor. Incidentally, to receive sunlight inside the room you can also use a desk-top version of the stand **S** placed on the table. Move the stand along the table and move the slider with a prism-holder up and down, until the prism will intercept the beam. You can set the prism either horizontally or vertically (attaching it to the slider).

3. Immutability of color

Mary. Thus, according to Newton, the difference between white light and colored light is that white light is compound. How do we know that colored light is not compound?

Teacher. We don't. That was exactly Newton's second point: to **demonstrate** that colors of the spectrum are simple, or indecomposable. Let us repeat his experiment shown in Fig. 7.7. We have to

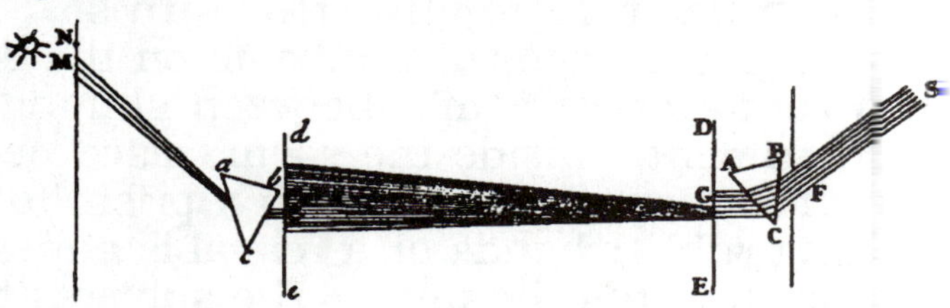

Fig. 7.7. Newton's proof of the immutability of light. From his *Opticks, Bk.1*.

separate light of a specific color from the whole spectrum and check whether it will not change its color after being refracted by another prism. When experimenting outdoors, set the prism $\mathbf{P_1}$ and $\mathbf{P_2}$ on two stands and add a slider **A** to the second stand to be able to move the prism $\mathbf{P_2}$ up and down (Fig.7.8). Intercept a part of the spectrum with the paper shield **S** and compare the colors seen on the paper with those projected on the screen **Q** on the ground. Make a 3 mm wide slit in **S** to select a narrow part of the spectrum. Move the stands so as to project the spectrum created by the first prism on the second one. Adjust the prisms so as to illuminate the slit with light of a single color, for instance, red. What color is the image projected on the screen **Q** by the second prism? Compare it to the color seen on the shield **S**.

lab

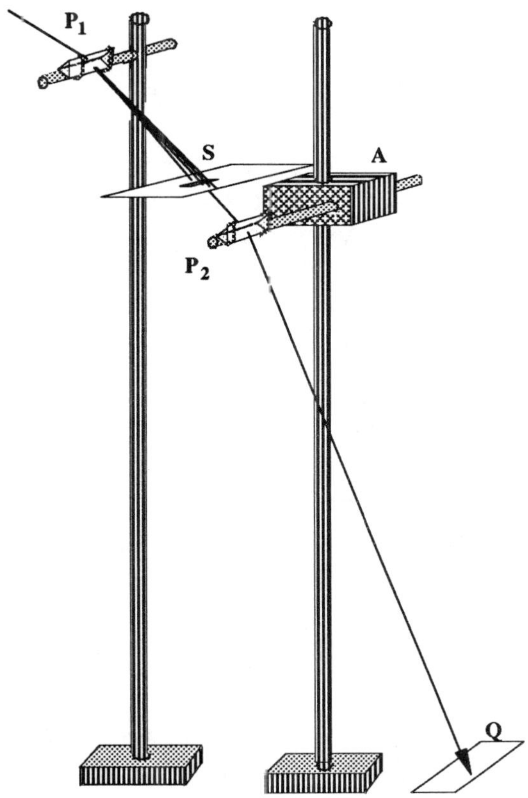

Fig. 7.8. Testing the immutability of light .

Dorothy. The image made by the prism $\mathbf{P_2}$ is a red rectangle. This red color appears to be the same as the one in the spectrum seen on the shield **S**.

John. We saw the image as mostly red but a part of it was orange.

David. We observed the same thing and found that by moving the second prism farther from the first one we were able to remove orange color. The closer the prisms to one another, the narrower each simple color and the greater the possibility that some of the neighboring colors will also pass through the slit. This is difficult to see when looking at the slit, but the second prism increases the separation of colors, and the mixture becomes noticeable. To avoid mixing several colors, the slit should let through only a **central** part of a simple color.

Teacher. Thus, we have found that refraction does not change the color of a simple (or *monochromatic*) light. How about reflection?

Mary. The red image we see on both screens is the result of reflection. However, since we cannot look directly through a prism at the sun, we cannot compare the reflected light to the transmitted one. The only thing we can do is to try multiple reflections. We can, for instance, intercept the light emerging from the second prism with a mirror and reflect it on a white screen.

good point!

Michael. We've just tried this and found that the second reflection didn't change the color.

4. Synthesizing white light

lab

David. Obtaining colors from white light does not necessarily prove that white is a mixture of colors. For instance, you can make carbon dioxide by dissolving soda in water, but this does not mean that soda is an ingredient of carbon dioxide. If white light is a mixture of colors, we have to obtain it from colors.

Teacher. Very good! That was exactly Newton's third point, to the proof of which he devoted a great deal of effort. He offered several demonstrations, but we will repeat only the one which involves a converging lens (Fig. 7.9). Place a prism on a stand and project the spectrum on a white screen laying on the ground. Move the lens between the prism and the screen so as to intercept light and watch the spectrum. What do you see?

Ruth. I saw the spectrum flip: the red and violet exchanged places.

Teacher. Now move the lens slowly to catch the moment of the flip. What do you see?

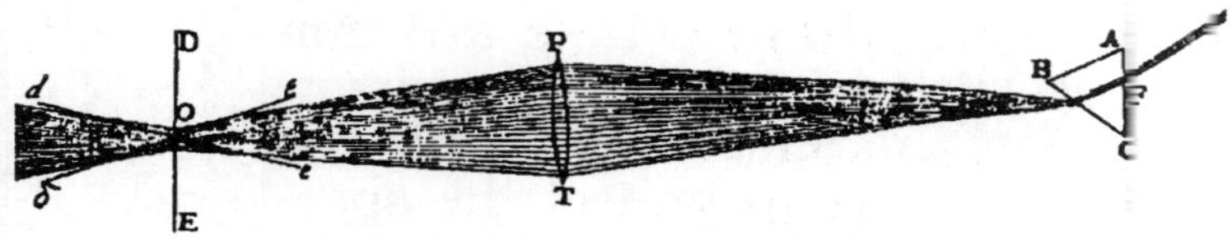

Fig. 7.9. Newton's proof that white light is a combination of spectral colors. From Newton, *Opticks*, Bk.1.

Michael. The image is small, bright, and **white**!

wow!

Mary. I know why it is white! Since the image is small, this is a focal spot, which means that rays of all colors come together at this spot. Thus, we've proved that adding together all spectral colors makes white light.

5. Seeing the world in color

Background

Ruth. We know that reflection doesn't change the color of light, and that we see bodies because they reflect light. If so, why does each body always appear to be of the same specific color (for instance the leaves always appear green) although all of them reflect white light?

Preliminary experiments

investigation

Teacher. Are you sure that your sweater always look blue?

Ruth. Of course!

Teacher. Well, let us do an experiment. Don't worry, we won't use your sweater! Take these sheets of typing paper with strips of colored paper/plastic about 10 cm long glued to them and spaced by about 1 cm. You are going to study the colors of these strips when illuminated by light from different parts of the solar spectrum. In a dark room, set a prism on a stand and project a spectrum on the table. Place the red strip across the spectrum and write down how this strip appears in the red, yellow, green, blue, and violet parts of the spectrum. What are the results?

Dorothy. The red strip looks the brightest in the red part of the spectrum and the darkest in the green and blue parts, it is also visible in the violet region. This means that red paper reflects plenty of red light, some yellow, and a little of violet.

Formulating a problem

David. How does the color of a body relate to the color of light it reflects?

Hypothesis

Mary. Each body reflects light of different colors but to a different extent, so that the light of its own color (that is, its color in daylight) dominates the rest.

Test

Ruth. Let us do the same experiment with yellow, blue, and other strips.

Michael. The yellow strip reflects more light in the yellow-green area (the yellow is very narrow), it also reflects some red-orange light, and very little of violet.

John. A blue strip reflects primarily blue light and also some violet in the red-orange region.

Dorothy. We found that a green paper reflects mostly green. I must also note that white paper reflects well light of all colors.

Teacher. Fine. It may be useful to put your results into a table (Table 7.1), using the scale for an estimated intensity from 0 to 3.

Table 7.1.

spectral colors	colors of paper							
	purple	red	orange	yellow	green	blue	indigo	violet
red orange yellow green blue indigo violet								

general conclusion ➡ **Each body shows the color of light that it reflects the most; a body is white if it reflects well light of all colors.**

Discussion

Teacher. How about the so-called "natural" colors of bodies, or colors seen in daylight?

Michael. If white light consists of all colors, red bodies reflect only the red part of it and absorb the rest, green bodies reflect the green component and absorb others, and so on.

John. We know now that opaque bodies display the color of light they reflect most intensely. How about transparent bodies?

Teacher. Let us conduct an experiment. Here I have a piece of a clear plastic. What color is it?

Michael. Red.

Teacher. I am now placing it on the overhead projector. Look at its image and tell me how the color of this plastic is connected with the color of transmitted light.

Dorothy. Red plastic transmits red light.

Hypothesis

Transparent bodies transmit light of their own color.

Test

Ruth. We repeated this experiment with blue, green, and yellow plastics and confirmed the hypothesis.

general conclusion ➡ **Clear bodies are colored according to the light they transmit.**

?

Mary. Wait a second! A red filter is red because it reflects red light, right? But how can a body simultaneously reflect and transmit the same light?

Teacher. This is an excellent question. In fact, how do you know that a red filter reflects red light?

John. Don't we all see it as red?

David. That does not answer the question: John believes that the light we see is **reflected** from the filter. Instead of presuming this, let us verify it in a direct experiment. I will use this filter as a mirror to reflect sunlight onto a white screen. We see both the transmitted and reflected images: the transmitted image is red, but the reflected one is **white**! I don't believe my eyes!

wow!

Teacher. You will have an opportunity to solve this mystery at home. In the meantime, I'd like to note that some seventeenth-century scientists used this experiment to refute the theory of Aristotle and his followers about light picking up the color off a body.

Mary. Indeed. Well, it looks that we have finally established the truth of Newton's theory of colors.

John. Certainly. Now it seems so obvious. I bet Newton's contemporaries accepted it without objections.

Teacher. Not at all!

6. Response to Newton's theory

history and discussion

Teacher. It is difficult to understand the merits and the flaws of Newton's theory of light and color without knowing how it was received. The general impression was that it was an "**extraordinary hypothesis, which...overturns the very basis of dioptrics.**" (*Newton's Papers,* 87) The objections to it were of several types. First, Newton's attempt to establish the concepts of refrangibility and color as inherent properties of light was considered to be a step back from mechanical philosophy to Aristotelian "innate qualities." Second, the theory was judged to be too complicated compared to other theories of color, since it utilized an infinite number of properties of light (colors). Third, some scientists suspected that the oblong shape of the spectrum could be accounted for without the concept of different refrangibility, by such factors as unequal incidence of light or the presence of clouds near the sun (Pardies, Linus, Gascoin). Fourth, many believed that whiteness could be produced with fewer than seven colors. Hooke and Huygens, for instance, claimed to have used only two. Finally, some considered Newton's theory of color to be no better than those of Descartes, Hooke, Grimaldi, and others who asserted that colors result from modifications of light by bodies, in particular, when light rays are split, diffused, expanded, or condensed.

Michael. Had Newton answered these objections?

Teacher. He had. Perhaps, he was too successful in defending himself. When A. Lucas obtained an unusually short spectrum, Newton accused him of not following his instructions. However, Lucas was right, and Newton's mistake delayed the discovery of achromatic lenses. Where Newton didn't succeed it was with proving that **seven** colors are necessary to produce white. He also had difficulties with demonstrating that his theory was the only correct one. Newton claimed that his theory was independent of any hypotheses on the nature of light and could be reconciled with both the emission (corpuscular) hypothesis and wave hypothesis by attributing the difference in refrangibility to a different size (or mass) of light particles in the former and to the difference in the wavelength in the latter. He believed that other theories of light and color were **speculations**, while his theory was proven by experiment. He explained his scientific method as follows: "**the best and safest method of philosophizing seems to be, first to inquire diligently into the properties of things, and establishing those properties by experiments and then to proceed more slowly to hypotheses for the explanation of them. For hypotheses should be subservient only in explaining the properties of things, but not assumed in determining them; unless so far as they may furnish experiments. For if the possibility of hypotheses is to be the test of the truth and reality of things, I see not how certainty can be obtained in any science; since numerous hypotheses may be devised, which shall seem to overcome new difficulties.**" (*Newton's Papers,* 106)

Hooke pointed out, however, that Newton did base his theory on a hypothesis, because "why there is a necessity, that all those motions, or whatever else it be that makes colours, should be originally in the simple rays of light, I do not yet understand the necessity of, no more than that all those sounds must be in the air of the bellows, which are afterwards heard to issue from the organ-pipes" (*Newton's Papers,* p.111). This question may be paraphrased in the following way: does an apparatus

interesting!

(organ-pipe or a prism) **produce** simple tones from noise ("white noise") and colors from white light, or does it only **separate** those tones and colors that existed before in a **mixture**? Newton could not answer this question and he never attempted to, although Hooke's approach to the problem was no less logical than Newton's. Two centuries after their dispute, physicists came to the conclusion that Newton's view did not suit them any longer, and Hooke's idea came to dominate optics.

Dorothy. Well, if Newton's theory turned out to be false, why do we need it?

Teacher. At his time (and even much later), Newton's approach to colors was better than Hooke's because it was simpler both physically and mathematically. Newton's idea was sufficient for transforming physical optics into a quantitative experimental science. Introducing into optics monochromatic light was one of Newton's most important achievements, because all optical phenomena appeared much simpler when viewed in monochromatic light. Thus, we study Newton's theory because it offers a simpler explanation of many phenomena of colors. In high school physics, it covers all our needs.

important

Mary. What was the cause of Newton's error about the possibility of achromatic lenses?

7. Achromatic lenses

history

Teacher. The reason for Newton's mistake was his unfortunate selection of the law of dispersion. *Dispersion* refers to the difference in refraction for light of any two colors of the spectrum, or, to the difference in the length of different parts of the spectrum. The law of dispersion is linear if the spaces occupied by different colors in the spectrum retain the same ratio for any material of the prism. With some of his experiments conforming to the linear law and others, to the quadratic law, Newton's choice of the linear law was probably affected by theoretical considerations. To preserve one-to-one correspondence between color and degree of refrangibility, the latter had to be a property of light only and not of matter. His favorite analogy between colors and musical tones required the same: as the ratio of the string's lengths producing two particular tones does not depend on the string's material, so the ratio of the spaces occupied in a spectrum by two given colors should be independent on the substance of the prism.

interesting!

Since the linear law of dispersion left no hope for correcting chromatic aberration of lenses, Newton insisted that the future of astronomy is in improving reflecting telescopes, which are free of chromatic aberration. In fact, soon after Newton's death, Chester More Hall constructed the first achromatic lens. However, Hall neither published nor commercially developed his invention, and it remained unknown. Euler challenged Newton by referring to the human eye as an example of an achromatic optical system (later, this analogy turned out to be mistaken). In 1747, he used Newton's dispersion law to show that a glass-water-glass compound lens could be free of chromatic aberration. In 1754, Samuel Klingenstierna further demonstrated that Newton's dispersion law contradicted his own experiments. Finally, ***John Dollond*** (1706-1761),a silk-weaver turned an instrument maker, settled the dispute in 1758 by constructing a system of crown glass and flint glass lenses in which chromatic aberration could be considerably reduced without increasing spherical aberration. This discovery had a great impact on the development of refracting telescopes and microscopes.

8. Rainbow

a. reflection or refraction?

history and discussion

Teacher. One of the successes of Newton's theory was his explanation of the rainbow. What have you noticed about a rainbow?

Dorothy. It appears sometimes after or during a rain when the sun is behind an observer. A sprinkler can also make one.

Teacher. Can reflection or refraction be involved in this phenomenon? What could be the origin of a rainbow?

John. As the case of a sprinkler shows, a rainbow must be associated with water drops. Perhaps reflection is involved because the sun is behind an observer.

Mary. A reflection does not create colors, thus it must be refraction.

Teacher. Well, Aristotle favored reflection.

Ruth. But this is impossible! Did he ever experiment?

Teacher. You already know from the previous discussions that he did. Incidentally, he probably knew more about rainbows than you do. For instance, he was aware that the size of the bow decreases when the sun rises higher in the sky, so that no rainbow could be seen at noon in Greece (and in the Twin Cities too!). He also knew that the moon also produces rainbows, and that a rainbow can be made by sprinkling water. Moreover, his argument for reflection **was** based on an observation: after seeing very small pieces of a mirror producing colors he compared water drops to tiny mirrors.

interesting!

David. He probably mistook refraction for reflection. Beveled mirrors also make colors, but their cause is refraction in a prismatic edge of the mirror. A very small mirror can produce a similar effect at its edge.

Teacher. Very good! That was indeed a misinterpretation of an experimental result, which happens in science quite often. The first correct qualitative description of the rainbow was given by Theodoric of Freiberg (ca.1250-ca.1310) who stated that the primary rainbow is formed by the rays which experience two refractions and one reflection in each water drop, while the secondary rainbow is formed by the rays which are twice refracted and twice reflected. Theodoric's explanation of colors of a rainbow was Aristotelian: the red refracted rays are produced by the incident rays which have the shortest path within a drop and thus are absorbed the least, while the blue refracted rays are produced by the incident rays which have the longest path inside the drop and are absorbed the most. Theodoric's' ideas were not understood and the correct theory of the rainbow came to us from Descartes. In Descartes' diagram (Fig.7.10), **ABCDE** is the path of light responsible for the primary

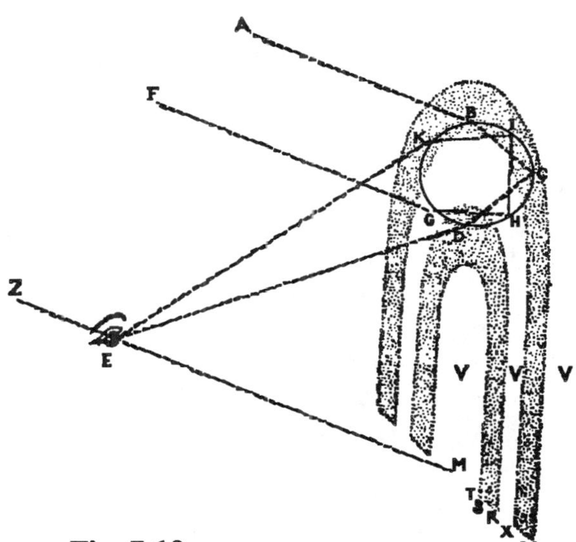

Fig. 7.10. Descartes' explanation of a rainbow. From Descartes, *Discourse on Method* [1637].

rainbow, and **FGHIKE** is the path of light which makes the secondary bow. Descartes found mathematically that the primary rainbow and secondary bows are seen at the angles about 42° and 52°, respectively, between the refracted rays and the direction of sunlight **ZEM** (Fig.7.10) He verified his theory by experiments with a glass globe filled with water. Let us now repeat this experiment. The simplest and the most spectacular way of doing it is by using sunlight outdoors.

b. a rainbow in a single drop

lab

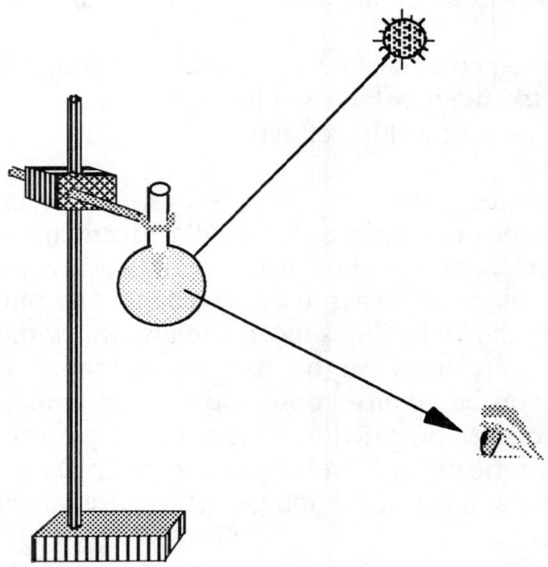

Fig. 7.11 Testing the theory of rainbow.

Teacher. Fill a round-bottom flask with water and hold it before your eyes with the sun behind you (Fig. 7.11). Move the flask slowly up and down and watch its bottom part. At a particular angle, you will see a very bright small spot. By slightly moving your eyes you will see the spot showing different colors. This is a model of an internal bow: the flask corresponds to a single raindrop. In a similar way you will discover a bright spot at the top of the flask, which represents the external bow. Have you noticed a difference in their brightness? Do you know the reason for it?

Mary. The upper spot is darker than the lower one, which agrees with the fact that the secondary bow as weaker than the primary one. The reason is an extra reflection (Fig. 7.10), because at each reflection light loses some of its intensity.

Teacher. If one drop makes only one brilliant point, how is the whole bow formed?

John.. Many water drops participate in this, to be specific, those from which refracted rays enter the eye forming the same angle with the direction to the sun. That means that the bow is a base of a cone (a circle), the apex of which is the eye, and the axis is the direction of incident light. Is that right?

Teacher. Very good! The figures derived by Descartes for the angular size of a rainbow referred to it as a whole. It was Newton who first explained the coloration of both rainbows and derived theoretically the angular size of each colored ring. His explanation was based on a correction of the law of refraction, according to which the index of refraction **n** is constant only for light of a specific refrangibility and differs for light of different colors.

III. EXPERIMENTING WITH COLORS

(student instruction sheets)

1. Sprinkler

home

Equipment. A sprinkler.
Procedure. On a sunny day, turn a sprinkler on and position yourself so as to see a rainbow.

Investigate.

1. Where should the sun be relative to you?
2. What is the sequence of the colors you see (make a drawing)?
3. Does the rainbow change if youchange the pressure of water? Can you suggest a reason for this change?
4. Does the rainbow change if seen from the second floor? from a hill? from the ground?
5. Is it possible to see a rainbow any time of the day?
6. Can you see the secondary rainbow? If yes, show its position relative to the primary bow.

Remember: while changing one variable (position, time, etc.) keep everything else the same.

Conclusion. List the conditions necessary to see a rainbow, which: 1) has several bright clear colors; 2) is a larger part of a circle.

2. Soap bubbles

lab
home

Equipment. A detergent ("Joy," for instance), a yogurt container or a Styrofoam cup, a cocktail straw.
Procedure. Pour a tablespoon of a much diluted detergent into a container and moisten all the walls and the rim. Make a bubble with a cocktail straw, remove the straw, and let the bubble sit on the rim.

Investigate.

1. What is the form of colored fringes and their sequence from the top?
2. Do colors change with time? Watch the same spot on the bubble, for instance, the reflection of a window.
3. Why do colors change? Why do bubbles break? Are the two causes connected?
4. Are reflection or refraction involved in producing these colors?

Conclusion. Can you prove that the thickness of the film is important for making colors in this experiment?

3. Two glass plates

lab
home

Equipment. Two microscopic slides or other glass plates.
Procedure. Tape the plates together at two opposite sides. Hold them against a dark background and look at the upper plate searching for colors. If you don't see any, hold them with fingers of both hands and press either perpendicularly to the surface, or parallel to it until you see colored fringes.

Investigate.

1. Describe the shape and order of the colors.
2. Do the fringes change when you vary the pressure on the plates?
3. Do the fringes change when you look at the plates at a different angle? (Watch the same spot and keep the pressure the same.) Describe the changes.

Conclusion. Can you produce these colors with a single glass plate?

Note. Robert Boyle was the first to study soap bubbles; Hooke (1665) and Newton (1675) examined them and the colors of double glass plates in detail and showed that they are made by thin films. In the nineteenth century, this phenomenon came to be known as *interference of light*. Do you see any proof in your experiments that colors are made by **thin** films?

Teacher's note. These colors are produced by **thin** films only, in this case, by an air film between the plates. The color seen at a particular place depends on the film's thickness at this place. That is why pressure affects the colors.

4. Hair

lab
home

a. experimenting with sunlight

Equipment. Human hair and a wire mounted on a slide frame, floor stand, white screen.
Procedure. Darken the room and let sunlight in through a pinhole. Place a white screen on the floor where you see the sun's image. Tape a human hair to a slide frame and fix the frame to a tall stand. Move the stand so that the hair intersects the beam of light. The hair's shadow will cross the bright circle on the floor. Observe this image.

Investigate.

1. Do you see fringes inside the shadow?
2. Do you see colored fringes outside it? If you don't see them with a naked eye, use a magnifier.

3. How do the fringes change when you move the screen to and from the hair?
4. How does the shadow produced by a wire of 0.5-1.0 mm in diameter differ from that of a hair?

Conclusion What difference between the hair and the wire could be responsible for the differences in their shadows? Can you prove it?

Note. It was the Italian astronomer Father ***Francesco-Maria Grimaldi*** (1618-1663) who discovered these colored fringes. He insisted that it was a new phenomenon which has nothing to do with either reflection or refraction, and he called it *diffraction of light*.

Teacher's note. The external fringes may show some coloration, which is easier to see with a hair. You can see up to three orders of colored fringes with the red outside and blue inside. If using a magnifier, take a semi-transparent material (a ground glass, for instance) for a screen and place the eye **behind** the screen. A wire produces also black and white internal fringes (all fringes run along the length of the diffracting body). A hair shows only one white band in the middle of its shadow. The width of the fringes increases with the distance between the screen and the hair, thus the fringes are visible only at a considerable distance from the body.

b. experimenting with candle light

Equipment. Hair mounted on a slide frame by means of scotch tape, candle, wire.

lab home

Procedure. Darken the room. Place before the eye a slide frame with a hair mounted on it vertically and look at a distant candle flame. Move the slide slowly across the eye until the hair is exactly before the pupil. At this point, you will see the flame surrounded at both sides with vertical colored fringes.

Investigate.

1. Describe the sequence of colors beginning from the flame.
2. Do colored fringes change in size, the interval between them, brightness, etc. if you move the hair to or from the candle)?
3. Does a wire produce the same result as a hair (mount a wire on the same frame as the hair)?
4. How would you check whether this phenomenon is due to refraction or reflection?

Conclusion. What is different in the hair and the wire that could be responsible for the appearance or disappearance of colors? Can you support your hypothesis by experiment?

Note. These colors are also produced by diffraction; they were discovered in 1802 by the English physicist ***Thomas Young*** (1773-1829).

Teacher's note. Each fringe is an image of the light source. There are several orders of fringes located symmetrically about the candle, in each order the outermost color is red. The spacing between the fringes increases with the distance between the eye and the light source, which suggests that each fringe is always seen at a specific angle. A wire does not produce colored fringes, which implies that only very **thin** objects can create such fringes.

BIBLIOGRAPHY

Aristotle VII, Meteorologica, in Loeb Classic Library (Cambridge, MA.: Harvard Univ. Press, 1952), pp. 245-284.
Aristotle XIV, Minor Works, in Loeb Classic Library (Cambridge, MA.: Harvard Univ. Press, 1980).
R. Descartes, *Discourse,* pp. 332-345.
R. Boyle, *Experiments and Considerations Touching Colours* [1664] (facsimile; New York: Dover, 1964), part 3.
R. Hooke, *Micrographia* [1665] (facsimile; New York: Dover, 1961), pp. 65-66.
I. Newton, "New Theory about Light and Colours," *Phil. Trans.*, 6 (1671/2), 3075-87; also in I. Bernard Cohen (ed.), *Isaac Newton's Papers and Letters on Natural Philosophy* (Cambridge, MA., 1978), pp.47-59.
I. Newton, *Opticks* [1704] (reprint of the 1730 edition; New York: Dover Publ., 1952), book I.
J. Priestley, *History of Light*, pp. 48-54, 256-279.
T. Young, *Miscellaneous Works* 2 vols (London, 1855), I: 170-178.
H. Helmholtz, *Treatise on Physiological Optics,* 3 vols. (facsimile; New York: Dover, 1962), II: 120-65.
M. Roberts and E. Thomas, *Newton and the Origin of Colours* (London, 1934).
E. Grant (ed.), *Source Book*, pp. 435-441.
Alan Shapiro, "Newton's 'Achromatic' Dispersion Law," *Archive for History of Exact Sciences 21* (1979), 91-128.

Chapter 8

THE EYE

I. ANATOMY

1. The eyeball

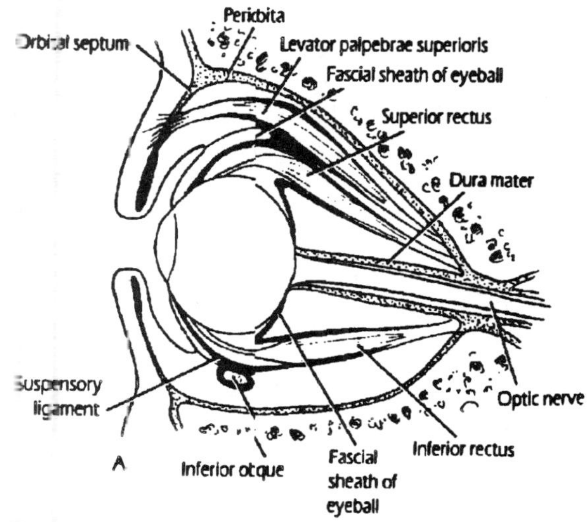

Fig. 8.1 Extraocular muscles. From R.Snell&M.Lemp, *Clinical Anatomy of The Eye* (Boston, 1989), p. 118; reprinted by permission of Blackwell Scientific Publications..

The eyeball consists of two spherical segments: the anterior segment is about one-sixth of the eyeball, its radius of curvature is 8 mm; the posterior segment is opaque, its radius of curvature is 12 mm. The eyeball is enveloped by a thin membrane, the *fascial sheath*, which separates it from the orbital fat. The fascial sheath is pierced by the tendons of six extrinsic muscles: two oblique muscles and four direct muscles (*recti*) These muscles move the eyeball and the sheath together on a bed of orbital fat; they move it upward, downward, medially and laterally (Fig. 8.1).

The globe of the eye consists of transparent media covered by three coats (Fig. 8.2) The outermost coat is fibrous, its posterior part - sclera - occupies five-sixths of the eyeball and is opaque, and its anterior part - cornea - is transparent. The main function of the sclera is to protect the eye. Its anterior part is visible as the "white of the eye," because in the adult the sclera is white. In children, the sclera is thin enough to show through the underlying choroid, which gives it a bluish tint. From the front, the cornea is convex and slightly elliptical, while

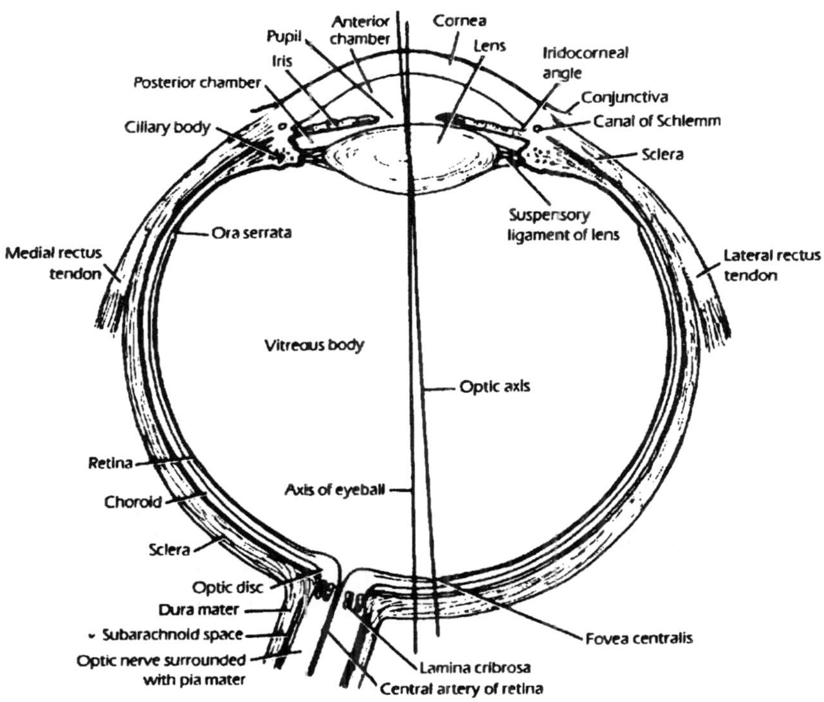

Fig. 8.2. The eyeball. From R.Snell & M.Lemp, *Clinical Anatomy of The Eye* (Boston, 1989), p. 122; reprinted by permission of Blackwell Scientific Publications.

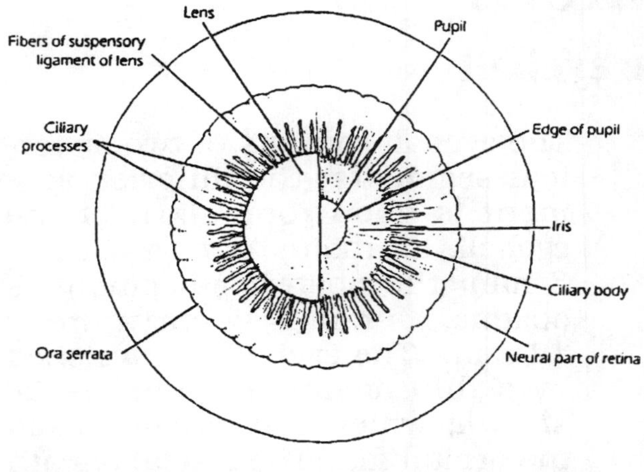

Fig. 8.3. From R.Snell & M.Lemp, *Clinical Anatomy of The Eye* (Boston, 1989), p. 146; reprinted by permission of Blackwell Scientific Publications.

from the back it is concave and circular, with the diameter of 11.7 mm. At the center the cornea is 0.5-0.6 mm thick and its thickness reaches 1.2 mm at the periphery. With its index of refraction 1.38, the cornea is the main refractive part of the eye, and most of the refraction occurs at its anterior surface where the difference in the index of refraction of the cornea and the adjacent medium (air) is the greatest. Sometimes, the cornea is more curved in a vertical than in a horizontal plane, which produces astigmatism.

The middle coat is vascular and pigmented, it consists of the *choroid*, the *ciliary body*, and the *iris* (Fig. 8.3). The main function of the choroid is to nourish with blood the outer layers of the retina, and it also conducts blood vessels to the anterior areas of the eye. The ciliary body extends from the choroid to the iris. It is a ring of about 6 mm wide, triangular in its cross-section. Its anterior part gives rise to the *ciliary processes*, between which the fibers of the *zonule* (suspensory ligament) of the lens pass to attach to the posterior surface (Fig. 8.4). The ciliary body can be divided into: 1) ciliary epithelium, 2) the ciliary stroma, and 3) the ciliary muscle. The ciliary epithelium consists of two layers of pigmented and non-pigmented cells. The ciliary stroma consists of a connective tissue rich in blood vessels and pigmented cells, which extends into the ciliary processes. The ciliary muscle consists of the longitudinal, the radial, and the circular muscle fibers. The contraction of the longitudinal and circular fibers pulls the ciliary body forward during the accommodation.

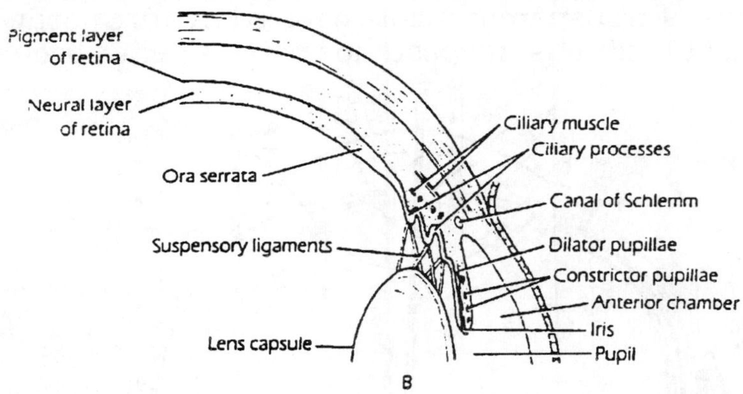

Fig. 8.4 From R.Snell & M.Lemp, *Clinical Anatomy of The Eye* (Boston, 1989), p. 10; reprinted by permission of Blackwell Scientific Publications.

The iris is the most anterior part of the vascular coat. It is a thin pigmented circular disk perforated near its center by a circular aperture called *pupil*. It resembles a cone whose apex is cut off at the pupil. The iris muscle consists of straight fibers *(dilator)* and circular fibers *(sphinc-*

ter) which join near the edge of the pupil (Fig. 8 5). The dilator originates from the posterior membrane. When it contracts it draws the pupillary margin towards its origin and thus dilates the pupil. The sphincter is a flat bar of intertwining plain muscle fibers in the stroma forming a ring around the pupillary margin. When it contracts it reduces the size of the pupil. The pupil varies in size from 1 to 8 mm to control the amount of light entering the eye. Also it contracts to accommodate the eye to different distances. The color of the iris changes with age. When babies are born with blue eyes it is because the dark pigment on the posterior epithelium seen through the transparent stroma (which initially has no pigment of its own) appears blue, in the same way as veins look blue through the skin. With age a number of pigment cells in the anterior limiting layer and the stroma increases, and this changes color: if there are few pigment cells, the eye looks blue or gray; if there are many pigment cells, the eyes become brown.

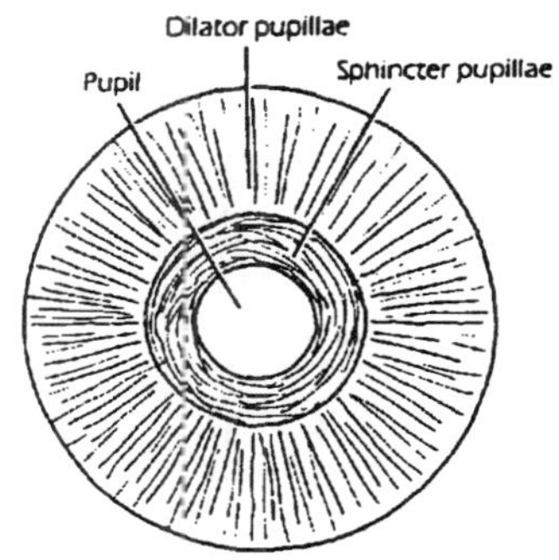

Fig. 8.5 From R.Snell & M.Lemp, *Clinical Anatomy of The Eye* (Boston, 1989), p. 153; reprinted by permission of Blackwell Scientific Publications.

The innermost coat is the *retina*, which contains the nervous elements responsible for vision, the *rods* and *cones*; and it continues forward over the ciliary body as the *ciliary epithelium*.. Most photoreceptors are concentrated in the area called *fovea centralis*.. There are no photoreceptors where the optic nerve exits at the *optical disk*, and thus this place is insensitive to light (the *blind spot*).

2. The lens

The eye is a compound lens which consists of the cornea *and* the *crystalline lens*. The spaces between the cornea and the lens - the *anterior and posterior chambers* - are filled with a clear fluid called *aqueous humour*. The space between the lens and the retina is filled with a jelly called the *vitreous body*. Since the index of refraction of the lens (1.339) is only slightly higher than that of the surrounding media (the aqueous, 1.336; and the vitreous, 1.334), it contributes only 17 diopters of refractive power as compared to 44 diopters of the cornea. However, unlike the cornea, the lens can change its focal distance and adjust the eye for seeing at different distances. Its index of refraction increases from 1.36 on the periphery to 1.40 at the center. The diameter of the lens is 9-10mm and its thickness is 4-5mm (at birth, it is spherical). The anterior surface is a segment of a sphere the radius of which is 9mm, its center is known as the *anterior pole*. The posterior surface is a segment of a sphere with the radius of 5.5mm. The two surfaces join at the equator. The lens contains 66% of water and 33% of proteins. Like the cornea and the vitreous, the lens has no blood vessels: it obtains nutrients from the adjacent aqueous and vitreous fluids. The lens consists of: 1) an

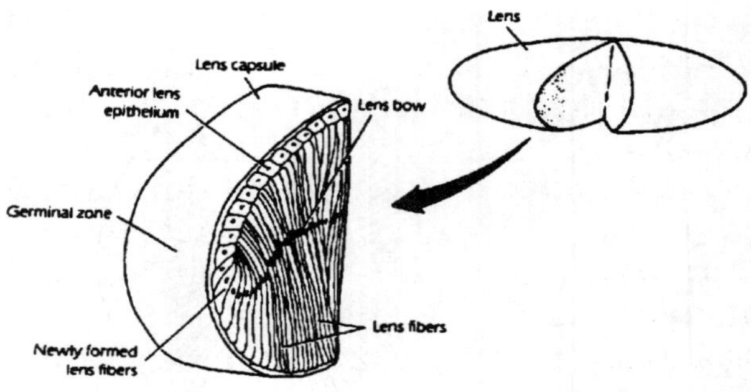

Fig. 8.6. The lens. From R.Snell & M.Lemp, *Clinical Anatomy of The Eye* (Boston, 1989), p. 8; reprinted by permission of Blackwell Scientific Publications.

elastic capsule, 2) the anterior epithelium, 3) the nucleus, and 4) the lens cells or "fibers" (Fig. 8.6). The capsule is a transparent and very elastic envelope. It is thicker in the periphery than in the center, which gives its anterior surface a hyperbolic form during accommodation. The various elements forming the lens are bound together by an amorphous substance. Each lens fiber is a long prismatic six-sided band. The lens grows throughout a person's life, and the newest fibers are laid on externally on the older deeper ones, thus the lens has a laminated structure. Near the equator the lamina are cut transversally while the anterior- posterior section has concentric layers of fibers. In young people the lens is colorless, after the age of 40 it acquires a yellow tint which becomes deeper and deeper with age. It had been suggested that the yellow pigment in the lens absorbs the ultraviolet and prevents the retina from being damaged.

The lens is held in position by a number of delicate, radially arranged fibers known as *suspensory ligament* or *zonule* (Fig.8.4). The zonule fibers arise from the epithelium of the ciliary processes and run toward the equator of the lens. The zonule enables the ciliary muscle to act on the lens. The action of both longitudinal and circular fibers of the ciliary muscle is to slacken the suspensory ligament of the lens. The circular fibers act directly as a sphincter diminishing the circumference of the ring formed by the ciliary body. The longitudinal fibers pull on the epi-choroidal tissue and drag the ciliary body forward.

According to Helmholtz's theory of accommodation, when the eye is in its passive state (focused at infinity), the ciliary muscle relaxes and reaches its maximal diameter. When this muscle is extended it stretches the zonule, which in turn pulls on the lens. This flattens the front and the back of the lens and makes it accommodate for the distant objects. Under these conditions, the lens substance is in its undistorted form and the capsule exerts no influence on it. When the eye accommodates to a close object the ciliary muscle contracts, shrinking the ciliary ring in diameter and also moving the muscle slightly forward. This reduces the tension of the lens suspensions, the zonule decreases its stretch and the elastic capsule presses upon the liquid inside it and shifts it so as to make the lens more spherical.

This theory is still in use, however some modifications have been made. In particular, it was found that the capsule has no liquid inside it. E. Fincham emphasized the role of the capsule's elasticity and of its varying thickness. In particular, he suggested that during accommodation the thicker ring of the anterior capsule, surrounding the central region, contracts; while the thinner central capsule bulges forward in a more pronounced fashion. Weale and Fisher pointed out that the lens substance has elasticity of its own that is mechanically anisotropic. J. Koretz and G. Handelman discovered that during a contraction of the muscle the zonules not only become less taut but also move slightly which reduces the compression in the lens. They built a mathematical model to explain the loss of accommodation with age (from 14 diopters at the age of 10 to 2 diopters at 45 and zero at 70). The model explained why the older the lens the more it is curved to achieve the same degree of accommodation (Brown): the extra curvature is necessary to compensate for the reduction of the index of refraction. It also predicted that in people older than 45 the front of the lens cannot change its shape. Perhaps due to its growth the lens advances so far from the ciliary muscle that it pulls the zonules taut, and they become unable to relax.

The possibility of seeing objects clearly at different distance is not entirely due to accommodation. For instance, when the eye is focused at 1 m (the diameter of the pupil is 4mm), all the objects between 0.97 m and 1.03 m will be seen distinctly. This is called the *depth of field*, and it is caused by the finite size of photoreceptors. To activate a receptor the image may be of a finite size (a blur), which allow the object some freedom of movement to and from the eye relative the place where it would have produced a point image. The depth of field increases with the contraction of the pupil: in the previous example, reducing the pupil to 2 mm would make the objects clearly visible between 0.94 m and 1.06 m.

Some experiments showed that the stimulus to accommodation is provided by the blur: when the circle on the retina reaches a specific size the eye starts to accommodate. Other authors see the stimulus in chromatic aberration: when the object is very close its image on the retina is surrounded by a red fringe, while if the object is very far its image has a blue fringe. If this is the case, it would have been difficult to accommodate in monochromatic light, which conforms to experiment (Fincham).

3. The retina

Traditionally, the retina has been divided into several layers (Fig. 8.7): 1) the pigment epithelium, 2) the photoreceptors: rods and cones, 3) the external limiting membrane, 4) the outer nuclear layer, which contains the cell bodies of the photoreceptors; 5) the outer plexiform layer provides the junction of the first- and second-order neurons: rod spherules and cone pedicles connect with dendrites of the *bipolar* and *horizontal cells*; 6) the inner nuclear layer contains 8 to 12 rows of nuclei belonging to bipolar, horizontal, *amacrine* and *Müller cells*; 7) the inner plexiform layer provides a junction of the second (intermediate) and the third (cerebral) neurons: it is composed of axons of the bipolar and amacrine cells and their synapses with the dendrites of the *ganglion cells*; 8) the ganglion cell layer contains the bodies of the second (cerebral) neuron; 9)

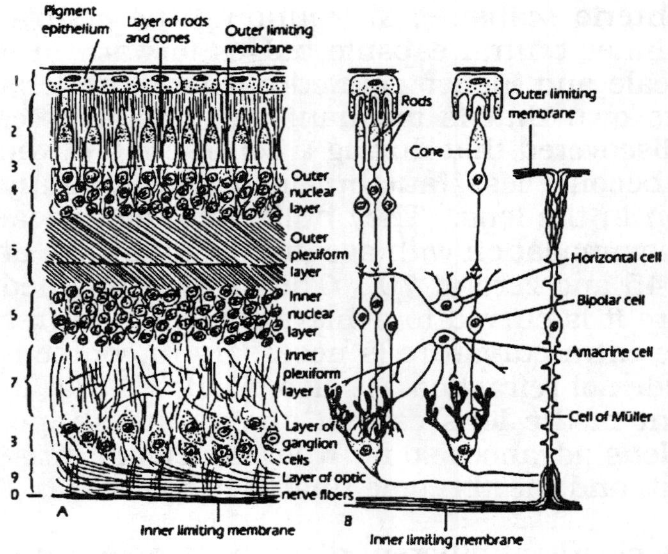

Fig. 8.7 The retina. From R.Snell & M.Lemp, *Clinical Anatomy of The Eye* (Boston, 1989), p. 161; reprinted by permission of Blackwell Scientific Publications.

the nerve fiber layer is composed of the unmedullated axons of the ganglion cells and runs parallel to the retinal surface; and 10) the internal limiting layer is a basement membrane derived largely from the retina and partly from the peripheral vitreous.

Each rod and cone cell consists of the *end organ*, the *cell body* with the nucleus, and the *inner terminal fiber* (Fig. 8.8). The end organ (40-60 mm long) is divided into the outer segment consisting of a stack of membranous disks (2μ in diameter) and the inner segment with outer ellipsoid and inner myoid areas. The disks contain light-sensitive pigments, rhodopsin in rods and possibly iodopsin in cones. The rods are responsible for the vision in dim light and black- and-white images and for registering movement at the periphery of the field of vision. The cones are adapted to bright light and can read the fine details and colored images. The retina contains about 120 million rods and 6.5 million cones. The rods are absent in the fovea where the cones are most dense. The outer segment rests on the pigment epithelium; the other end is called the *synaptic body*, through which the effect of light is transmitted to the bipolar or horizontal cell. In the rods this end is called rod spherule and in the cones, the cone pedicle.

The bipolar cells have a radial orientation. One or more dendrites of a bipolar cell pass outward to synapse with the cone or rod terminals. The single axon is directed inward to synapse with ganglion cells and amacrine cells. There are several types of bipolar cells. Rod bipolar cells connect several rods to from one to four ganglion cells. Diffuse bipolar cells connect many cones with many ganglion cells. Midget bipolar cells connect a single cone with a single midget ganglion cell.
The ganglion cells are the second neurons in the visual pathway. Most of them are small (midgets) and multipolar, their dendrites synapsing with the axons of bipolar

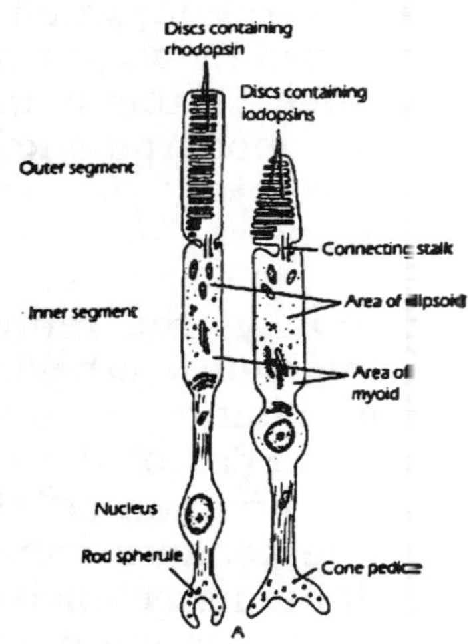

Fig. 8.8 The photoreceptors. From R.Snell & M.Lemp, *Clinical Anatomy of The Eye* (Boston, 1989), p. 163; reprinted by permission of Blackwell Scientific Publications.

cells and amacrine cells. Horizontal cells are situated close to the terminal expansions of the rods and cones. They are multipolar too and have one long and several short processes which run horizontally parallel to the retinal surface. Some horizontal cells synapse with seven cone pedicles, others synapse with 10 to 12 rod spherules. The horizontal cells respond to the neurotransmitter released when the rods and cones are excited by light. Apparently, they liberate an inhibitory transmitter which stops the activity of the bipolar cells some distance away. Amacrine cells are stimulated by the bipolar cells, and they in turn excite the ganglion cells.

II. THE GREAT DEBATE

The investigation of the eye has a long history, and here we will discuss only some important findings of the 17th and 18th centuries, which provide interesting examples of interactions between physics and biology and also between experiment and theory.

1. The seat of vision

Kepler's idea that the retina is the true seat of vision had a long way to go before it was generally accepted. A Jesuit mathematician and astronomer ***Christoph Scheiner*** (1575-1650), discovered in 1625 that after removing the coat from the back of the eyes of sheep and oxen, he could see on the retina sharp images of the bodies in front of the eye. He also found that the index of refraction of the aqueous humour was close to that of water, that the crystalline's refraction was close to that of glass, and that refraction of the vitreous humour was in between the two.

history

Edme Mariotte (1620-1684), member of Paris Academy of Sciences, known mostly as a co-author of the Boyle-Mariotte law, was curious to know why the optical nerve enters the eye not exactly opposite the pupil. In 1668, he conducted the following experiment. He attached to a dark wall two small pieces of white paper, one directly against his eye and another about two feet to the right and somewhat lower than the first one. With his left eye shut and his right eye fixed on the first paper, he backed away from the wall and noticed that when he was ten feet from it, the second paper disappeared. Thus, he found that the retina has a blind spot. The discovery of the blind spot revived the controversy about the seat of vision. According to Mariotte, it was impossible that a part of the seat of vision could be insensible to light. Besides, he said, since the retina is transparent, it is made to **transmit** light farther rather than to stop it. In Mariotte's view, the seat of vision should be in the choroid because it is black and sensitive to light and since the iris — its continuation — contracts or extends the pupil according to the brightness of objects.

?

Jean Pecquet (ca.1600-1674), a physician and a member of Paris Academy of Sciences, objected to Mariotte that the retina is not fully transparent, and thus light coming through it cannot produce clear pictures on the choroid. He suggested that the loss of vision in the optical disk resulted from an abundance of blood vessels in the area. Moreover, he said, perhaps the loss of image takes place not in the retina but behind it, in the choroid, and it is due to an interception of light by blood vessels. Mariotte replied that the transparency of the retina was proven, for the eye of a dog, which has a white choroid, reflected a clear image of a candle flame. Moreover, he said, it is impossible to explain how the retina commands the pupil to contract when too much light enters the eye. On the other hand, if the choroid is sensitive to light, it can contract its fibers extending to the iris and reduce the pupil's diameter.

Claude-Nicolas Lecat (1700-1762), a physician from Rouen, supported Mariotte because he found that old people who do not see distinctly, have a less black choroid than the young ones. ***Jean Mery*** (1645-1722), a Paris physician and a member of Paris Academy of Sciences, also sided with Mariotte. In his view, the choroid reflects light back to the retina where it affects the nervous fibers. In those places where the choroid does not reflect because it is pierced, the retina has no sensitivity (the blind spot). Also, when a cat's head is immersed in water, and sunlight comes into the eye through a widely open pupil, the choroid is clearly seen while the retina is invisible. But if the retina does not stop enough light, Mery said, it cannot be the chief seat of vision. Astronomer ***John Michell*** (ca. 1724-1793) argued that if the images were formed either on the anterior or the posterior surface of the retina, they must be confused, since each point of the retina can be reached not only by direct rays but also by the rays reflected from the choroid. For this reason, Michell concluded that the seat of vision must be in the choroid .

interesting!

Thus, the facts then known (the blind spot, the transparency of the retina, etc.) were insufficient to resolve this controversy, and the necessary direct anatomical evidence about the retina was not yet available. The debate ended only when the improvement of microscopes in the 19th century made possible the discovery of rods and cones.

2. Accommodation

a. early theories

history

The first theories of accommodation attributed its cause to changing the shape of either the eye globe or the lens. Kepler supposed that the ciliary processes could change the distance between the crystalline and the retina. Descartes was the first to involve the crystalline lens. In his view, the ciliary processes, being the tendons of the lens, could make the latter to contract and increase its curvature. Musschenbroek speculated that a part of the change of the curvature could be due to a contraction of its capsule.

interesting!

?

Other scientists had been seeking the cause of the accommodation in the change of the pupil's size. ***Philippe de Lahire*** (1640-1718), a Paris architect and astronomer, member of Paris Academy of Sciences, denied the existence of any accommodation of the eye. To prove this, he made two holes in a card the distance between which was smaller than the pupil's diameter. He shut his left eye and looked through the holes with his right eye at a small black spot on a white background. Lahire found that by properly adjusting his eye he could see two images instead of one. He explained this result with the following diagram (Fig. 8.9). The beams of light **OH** and **Oh** passing through the holes **H** and **h** are refracted in the eye and intersect at the point **o**. If the eye is accommodated to **O**, then the intersection takes place on the retina **RR**, and **o** will be the single image of the object **O**. If the eye is not accommodated to **O**, and the image is received not on the retina but at **AB** or **CD**, one will see two images **xx** or **cc**. Lahire argued that if he saw an object clearly at a distance of 1 foot, then, after moving 4 feet away from it, he had to readjust his eye. However, when looking at the object through two holes he saw two images,

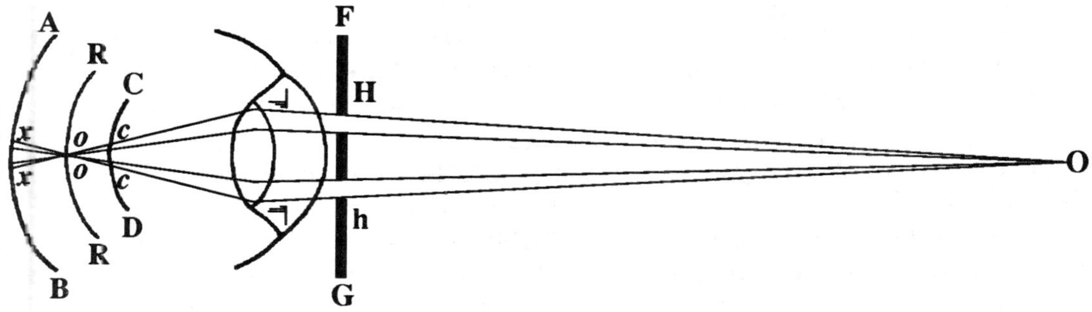

Fig. 8.9. From W. Porterfield, *Treatise on the Eye* (1759). Courtesy of Wangensteen Collection, University of Minnesota

which means that accommodation did not occur. He also pointed out that the only known change in the eye is the change of its pupil, and that it had been observed that when the eye is focused on a nearby object the pupil contracts, and when the object moves away the pupil dilates. Thus, Lahire concluded, the true organ of accommodation is the pupil.

Charles Le Roy (1726-1779), professor of medicine at the University of Montpelier, sided with Lahire. According to Le Roy, the lens could not be moved by means of the ciliary processes: neither the strength nor the location of these fibers (he had doubts even regarding their muscular nature) were adequate for this purpose. In 1755, Le Roy compared the eye to the camera obscura in which reducing the aperture increases the range of distances at which the objects are seen clearly. He also found that if an object was too close to the eye to be seen distinctly, its sharpness could be improved by looking at it through a small hole in a card. According to Le Roy, these facts supported the pupil hypothesis. In particular, the hole in the card played the role of a very small artificial pupil, which extended the range of accommodation. Joseph Priestley noted in 1772 that actually the small hole improved the image by reducing the eye's spherical aberration, because it stopped all the rays except for the paraxial ones.

good point!

b. Porterfield

In 1759, ***William Porterfield*** (d. 1771), a physician in Edinburgh, offered a direct experimental proof that the eye adjusts itself to a clear vision for each position of an object within some range. He used the idea first expressed in 1619 by Christoph Scheiner. Scheiner made two small holes in a card so as to have the distance between them less than the pupil's diameter. He placed the card near one eye and closed the other. A pin held about 30 cm from the card appeared double when the eye was focused at a distant object; as soon as the focus was shifted to the pin, the pin appeared single, while the distant object doubled.

history

Porterfield bases his reasoning on the following facts which he calls "axioms".

Axiom 1. *When an object, seen with both eyes, appears double, by reason that its distance is less than that to which the eyes are directed, upon covering either of the eyes, the appearance that is on the contrary side will vanish; and if it appears double, because its distance is greater than that to which the eyes are directed, upon covering either of the eyes, the appearance that is on the same side will vanish.* (Porterfield, 408)

Axiom 2. ***When an object appears double, from its being seen with one eye through two small holes made in a card..., if its distance be greater than that to which the eye is accommodated, upon covering either of the holes, the appearance that is on the same side will be made to vanish; and if its distance be less than that to which the eye is accommodated, upon covering either of the holes, the appearance that is on the contrary side will be made to vanish.*** (ibid., 410)

The truth of the second axiom can be seen from Fig. 8.9. Let the eye be focused on **O**, its image **o** being on the retina **RR**. If the object is farther from the eye than **O**, its image will be before the retina **RR**, that is in the position **CD**. In this case, if we close the left hole, the left image disappears, in agreement with **Axiom 2**. If the object is nearer than **O**, its image moves beyond the retina **RR**, in the position **AB**. If we now close the left hole, the right image disappears, again confirming **Axiom 2**.

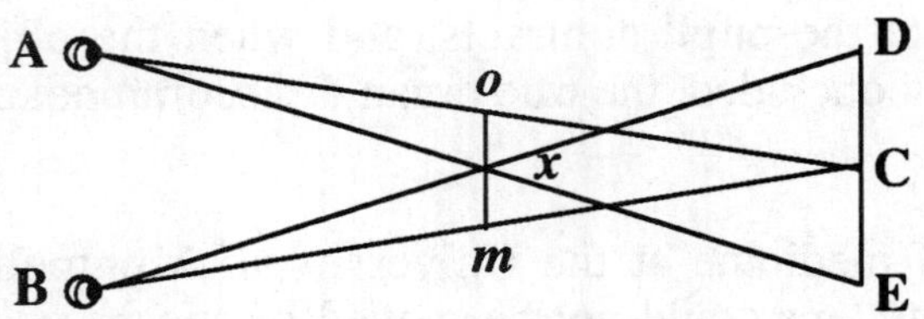

Fig. 8.10. From W. Porterfield, *Treatise on the Eye* (1759). Courtesy of Wangensteen Collection, University of Minnesota.

Next, Porterfield proves **Axiom 1**. Let **A** and **B** be the eyes (Fig. 8.10) and **x** the object that is closer to the eyes than the point **C**, at which the eyes are focused. The right eye **B** will see this object at **D**, and the left eye **A** will see it at **E**. Thus, if we cover the left eye, the image **E** , which is on the contrary side, will vanish. In a similar way, let **C** be the object and x the point to which the eyes are directed (**C** is farther away than **x**). Then the left eye will see the image **O** and the right eye will see the image **m**. If we now cover the left eye the image to vanish will be on the same side. It is easily to see all of this experimentally.

important

Porterfield applies these "axioms" to determine at which point his eye is actually focused, and also whether he is able to refocus it at different distances. He took a small metal plate **IK** (Fig. 8.11) provided with two narrow parallel slits (a slit is better than a hole because it transmits more light), the distance between which was less than the pupil's diameter. While holding the slits vertically near his right eye **B** (his left eye being shut) he looked at a small bright vertical object **O** (a narrow slit which covered a lantern). This object was kept at such a distance as to appear single. When he opened both eyes and directed them to a distant point **P,** he observed three images **a**, **b**, and **C**. When continuing to look at **P** he covered his left eye **A**, the image **a** disappeared, but when he covered the right eye **B**, the images **b** and **C** disappeared. From this he deduced

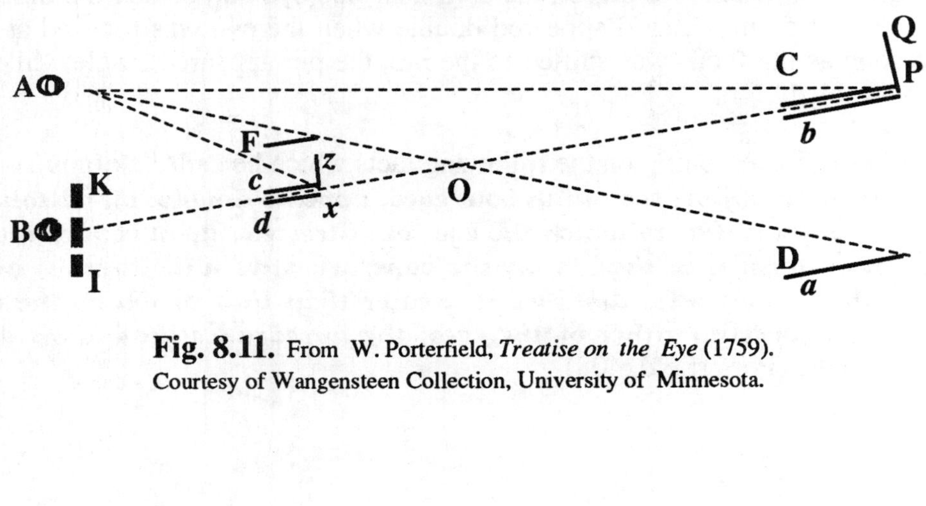

Fig. 8.11. From W. Porterfield, *Treatise on the Eye* (1759). Courtesy of Wangensteen Collection, University of Minnesota.

(Axiom 1) that the focus was closer to the eye than **P**. He confirmed this by covering one of the slits: of the two images **b** and **C** the one that disappeared was on the opposite side (Axiom 2). Then he turned his eyes to the point **x** and again observed three images **d**, **e**, and **F**, but this time **F** which belonged to the left eye **A** was on the left side, which meant that **x** was closer than **O** (Axiom 1). He obtained the same result by covering one of the two slits (Axiom 2). When the eye was focused on **P**, the object **O** appeared double, and the distance between the two images decreased when **P** moved towards **O**. If the points **P** and **O** coincided, only a single image was seen. And finally, when the point of accommodation **P** was closer than **O** the object again appeared double, and the distance between the images increased when **P** moved towards **x**.

Searching for the explanation of accommodation, Porterfield critically reviews various theories involving a change of the eye's shape. Keil, for instance, supposed that since the aqueous humour easily changes its shape, either ligamentum ciliare or the oblique muscles squeeze the middle of the eye globe and make it oblong, which helps to see at a short distance; meanwhile Boerhaave attributed the same role to the four straight muscles. In Porterfield's view, the oblique muscles do not embrace the globe as a ring and thus cannot perform such an action; as to the straight muscles, their action would draw the eye inward and increase the inner pressure which will expand the eye sideways and thus keep the initial shape. He also notes that compressing the eye would make the objects distinct only for particular distances or directions, and it is not applicable for fish and birds with bony eyes.

Porterfield is certain that the lens must be somehow involved in accommodation because he knows that the eye loses its ability to accommodate after couching the cataract, when the lens is removed from the eye. However, he offers several objections to changing the lens' curvature. First, he says, it is not necessary to change the focal distance of the lens because a camera obscura provided with a lens makes equally sharp images of the objects at very different distances from it. Second, the ciliary processes, which are supposed to extend the lens, act in a direction different from the natural transverse axis of the eye. Third, the lens is softer outside and much harder at the center, but this softness serves not to facilitate the squeezing of the lens but to reduce its spherical aberration. Fourth, the lens cannot be contracted by its internal muscles, because the lens fibers intertwined in a way that does not help changing the figure of the lens, and it remains to be proven that these fibers are muscular. Fifth, the lens is not attached to any other parts of the body, and is kept in place by its capsule, with which it has no firm connection either.

good point!

?

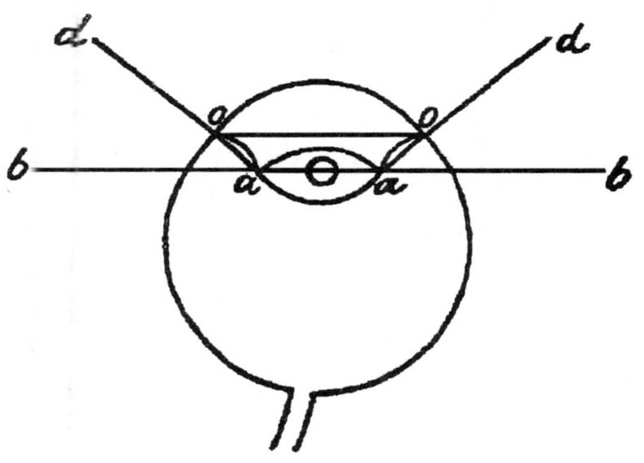

Fig. 8.12. From W. Porterfield, *Treatise on the Eye* (1759). Courtesy of Wangensteen Collection, University of Minnesota.

Like Kepler, Porterfield believes that accommodation is due to the change of the distance between the lens and the retina, and the ciliary processes are well adapted for moving the lens back and forth (Fig. 8.12). Besides, he notes, when the lens is brought forward the cornea itself becomes more convex, which provides an additional refraction. To scientists denying the muscular nature of the ciliary processes on the basis that their fibers are not red, Porterfield points out that muscles of the intestines, stomach, and the pupil are not red either. Some scientists calculated, however, that a small shift of the crystalline would be insufficient to explain a considerable difference in refraction of light from close and distant objects. For this reason, Jurin offered another solution: a contraction of ciliary muscle draws a part of the surface of the anterior chamber a little forward and outward,which produces a redistribution of the aqueous liquid, and the whole anterior surface becomes flatter. This, according to Jurin's calculations, provided for a distinct vision from 15 inches to 14 feet without any movement of the crystalline. In Porterfield's judgement, all these calculations were inconclusive, because the data about the dimensions and refractive power of different parts of the eye were inexact.

Thus, Porterfield proved that the eye accommodates to different distances. On the basis of his experiments, he developed an instrument, called *optometer*, for measuring the distances of the near and far points of distinct vision, This instrument became an important factor for Thomas Young's involvement in the debate.

c. Thomas Young

history

In 1793, Thomas Young, who was then only 20 years old, once again revived Descartes' hypothesis and gave it a new experimental support. When examining through a magnifier the crystalline of an ox, Young discovered that each coat of the lens consisted of six muscles attached to six tendons meeting at the axis of the crystalline (Fig. 8.13). In Young's view, these muscles were perfectly arranged to contract and make the lens spherical. He calculated a possible change in the focal distance of the eye due to such a contraction and found it to be reasonable. In 1793, Young published these findings in the *Philosophical Transactions of the Royal Society of London*, and next year he was elected a member of this body.

At the same time ***John Hunter*** (1728-1793), a famous London surgeon and anatomist, found the periphery of the lens of a fish laminated and interpreted the fibers as muscular. For a long time, he had had an idea that the lens accommodates by the action of its own muscles, and there was an apparent proof of his hypothesis. However, Hunter died before publishing his discovery, and it was presented to the Royal Society by his assistant and brother-in-law ***Everard Home*** (1756-1832) who succeeded Hunter as a surgeon. For a short time, rumors circulated in London that Young's work was not original because he somehow learned about Hunter's idea. However, very soon Home, the champion of Hunter's priority, decided that the discovery was unworthy of defense. On the basis of his experiments, he suggested (1794) that accommodation is produced by four straight muscles of the eye which compress lateral and posterior parts of the eye and by forcing the aqueous humour forwards against the center of the cornea, shorten the radius of cur-

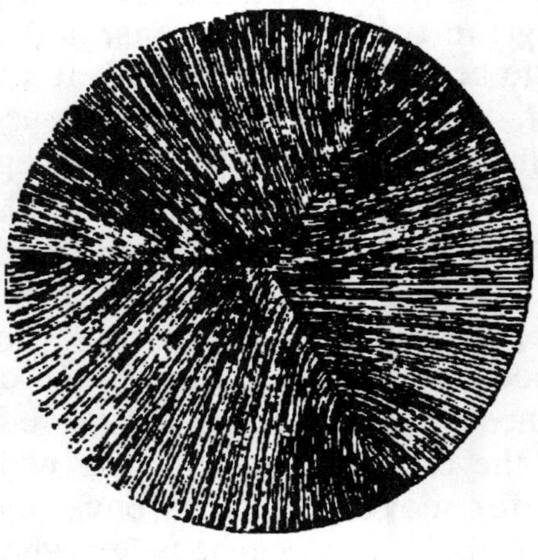

Fig. 8.13. From T. Young,"Observations on Vision," *Phil. Trans.*, 1793.

vature of the cornea and increase its distance from the retina. As to the lens' laminae, following the suggestion of ***Jesse Ramsden*** (1735-1800), the famous instrument maker, Home changed their role from deforming the lens to reducing reflection and correcting spherical aberration. To prove that the lens plays no role in accommodation, Home and Ramsden measured the depth of the visual field on a patient with one lens removed and the other eye normal. They found that with a proper lens the operated eye accommodated at close distances (1 to 7 inches), which confirmed their hypothesis. To test whether the cornea changes its curvature Ramsden and Home observed through a microscope an image reflected by the cornea and found some movement of it during accommodation. They interpreted (1795) their results as proving that the deformation of the cornea is responsible for one third of the total accommodation, the rest being the result of the elongation of the eye and a forward movement of the lens produced by ciliary processes.

?

Young's first reaction to the claim that the eye can accommodate even without a lens was to acknowledge his defeat. However, after he became familiar with Porterfield's optometer and made his own version of this instrument, he resumed his research. In 1800, Young presented his objections to Ramsden and Home. First, he denied that a gradual change in the index of refraction from the lens' periphery to its center could reduce spherical aberration. Secondly, his experiments showed no change in the cornea during accommodation. Thirdly, Young pointed out that the distinct vision by the operated eye had nothing to do with accommodation and resulted simply from the introduction of a small aperture. Aided by his optometer, Young also tested people with their lenses removed and demonstrated that they could focus only at a single distance. Finally, he showed experimentally that there is no elongation of the eye and concluded that the only way to provide accommodation is by changing the lens' curvature. Young did not know the mechanism of this process, although he came close in his hypotheses to some parts of the future Helmholtz's theory. In particular, he said that "the capsule is highly elastic; and, since it is laterally fixed to the ciliary zone, it must cooperate in restoring the lens to its flattest point." (1801:57)

good point!

Home met Young's challenge and applied an optometer to measure the focusing ability of one of his patients with his lens removed. The results were positive: the patient could focus his eye from 8.3 to 13.3 inches. Home remarked that the person saw everything well with his naked eye, and his vision was best in bright light. Perhaps, it was this bright light which accounted for Home's result: a strongly contracted pupil made vision more distinct, as in the previous Home's experiment. Moreover, the optometer itself had a narrow slit which also could increase the depth of vision. Young did not reply, perhaps because his interests shifted to other subjects, in particular to his theory of light.

Thus, the only thing proven by the early 19th century about accommodation was that it existed. While opticians suggested several possible mechanisms, anatomists did not have sufficient evidence to make a choice. The contemporary microscopy could not decide, for instance, whether ciliary processes contained muscular fibers, and there was no instruments to observe the inside of of the eye. Those improvements came later, and they gave Helmholtz an opportunity to create his theory of accommodation. It is of interest that some ideas on the nature of accommodation which had been considered mutually exclusive in the 18th century (a change of the pupil and a change of the lens' curvature) became complementary in the modern theory.

important

III. EXPERIMENTS

1. Model of the eye

lab

Purpose. In this experiment you will work with a simplified dioptric model of the eye which has two refracting media, the “cornea” and the “lens.” The lens can change its curvature and accommodate the eye to close objects. The purpose of this lab is to show both the accommodation and the correction of the eye.

Equipment. The “eye globe” (Fig. 8.14) consists of two parts cut out of a 2 liter pop bottle: the "cornea" is made of a clear front part, while the "sclera" is made of an opaque back part of the same bottle. The “cornea part" has an opening in the front for a lens which refracts light as the cornea should. This lens is taped to the eyeball from outside and it is covered with a black paper disk that has a small opening in the center - the "pupil." The back part of the eye has an opening at its center to let in a movable screen which receives the image produced in the eye and thus allows to show that sometimes the sharp image may be obtained, depending on the shape of the eye, before the retina or behind it. The "crystalline lens" was designed by Linda Baloun of Highmore High School (Highmore, SD) and Dr. Albert Kuhfeld of The Bakken. It is mounted in the back of the cornea part and is made of Plexiglas. One wall of the lens is made of a clear plastic wrap (Fig. 8.15). The lens is filled with water by means of a syringe, and by changing the amount of water the flexible surface can be made more or less curved.

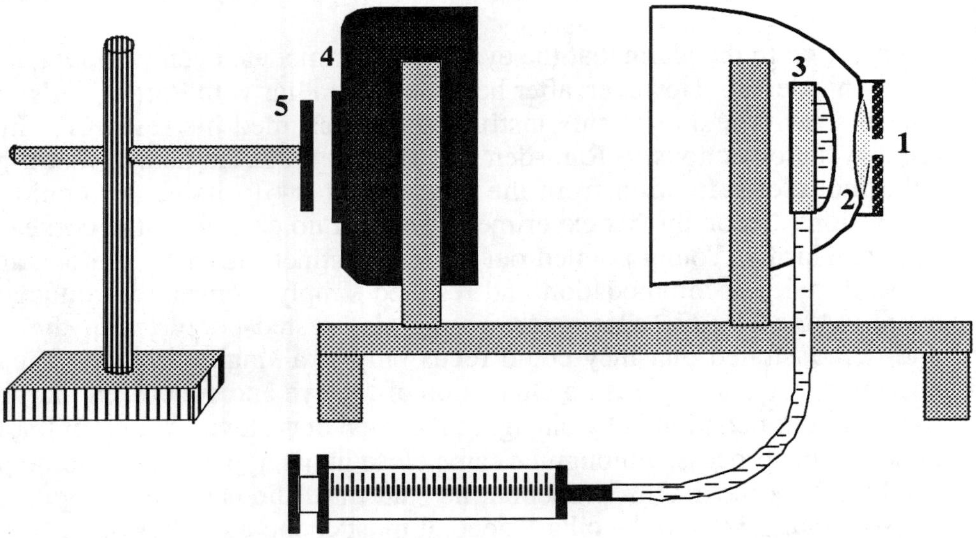

Fig. 8.14. Model of the eye: 1) the "pupil"; 2) the "cornea"; 3) the flexible "lens"; 4) position of the "retina"; 5) movable screen.

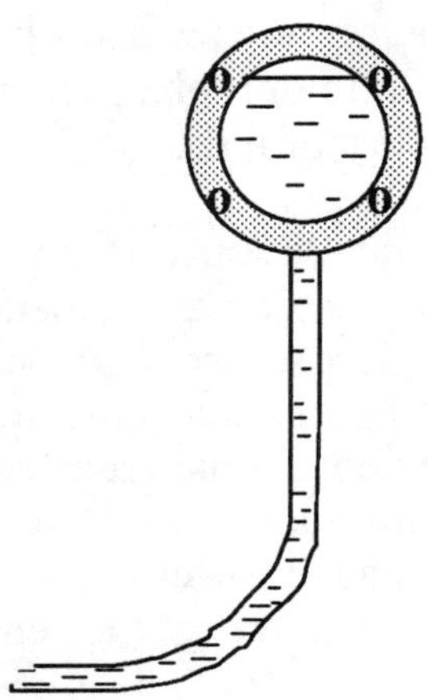

Fig. 8.15. Flexible lens.

a. accommodating the eye

Procedure.
Reduce the amount of water in the lens to a minimum that still allows one to see a clear image on the screen (water must fill the whole chamber and stretch the wrap so much as to leave no wrinkles). Place a candle flame 70-80 cm from the eye, and move the screen until you obtain a sharp image, after which do not move the screen. Then place the candle at 25 cm from the eye: the image becomes indistinct. Slowly add water into the lens to increase its curvature until you see again a sharp image

Investigate.
Repeat the experiment with the candle at different distances to find the range of distances at which your lens is capable of accommodating the eye.

b. correcting the eye

Theory. In the *myopic* (nearsighted) eye the image is located before the retina, which is corrected with a diverging lens (Fig. 8.16a). In the *hyperopic* (farsighted) eye the image is located beyond the retina, and the correction requires a converging (positive) lens (Fig. 8.16b).
Equipment. The model of the eye made either myopic or hyperopic by changing the "cornea", an assortment of lenses, a lens-holder.

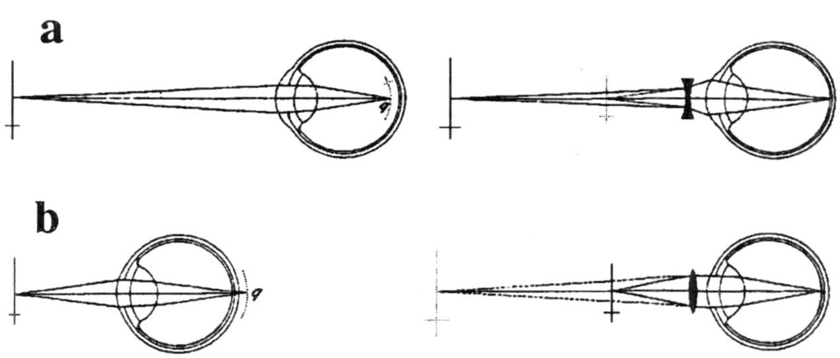

Fig. 8.16. Correcting the eye. correction.

Procedure. First establish the character of the eye defect. Place a candle at 40 cm from the eye and move the screen until you obtain a sharp image. If the screen is inside the eye, the eye is myopic; if the screen is outside the eye, the eye is hyperopic. Select a converging or diverging lens, depending on the defect, place it in a holder close to the eye and check the new position of the image. Change lenses until you find one which moves the image to the retina (the screen being near the back opening).
Investigate.
Does your selection depend on the distance between the object and the eye?

2. Accommodation

a. Scheiner's experiment

Equipment. An index card, a pin.

lab home

Procedure. Using a pin make two holes in an index card 2 mm apart. Move the card across the eye until both holes will be against the pupil. When holding a pin about 30 cm from the card (Fig. 8. 17) look through the holes at a distant vertical **narrow** object such as a chalk line on the board or a line drawn on white paper with a dark marker (it should be brightly lit). Do you see the pin double? Be sure you **focus** on the line.

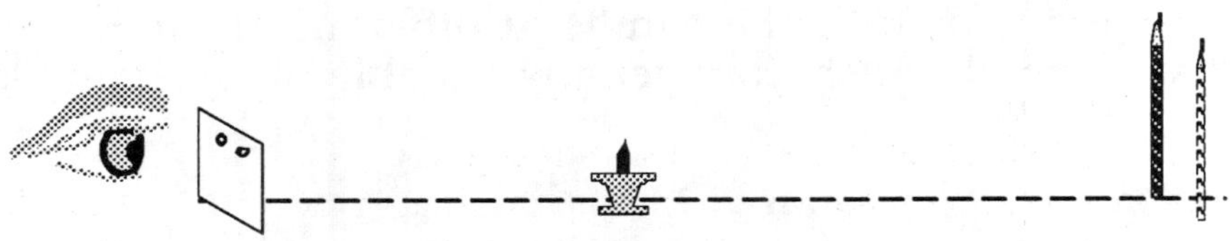

Fig. 8.17. Scheiner's experiment.

Now, look intently at the pin. Do you now see the line double? Repeat several times refocusing your eye from the pin to the line and back. If you wear eyeglasses, use them when it helps: farsighted people need them to look at the close object; and nearsighted people will benefit from them when focusing at the distant object.

b. Young's optometer

lab

Background. While Porterfield used a single object to focus on which he moved to or from the eye, Young opted for many objects placed at different distances from the eye at the same time. The objects were points belonging to a straight line: the point at which the eye was focused appeared single, the rest were doubles. Thus, when looking along a straight line drawn on paper the eye should see two intersecting lines.

Equipment. A meterstick 1 m long, index card, razor blade. To make a modified Young's optometer (Fig. 8.18), cut a paper strip 1 m long and 1-1.5 cm wide and attach it to the meterstick so as to leave the scale open . At each end of the strip mark a mid-point and draw a pencil line connecting the two. Cut a piece of an index card 1x6 cm and make two parallel slits 4 cm long in the middle of it, parallel

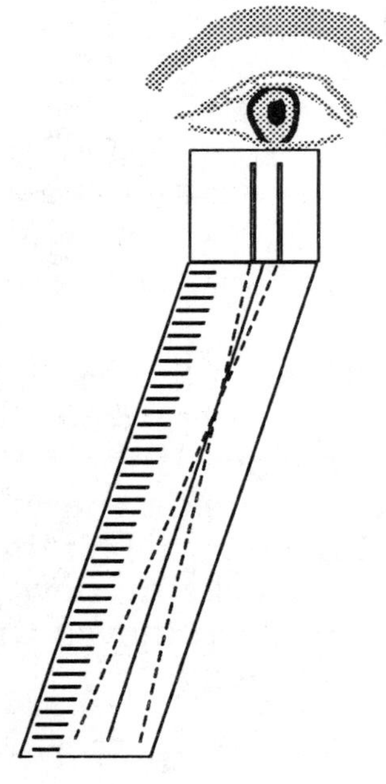

Fig. 8.18. A model of Young's optometer.

to the long side. The width of each slit should be about 1 mm and the distance between them should not exceed 2 mm. Attach the double-slit to the narrow side of the meterstick, so as to have the space between the slit projecting against the line.

Procedure. Look through the slits at the pencil line (your other eye should be screened from it) so as to follow its course from one end to another: you will see two crossing lines. If you concentrate on different numbers on the scale, you will see the point of intersection moving. The closest distance at which you can make the lines cross is the *near point* of your eye; the greatest distance you can do it is the *far point*. Repeat the experiment several times and take the average reading. Check the other eye. If you have reading glasses, you must have them on. If you found that your near point is very different from 25 cm, you need new eyeglasses. Place different lenses between your eye and the slits (remove your eyeglasses if you have any) until you find one that brings the near point close to 25 cm. Measure the focal distance of this lens and calculate its optical power using eq. (5.3). Then go to an optometrist and compare his prescription to your own findings.

BIBLIOGRAPHY

E. Wolf's Anatomy of the Eye and Orbit (London, 1968).

R.S. Snell & M.A. Lemp, *Clinical Anatomy of the Eye* (Boston: Blackwell Scientific Publ., 1989).

J.F. Koretz & G.H. Handelman, "How the Human Eye Focuses," *Scientific American* (July 1988), pp. 92-99.

Adler's Physiology of the Eye, Clinical Application (St. Louis: C.Mosby Co., 1987), pp. 291-306.

H. Davson (ed.) *The Eye*, 3rd ed., vol.Ia (Orlando, FL: Academic Press, 1984).

R. Descartes, *Discourse*, pp. 91-113.

J. Priestley, *History of Light*, pp. 112-118, 183-202, 630-688.

W. Porterfield, *Treatise on the Eye* , 2 vols (Edinburgh, 1759), I: 389-450.

E. Home, "The Croonian Lecture on muscular motion," *Phil. Trans.*, 1795, 1-23; 1796, 1-26.

E. Home, "On the power of the eye to adjust itself to different distances, when deprived of the crystalline lens," *Phil. Trans.*, 1802, 1-11.

T. Young, "Observations on vision," *Phil. Trans.*, 1793, 169ff; also in his *Misc. Works*, 3 vols (London, 1855), I:1-11.

T. Young, "On the mechanism of the eye," *Phil. Trans.*, 1801, 23ff; also in his *Misc. Works* I:12-61.

Chapter 9

COLOR VISION

I. PRIMARY COLORS

1. Mixing colors

a. studying color filters

Equipment. Pieces of clear colored plastic mounted in slide frames. ***lab***

Teacher. To understand the nature of color vision it is important to grasp the meaning of color mixing. We have already done some experiments on mixing colors of the solar spectrum and opaque bodies. Now, we will focus on color filters. In Ch.7, we've presumed that a filter transmits light of a single color. How can we prove this?

Dorothy. We need a light source which produces light of selected colors. Watching these colors one by one through, say, a violet filter, we can establish how well the violet filter transmits light of different wavelengths. We could obtain a monochromatic light from parts of the solar spectrum produced by a prism, or from the desklight that we've used before provided with different color filters.

Ruth. Wait a second! If we want to prove that a violet filter transmits only violet light we may not use it to **create** monochromatic violet light! As to sunlight, you cannot rely on it this time of the year.

good point!

Teacher. Ruth is right about filters. Perhaps this violet filter transmits other colors as well but the eye cannot distinguish them. There is an instrument that can separate a narrow part of the spectrum called *monochromator*. It uses a system of prisms, slits, and lenses. However, it is an expensive instrument, while for a lab we need something simple and cheap even if less precise. There are also color filters that are truly monochromatic *(interference filters);* a set of these is less costly than a monochromator but still too expensive for a lab. We can, however, make our plastic filters more monochromatic. Imagine, for instance, that our violet filter transmits green light the intensity of which is one half of the intensity of violet light. What would be the relative intensity of green light emerging from **two** violet filters?

David. One forth of the intensity of violet light, and with three filters we can make it one eighth. Thus, by applying several identical filters we can weed out all wavelengths but one. This is great, because we can make a monochromator for the price of a few cents!

Teacher. This is not exactly so, for we have only a very limited selection of suitable plastic filters. Remember also that you cannot combine too many filters because the intensity of emerging light can become too low for making any estimates. Still, this is a reasonably good method for our purpose. Let us take five identical filters and look at the cloudy sky through a different number of them. How would you compare the colors you see?

Dorothy. With one red filter, I saw an orange color, three filters made it more red, and five filters produced a deep red color.

Michael. One blue filter transmitted a bluish white light, while five filters transformed the color into a deep blue. Apparently, five pieces of clear plastic is enough to create a monochromatic filter.

lab

Teacher. Now, make red, yellow, green, blue, and violet monochromatic filters by taping together five ordinary filters of each sort. We are going to study how an ordinary (single) filter of each type transmits red, yellow, green, blue, and violet light. For instance, you hold a red filter before your eye looking at a cloudy sky. Bring before this filter one by one different **monochromatic** filters, and describe the color you see. To prevent a confusion, avoid commercial terms and try to match the colors with some of the following: R (red), Y (yellow), G(green), B (blue), V (violet), P (purple), and Br (brown). Enter the symbols of colors into the Table 9.1 together with their estimated relative brightness expressed by numbers from 3 to 0 ("high," "medium," "low," and "none"). To make the comparison easier, hold two different

Table 9.1 (example)

light	filters				
	red	yellow	green	blue	violet
red	3R	3R	1P	0	2P
yellow	3R-0	3Y	3Y-G	3Y-G	2Y-Br
green	0	2G	2G	2G	1.5G
blue	0	2G	3B	3B	3B
violet	1.5P	1P	1B	2V	2V

monochromatic filters side by side so that you can see both through the filter under the investigation. What are your results?

John. The red filter transmitted a lot of red and orange, very little of green, no blue, and quite a bit of purple, which we expressed as 3R,3O, 0.5V, 0G, and 1.5P. The results about red and orange seem natural, but purple shouldn't be there, should it?

Mary. We also had a surprise: the blue filter transmitted a lot of yellow and blue, and enough of green and violet, the numbers being 0, 3Y-G, 2G, 3B, and 2V. Thus, a single piece of colored plastic can hardly be called a "filter" because it lets through a wide part of the spectrum.

Teacher. We can also interpret these results in terms of color mixing: red + violet = purple, blue + yellow = green, etc. We should be now able to predict the color of light transmitted by a combination of two filters. For instance, what color will produce a pair of red and blue filters?

David. After the red filter we'll have plenty of red and orange and some purple, but a blue filter doesn't let through any red: probably we'll see very little light at all.

Teacher. Very good! If we plot "intensity" versus "wavelength" (represented here by the color of our monochromatic filters as follows: red-

620 nm, yellow - 580 nm, green - 520 nm, blue - 480 nm, and violet - 400 nm), we'll obtain a *transmission spectrum* of our filter (Fig. 9.1a). It is similar to a graph obtained with a monochromator (Fig. 9.1b).

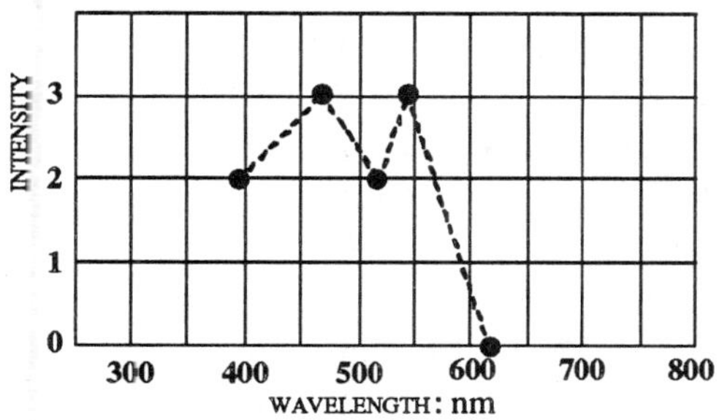

Fig. 9.1a. Transmission spectrum of a blue plastic filter.

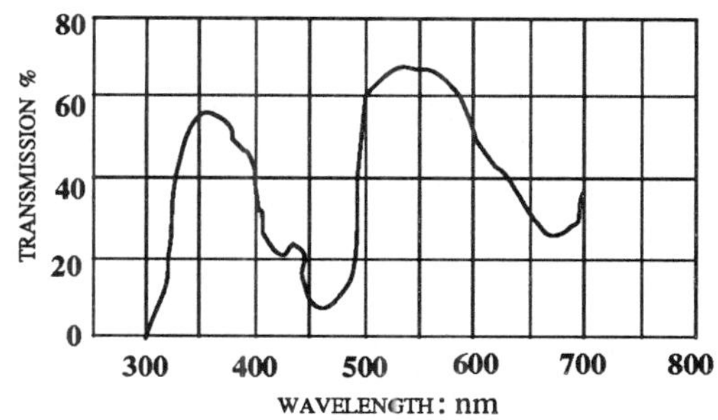

Fig. 9.1b. Transmission spectrum of the "pale green" filter. (GamColor).

Although the latter is more precise, the former is sufficiently good for our purpose.

b. mixing colors with filters

demo

Teacher. I am placing yellow and blue filters on the overhead projector. We see green color where the two overlap. Can you explain this result on the basis of Newton's theory?

Dorothy. We have just found (see Table 9.1) that a yellow filter transmits red, orange, yellow, and green, while a blue filter lets through green, blue, and violet. Thus, the light coming through both is green.

Teacher. Very good! This kind of mixing colors is called *subtractive*. There is also another procedure called *additive* mixing. Let us take two concave mirrors and tape a blue filter to one and a yellow filter to another. I will shine both images on the ceiling (or a wall) and move them until they cross. What color do you expect to see at their intersection and why?

Michael The two filters together will bring red, orange, yellow, green, blue, and violet, which is a complete spectrum. Thus, the mixture must be white.

Teacher. Fine! As you see, white it is. Two colors, such as blue and yellow, the mixture of which produces white are called *complementary*.

c. color-prism

lab

Equipment. A color-prism (Fig. 9.2) is made of 3 slide frames, carrying color filters (for instance, red, yellow, and blue). To make replacing a filter very easy, connect the frames with masking tape or "Tak." Make a white screen out of an index card, a little smaller than the frame. Attach it with paper clips to one frame inside the prism: it will serve for receiving and mixing colored images.

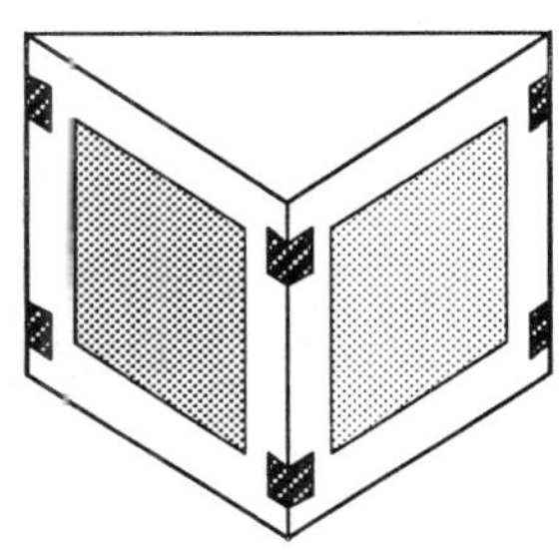

Fig. 9.2. The "color prism".

1. Additive mixture

lab

Teacher. Cover, for example, the red window of a color-prism, and direct the blue side of the prism to a bright window or to sunlight. Look at the screen through an open base of the prism: you will see it colored in blue. By means of a mirror shine light on the screen through the yellow window. What color do you see? Can you change it by manipulating the prism and the mirror? Try other pairs of filters.

John. The compound color was more or less white with a tint varying from yellowish to bluish, depending on the inclination of the windows and of the mirror. By changing these angles we make the yellow image brighter than the blue one, or vice versa. Apparently, a compound color depends on the relative intensities of its components: if the blue is stronger, we have a bluish tint, and if the yellow is stronger, the tint is yellowish.

Ruth. We obtained a sort of white with red and green filters, and we agree with John about the role of the relative intensity of the components. Apparently, to make a pure white one needs a proper ratio of intensities for the two components. What I don't understand, however, is why the tint of white was yellowish rather than reddish or greenish.

Teacher. According to Table 9.1, the amount of blue and violet light supplied by red and green filters is much smaller than their portion in the daylight spectrum, and lack of these colors leads to an excess of yellow.

2. Subtractive mixture

lab

Teacher. We will now perform another experiment with the color prism. Let sunlight pass through two windows of a color-prism and fall on a white screen. What color do you see? Then look on the screen through the two windows. Is the color the same? Repeat the experiment with different pairs of filters and compare the results with those from the first experiment. Explain the difference on the basis of Newton's theory.

David. When looking through a blue and a yellow filters placed together we saw the green color, just as when we projected sunlight through them. Since a yellow filter transmits red, orange, yellow, and green light, and a blue filter lets through the green, blue, and violet, only green light will emerge from both.

important

Mary. We tried a red and a green filters and obtained a yellowish color. A red filter transmits red, orange, a little of yellow, and a little of violet light; a green filter lets through yellow, green, and blue light. Thus, only yellow light should pass through both filters, which conforms to the theory. These results differ from the previous ones because now the mixture is subtractive rather than additive.

2. "The rays... are not coloured"

history and discussion

Teacher. Very good! Now, we can talk of the origin of colors. Although we have already established that color is connected with light, the nature of this connection is still unclear. First of all, where do colors originate from?

John. Colors are produced when white light meets a body. A red paper reflects red light, while a red filter transmits it. In either case, red light is separated out of white.

Mary. But was this light colored before reaching the eye, or did it create color after entering the eye?

Michael. I think that mixing colors proves that they exist outside the eye. If two colored bodies occupy the same place, as two pigments, or two color filters placed together, the colors mix there and we see the color of the mixture.

good point!

Teacher. In the additive mixture rays of each component reach the eye independently of one another and thus they can mix either inside the eye or outside it. If they mix outside, different rays either affect one another so as to change their refrangibility, or they don't. If the former were true, every intersection of rays of different colors would change their colors. Did this happen in your experiments?

Mary. No. After the rays from all parts of the solar spectrum crossed in a focus of a lens and separated again, we saw the same colors as before.

Teacher. Very good! Thus, rays of different refrangibility (or, wavelength) enter the eye without any change. This is an important point, and there was much confusion about it even as late as the 19th century. The question still unanswered is: do different rays bring their proper colors into the eye from outside or excite them in the eye after entering?

John. I think that colors are created outside the eye.

Dorothy. In my view, colors originate in the eye (or the brain).

Teacher. Logically, both answers are plausible, and it is up to experiment to decide which one is true. Some seventeenth-century scientists, such as Descartes and Boyle have already held that colors are produced by the brain. One of the arguments for this was the fact that some sick people saw colors in their dreams or with their eyes shut. Newton said that the very term "red light" is wrong and may be allowed for shortness sake only, the correct term being "red-making light," because

> **the Rays to speak properly are not coloured** [my emphasis - N.K.]. In them there is nothing else than a certain Power and Disposition to stir up a Sensation of this or that Colour. For as Sound in a Bell or musical String, or other sounding Body, is nothing but a trembling Motion, and in the air nothing but that Motion propagated from the Object, and in the Sensorium 'tis a Sense of that Motion under the Form of Sound; so Colours in the Object are nothing but a Disposition to reflect this or that sort of Rays more copiously than the rest; in the Rays they are nothing but their Dispositions to propagate this or that Motion into the Sensorium, and in the Sensorium they are Sensations of those Motions under the Forms of Colours. (*Opticks*, 124-25)

important

3. Mathematics of color I

history
and
discussion

Teacher. When assuming that color originates in the eye one has to adopt some procedure for mixing colors within the eye. In particular, it is necessary to decide on the number of different color receptors in the eye, which, in turn, is connected with the minimal number of primary colors sufficient to make any other color. Newton believed that seven spectral colors can do the job and devised a mathematical procedure for determining the result of such a mixture. He represented simple colors by seven arcs of a circumference (Fig. 9.3). The number of colors, the proportions of the arcs' length, and the coincidence of the red and violet extremes reflected Newton's belief in an analogy between colors and musical tones, in which a spectrum corresponded to an octave. At the center of each arc Newton drew a circle, the radius of which was proportional to the relative intensity of this color. The next step was to find the center of gravity of all circles. In the example displayed, Z is the center of gravity of a mixture, in which the intensities of violet, indigo, blue, green, yellow, orange, and red colors are as 1:1:2:3:5:6:10, respectively. Then Newton drew a radius OZ which crossed the circumference at Y. Since Y is located between E and F, the color of the mixture will be orange, closer to red. The distance of the center of gravity from the circumference determines how much white light is added to the color: if this point lies on the circumference itself, the color will be absolutely pure, if the point is at the center O, the color is white. Thus, in our example, the orange color will be considerably diluted with white.

interesting!

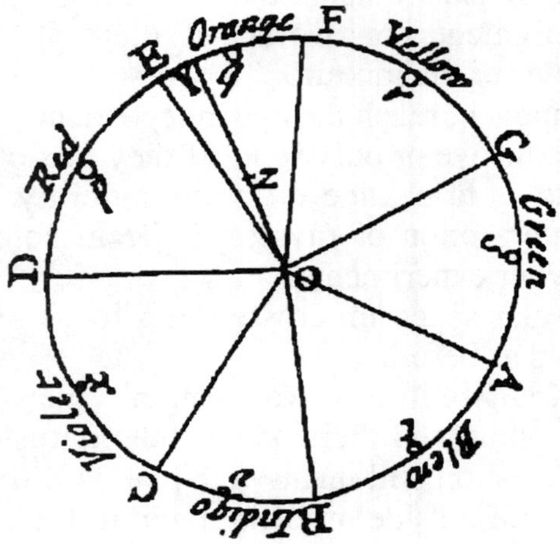

Fig. 9.3. Newton's circle. From I. Newton, *Opticks*, Bk. 1, pt. 2.

John. How did Newton justify the ratio of arcs representing different colors?

Teacher. The whole procedure, including this ratio, was empirical. It worked, and both practitioners and scientists were satisfied with it and used Newton's circle up to the middle of the nineteenth century.

II. NON-NEWTONIAN THEORIES OF COLOR

1. Three primary colors

Mary. Did someone try then to achieve the same with a fewer number of primary colors?

Teacher. On a qualitative level, yes. It had been known at the time from the experience in mixing pigments that any color can be made by mixing in different proportions red, yellow, and blue pigments. On this basis, a number of Newton's contemporaries argued that one should not multiply entities without necessity: if three primary colors were sufficient, one did not need seven. Moreover, Hooke and Huygens claimed that even two primary colors, red and blue, could do the job. In fact, they used different reds, the yellowish red and the bluish red, thus, actually, they also had three primary

history and discussion

colors. Some scientists believed that experiments with mixing two beams of colored light also contradicted Newton's theory. Boyle, for instance, reported that a yellowish flame appeared green through a deep blue glass. He also found that white paper illuminated by sunlight coming through a pair of red and yellow glasses appeared orange; it changed its color to purple when the glasses were red and blue, and to green when the glasses were blue and yellow. When viewing the solar spectrum through a blue glass, Boyle found its yellow part yellowish green. A similar effect took place when he received the solar spectrum on various colored bodies: the blue part of the spectrum showed a red surface as purple, and yellow light appeared green after being reflected from a blue body. Finally, Boyle superimposed two spectra produced by two prisms and saw a green color where the yellow part of one spectrum came upon the blue part of the other, and purple where red color became mixed with the blue one. (Boyle, 197, 222-27) In his view, these findings proved that all colors but red, yellow, and blue were compound.

Michael. Wait a second! According to Newton, to tell which color is simple and which is compound we have to use a prism rather than color filters.

Teacher. Right! And he insisted that a prism can find a difference where the eye cannot, as is the case with green simple and green compound light. However, those scientists who supported three primary colors questioned the propriety of using a prism to analyze light and suggested using colored filters instead. This view became quite popular in the early nineteenth century.

Dorothy. But shouldn't the result of the analysis be the same in both cases?

2. Overlapping spectra

history

Teacher. Apparently not, as ***Matthew Young*** (1750-1800), Professor of Natural Philosophy in Dublin, showed in 1800. From experiments similar to Boyle's, he deduced that there is no green part in the solar spectrum, and we see green due to a mixture of light from yellow and blue parts: in the same way he explained orange color. The indigo and violet were, in his view, a mixture of the blue and red. It was generally accepted then that the purple color was a mixture of the red and violet. Since the blue part of the spectrum appeared purple when seen through a red glass, M. Young asserted the existence of red color in it. As an intersection of two adjacent circles of different colors meant that light of the same degree of refrangibility produced different colors, M.Young concluded that Newton's concept of the solar spectrum as a sum of several colored images of the sun contradicted his idea of the one-to-one correspondence between the degree of refrangibility and color,

?

The matter became even more confusing when it was found that even the dispersion analysis of sunlight may show less than seven colors. William Wollaston, for instance,observed in 1802 only four simple colors in the solar spectrum: red, green, blue, and violet, with yellow being so narrow that he attributed it to a mixture of red and green. In 1808, ***Jean-Henri Hassenfratz*** (1755-1827) found that the number of colors in the solar spectrum varied from three to seven, depending on the time of day. He attributed this variation to selective absorption of sunlight in the atmosphere.

Like M. Young, ***Claude-Antoine Prieur*** (1763-1832) observed in1805 that the blue part of the solar spectrum appeared green through a green filter and violet through a violet filter, while the yellow part appeared red through the red glass and green

history

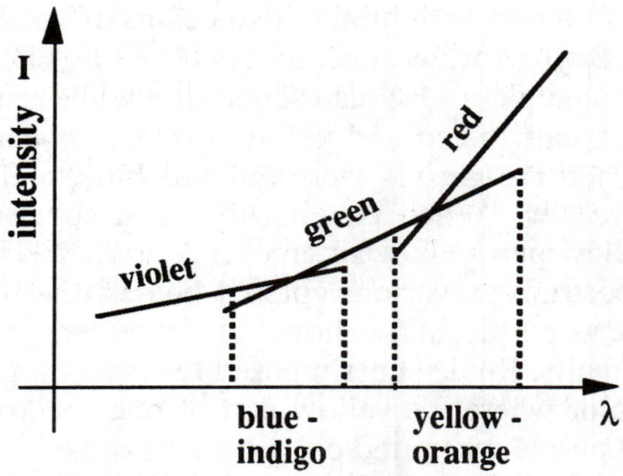

Fig. 9.4. Prieur's overlapping spectra. From *Annales de Chimie 54* (1805), 343.

through the green one. On this basis, Prieur claimed that the spectral blue light consists of green and violet light and the yellow is a mixture of red and green light. Accordingly, he represented the solar spectrum by three overlapping parts (Fig. 9.4) centered around the red, green, and violet (that was his choice of primary colors). Orange and yellow appeared at the intersection of red and green spectra, blue and indigo, at the intersection of green and violet parts. Thus, in Prieur's theory, like that of M. Young, light of the same degree of refrangibility produced more than one color.

A similar approach brought ***David Brewster*** (1781-1868) to one of the most bizarre theories of colors. Brewster was one of the most prominent opticians of the 19th century, best known for his investigations of polarized light. However, while a brilliant experimenter, he was not successful in his ventures into the realm of theory. In 1831, unaware of the works by Hassenfratz and Prieur, he revived the idea of overlapping spectra and developed it into a new theory of colors. To him, it was not a theory but a "fact" deduced from observations. The theory had four points: 1) white light consisted of three simple colors, red, yellow, and blue, by mixture of which all other colors are formed; 2) the solar spectrum, whether formed by refraction or by diffraction, consists of three spectra (red, yellow, and blue), which all begin and terminate at the same points (see his curves, intensity of light vs. refrangibility in Fig. 9.5); 3) all colors in the solar spectrum are compound, each of them consisting of red, yellow, and blue light in different proportions; 4) a part of white light is indecomposable by a prism; since its components have the same refrangibility, this part may be separated from colors. Brewster's statements about rays of the same refrangibility producing different colors, and about white light which cannot be decomposed by a prism con-

?

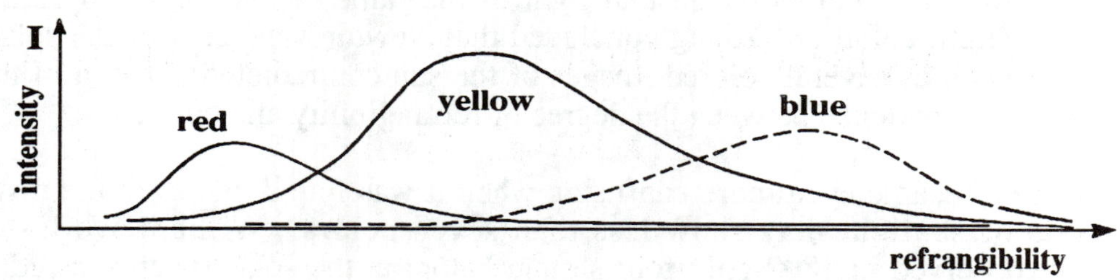

Fig. 9.5. Brewster's spectra. From D.Brewster, *Trans. Roy. Soc. Edinburgh 12* (1834), 123.

tradicted Newton's theory. He concluded (again!) that the only true way to separate colors from a compounded light is by absorption in colored filters. Here is an example of his reasoning that red color exists in all parts of the spectrum. First, Brewster said, it was generally accepted that red color is present in the red, orange, and violet sections. Second, the indigo and blue parts of the spectrum seen through some yellow liquids, displayed a clear violet tint, which implied to him that the red color existed there before. Brewster was so confident in the truth of his theory, that he even included it in his biography of Newton as a correction of Newton's theory of colors. Initially, Brewster's theory had some support, especially in Britain, but in the 1850s it was utterly refuted.

David. I guess I know what caused all this confusion: those scientists had no idea that some filters transmit light from different parts of the spectrum.

Teacher. Or, in other words, that a filter interfered in an unpredictable way with the process of analysis. Incidentally, ***Wolfgang Goethe*** (1749-1832), brought forth a similar objection even against a prism.

important

3. Poetry versus Science

Teacher. The famous German poet was also an amateur scientist, and sometimes he even attached more importance to his work on color than to his poems. In 1810, he published a book in which he criticized Newton's theory of colors and advanced a theory of his own. However, we cannot place Goethe among those critics of Newton that we have discussed above, for his objections were based on an entirely different approach to studying Nature. In Goethe's view, one must observe phenomena in the simplest possible way without using any instruments or mathematics. To Goethe, jamming light into a pinhole or passing it through a prism meant interference with Nature, which could only distort the reality of things.

Michael Did Goethe object to all experimentation?

Teacher. No, he did some experiments, but he viewed them as "less interfering" than Newton's although he did use a prism himself. According to him, light was a fundamental entity and thus any attempt to analyze it was meaningless. He considered color to be a property of bodies, referring to a coloration of some turgid liquids, which depends on whether it is observed along the direction of the transmitted light or perpendicularly to it. In Goethe's view, a single observation is sufficient to suggest a very general hypothesis or to refute a well established theory.

good point!

Ruth. Certainly, Goethe's approach to Nature was poetic rather than scientific. "Interference" apart, however, isn't it true that there was another important reason for so different interpretation of spectral analysis: apparently, M. Young and Prieur and Brewster all believed that color exists outside the eye?

III. TRICHROMATIC THEORY OF COLOR VISION

1. T.Young's idea

Teacher. You are absolutely right! Although Newton's idea of color originating inside the eye had been well known, even the Newtonians failed to see how to use it for compound colors. Thus, when Thomas Young offered in 1801 the correct idea of color vision, it was misunderstood and ignored for half a century. According to Young, each point of the retina contains three receptors, the particles of which vibrate in perfect unison with light of one of the three principal colors, red, yellow, and blue. At the same time, each of the three receptors also responds, although to a lesser degree, to a range of frequencies centered around the "resonant frequency." For instance, the waves of green light will affect both yellow and blue receptors and thus create the same impression as if yellow and blue light emitted by the same point

history

history

would have entered the eye independently of one another. In 1802, after William Wollaston had questioned the primary character of yellow light in the solar spectrum, Young changed the primary triad to red, green, and violet.

It is interesting that the existence of three types of receptors in the retina was first suggested by George Palmer in 1777. Apparently, Young was unaware of his work, but it wouldn't had helped him much even if he were, because the principal idea of Palmer's theory was very different from Young's. According to Palmer, each ray of white light consisted of three different rays (red, yellow, and blue) and each receptor was sensitive to only one of these rays: red receptor responded only to the red rays, etc. Thus, Palmer's theory, presumed that the decomposition of a compound light into primary components occurred **outside** the eye. Young did not offer any experiments to support his theory, that was accomplished half of a century later by ***Hermann von Helmholtz*** (1821-1894) and ***James Clerk Maxwell*** (1831-1879).

important

2. Experimental support

Teacher. In his youth, Helmholtz was interested in physics but he could not support himself through the university. Thus he turned to the department of medicine where he could receive government aid. In return for his scholarship, after graduation he had to serve for several years as a military surgeon. In 1849, at the age of 28 he was appointed Professor of Physiology at the University of Koenigsberg. It was the beginning of a brilliant career in science. Helmholtz's crowning achievement was the law of conservation of energy. He also made very important contributions to physiology, physiological optics, acoustics, physics, and mathematics.

In the case of Brewster's theory, Helmholtz demonstrated that the source for Brewster's "indecomposable" white light was a scattering of light by the irregularities in the prism and by multiple internal reflections in the prism, in the glass filters, and in the eye itself. As for the latter, Helmholtz showed by means of the ophtalmoscope which he invented in 1851, that some light from the illuminated part of the retina is reflected back to the cornea and then is reflected back to other portions of the retina.

While defending Newton's theory of colors in general, Helmholtz objected to Newton's method of obtaining white out of a mixture of colors. His point was that the theory of primary colors had not been demonstrated because the "true" mixture of colors can only be obtained with spectral light and not with pigments. Helmholtz was the first to mix two pure spectra (1852), which he let through a V-slit (Fig. 9.6), and observed through a telescope. Initially, he concluded that only one pair of simple colors, blue and yellow, can produce white. This result was criticized by ***Hermann Grassmann*** (1809-1877) who asserted that the existence of many pairs of complementary colors is a direct consequence of Newton's theory of mixing colors.

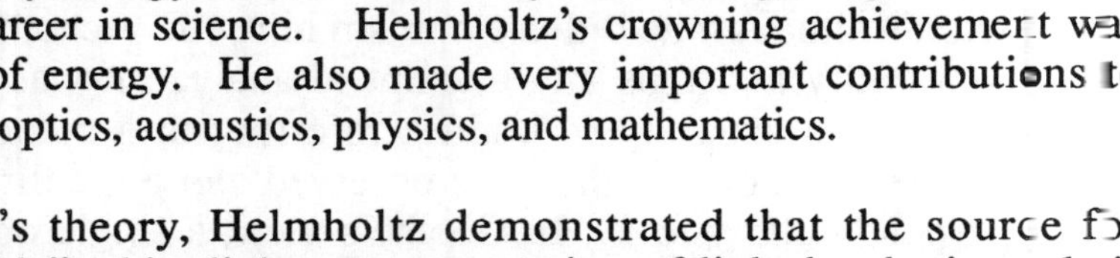

Fig. 9.6. From H.Helmholtz, *The Sensations of Vision*, v.1.

Grassmann revived two Newton's ideas of color perception: 1) that colors may be arranged in a circle in which violet is adjacent to red; and 2) that perception of color can be described by three variables, namely, the tint of light, its intensity, and a proportion of white light added to it. From these propositions he derived that for each given color there is a complementary color, which produces white when mixed with it. Grassmann showed that Helmholtz misinterpreted his own experimental data, for he actually observed more complementary colors than he thought he had. In response to this criticism, Helmholtz repeated his experiments with an improved apparatus. By means of two adjustable slits and a prism he combined two parts of a pure spectrum so as to obtain white image (Fig. 9.7), which was compared to the color of paper illuminated with daylight. He found that colors from red to green-yellow had their complementary colors between green-blue to violet, but in the area between green-yellow and green-blue a single color could be neutralized only by a mixture of two colors, red and violet. This proved, in his view, the existence of three primary colors. However, to reproduce some of the spectral colors, Helmholtz found a need to increase their number to five. For this reason, he at first rejected Young's theory and did not come to accept it until 1867.

history

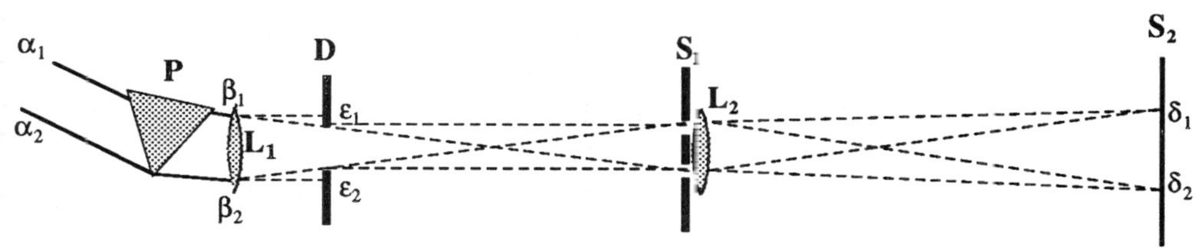

Fig. 9.7. From H. Helmholtz, *The Sensations of Vision*, v.1.

Mary. Are you implying that accepting **three** receptors was crucial for the new theory of color vision? Why was this number so important?

3. Mathematics of colors II

Teacher. Its significance was discovered in mathematics. Having concluded from experimental data that three and no more than three variables are necessary to describe a color, such as tint, intensity, and the proportion of white light, Grassmann offered a simple mathematical construction for mixing colors (Fig. 9.8). He assumed that each color may be represented by a vector the direction of which indicates its tint and the length, its intensity (initially, he ignored the presence of white light). If **a** and **a'** are complementary colors (for instance, indi-

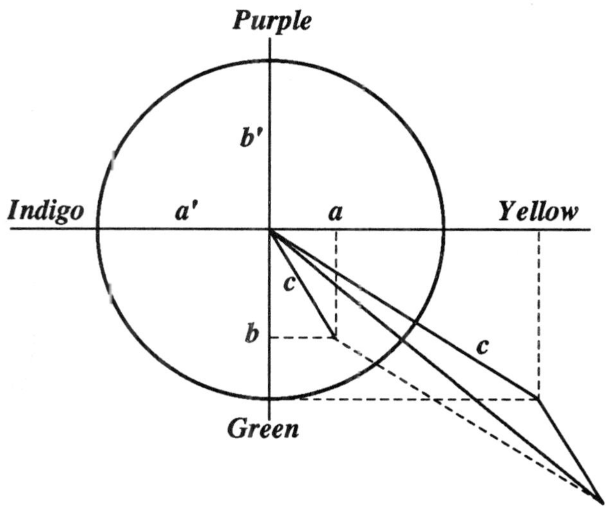

Fig. 9.8. From H. Grassmann, *Phil. Mag*. 7(1854), 264.

math is a friend !

history

go and yellow), they can be expressed by two equal and oppositely directed vectors. Let **b** and **b'** be another pair of complementary colors (for instance, green and purple) which produce in mixing the same amount of white light as **a** and **a'** do, and **a** and **b** make a rectangular basis. If two colors **e** and **e'** be expressed through this basis as $\alpha\mathbf{a} + \beta\mathbf{b}$ and $\alpha'\mathbf{a} + \beta'\mathbf{b}$ (**a'** and **b'** correspond to the negative magnitudes of α and β), their mixture can be given as $(\alpha + \alpha')\mathbf{a} + (\beta + \beta')\mathbf{b}$, or $(\alpha\mathbf{a} + \beta\mathbf{b}) + (\alpha'\mathbf{a} + \beta'\mathbf{b})$. This means that the result of a mixture of two colors can be found as a sum of two vectors representing the original colors. If white light is added to colors, the total intensity, in Grassmann's view, must be the sum of the amplitudes of colored and white light.

The next step was made by James Clerk Maxwell. Maxwell was born in Edinburgh in the family of a lawyer. As a boy of three, he was busy investigating how the kitchen bells were activated from different rooms, and at the age of 14 he published his first mathematical paper. James Clerk entered the University of Edinburgh at 16, and three years later went to Cambridge. Maxwell enriched science with a number of important discoveries in such areas as the nature of Saturn's ring (he proved that it can be neither liquid nor solid), kinetic theory of gases, statistical mechanics, thermodynamics, theory of elasticity, geometrical optics, and his crowning achievement - the electromagnetic theory. Although Maxwell is known as a theoretician, more than once he displayed a remarkable experimental skill, as exemplified by his research on colors.

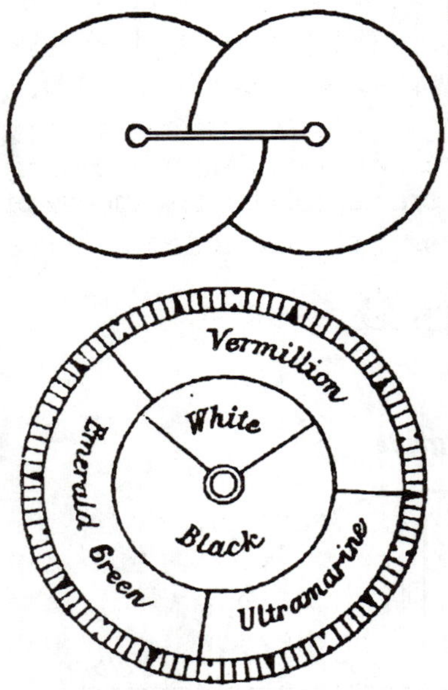

Fig. 9.9. From J.Maxwell, *Scientific Papers*, v.1. pl.1

Maxwell's interest in color mixing originated from his student years at Edinburgh where he assisted James Forbes in experiments with color tops. In 1852, he began experimenting with mixing colors of two spectra, but as he found himself anticipated in this by Helmholtz, Maxwell returned in 1854 to spinning tops. His improved top carried two concentric sets of disks of different diameter, the outer one consisting of three colored disks, and the inner set of white and black disks. In some experiments, the inner disks were also colored. The disks were arranged around the axis so as to expose each color as a sector of a variable size (Fig. 9. 9). The task of an observer was to change the intensities of the "basic" colors (by changing the size of the sectors) so as to equalize both the intensity and the hue of the two sets. Like Helmholtz, Maxwell also added spectral colors using a spectroscope with three slits of a variable width. An observer selected three parts of the spectrum and changed their intensities (by varying the width of the slits) until the mixture appeared white.

From these experiments, Maxwell obtained the "color equations" that express any color as a mixture of three basic colors and also of white and black. The basic colors in these equations were not at all the classical "primary colors": they were selected so as to produce as many compound colors as possible. To Maxwell the mathematician, the important factor was the **number of variables**, rather than their specific choice, because he knew that **in a linear system one triad of independent variables can be**

replaced with another. He deduced one set of variables from his experiments, it was "redness, blueness, and greenness." Like Grassmann, he characterized color by its *hue*, *shade* (intensity), and *tint* (the percentage of white light). Maxwell asserted that the two different methods of representing color are mutually convertible. In his view, the identical perception of two colors produced by light of different wavelength presumes an existence of three different receptors inside the eye rather than of three sorts of light outside it. Thus, Maxwell sided with Young against Brewster. Two principal concepts of his theory followed those of Young (and he was the first to uncover Young's theory) but in a more general form. First, Maxwell assumed

important

> the existence of three elementary sensations, by the combination of which all the actual sensations of colour are produced. It will be shewn that it is not necessary to specify any given colours as typical of these sensations. Young has called them red, green, and violet; but **any other three colours might have been chosen, provided that white resulted from their combination in proper proportions** [my emphasis - N.K.]. (Maxwell, *Scientific Papers*, I:121)

Second, Maxwell generalizes Young's idea that each elementary sensor responds to more than one sort of light: "**every ray** [my emphasis - N.K.] of the spectrum gives rise to all three sensations, though in different proportions."[ibid., 122]

important

He discovered that different observers supplied very consistent results, and the results obtained by observers with an ordinary vision agreed with those of color-blind people. Maxwell concluded that "the law of color vision is identical for all ordinary eyes." Following Newton, Maxwell gave a geometrical illustration of color mixing, with that difference that he used only three primary colors. He represented the three pure sensations (red, green and blue) by vertices of a triangle **R**(Red)**G**(Green)**B**(Blue) and their intensities by the weights suspended at these points. The center of gravity of two such masses represents the hue of a compound color, and the total mass, its intensity (Fig.9.10). For instance, this is how he interprets the following equation for a color-deficient person

$$0.19\,G + 0.05\,B + 0.76\,Bk = 1.00\,R$$

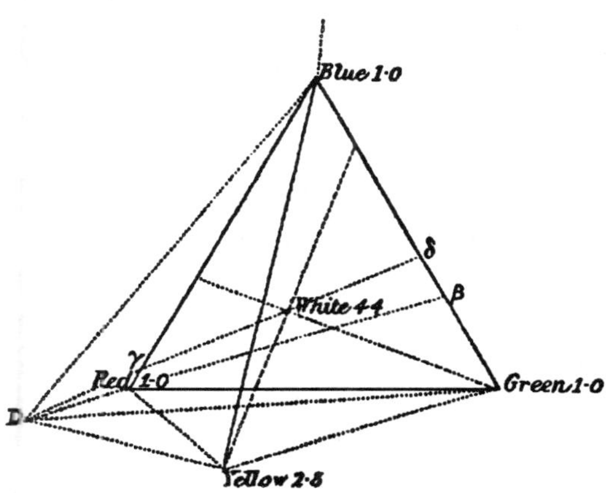

Fig. 9.10. From J.Maxwell, *Scientific Papers*, v.1. pl.1

First, he divides the line **BG** in the proportion 19 to 5 at the point b, this point representing 0.19 G + 0.05 B. To add to this 0.76 Bk he draws a line through **R** and **b** and finds the point **D**, representing black color (Bk), such that RD : Rb = 24: (100-24). Thus, Maxwell's color-deficient observer perceived red as a dark blue-green color. More exactly, he names green yellow, and to him the line **DW** (the point **W** stands for white) separates two types of colors: blue colors above it and yellow colors below it. Maxwell supposed that color-deficient people have one of their three receptors not functioning, and he found a strong confirmation of this idea in experiments. To give more support to his theory of three primary colors, Maxwell developed the concept of color photography (1855) and made the first color photograph (1861). In the same year, he also obtained the first experimental response curves (R, G, and B in Fig. 9.11) which he interpreted as the reaction of the primary receptors to light of different wavelengths.

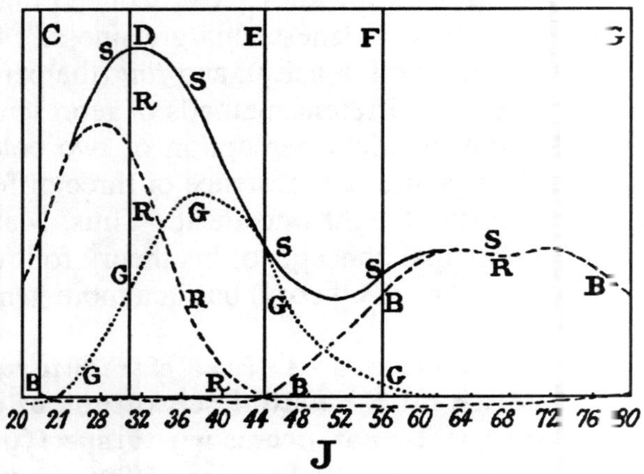

Fig. 9.11. From J.Maxwell, *Scientific Papers*, v 1. pl.1

IV. MODERN THEORIES

1. Psychological research

review

a. the response function

A number of phenomena, such as afterimages, simultaneous contrast phenomena, the peripheral color blindness of people with normal foveal color vision, and congenital color deficiencies support the view that there are three pairs of "opponent" colors: red/green, yellow/blue, and white/black. With a proper combination of intensities the opponent hues produce white. According to this theory, all hues can be divided in four groups each consisting of the mixture of two colors: red-yellow, yellow-green, green-blue, and blue-red. There are three pure hues: the unique yellow (580 nm) does not have any red or green; the unique green (500 nm) contains neither yellow nor blue, and the unique blue (475 nm), possesses neither green nor red (Fig. 9.12).

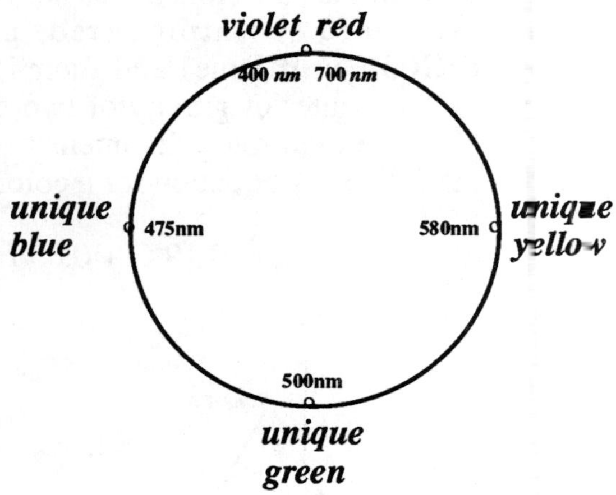

Fig. 9.12. The color circle.

To establish how the eye perceives light of different wavelengths is equivalent to finding the composition of reach color produced by a given wavelength. For instance, let us examine light of 510 nm, which, according to Fig. 9.12, produces a mixture of yellow and green colors. We take two monochromators (Fig. 9.13), set one at 510 nm and the other at 475 nm (the unique blue) and by means of a mirror produce an

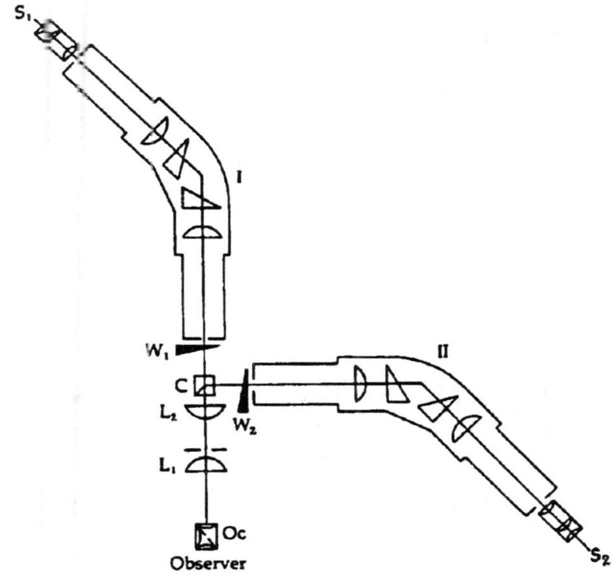

Fig. 9.13. From L.M. Hurvich, *Color Vision* (Sunderland, MA, 1981), p.56; reprinted by permission of Sinauer Associates, Inc.

additive mixture of the two lights in the eye of an observer. The intensity of each light can be controlled by an absorbing wedge. Since blue and yellow light add to white, we can produce as much blue light as necessary to neutralize the yellow component of 510 nm and make the field of view pure green. The measure of intensity of the blue light will show the intensity of yellow component as perceived in the 510 nm stimulus.

By cancelling the yellow part in both the green-yellow colors produced (500 to 580 nm) and in the red-yellow colors (580-700 nm) using 475 nm for the comparison light, we will obtain a curve called the *yellow response* (Fig. 9.14, the intensity is given on logarithmic scale). In a similar way, we can determine how much blue light is produced by wavelengths in the blue-red range 400-475 nm and in the blue-green range 475-500 nm using the unique yellow 580 nm as a comparison light. We will obtain the *blue response* curve (Fig. 9.15). We can also eliminate red from red-blue colors (400-475 nm) and from red-yellow colors (580-700 nm), by means of the unique green light of 500 nm and produce the *red response* (Fig. 9.16). Finally, we can cancel green in both blue-green colors (475-500 nm) and green-yellow colors (500-580 nm) by means of red light of 700 nm (Fig. 9.17). Let us now combine all four graphs in a single graph and represent, for convenience

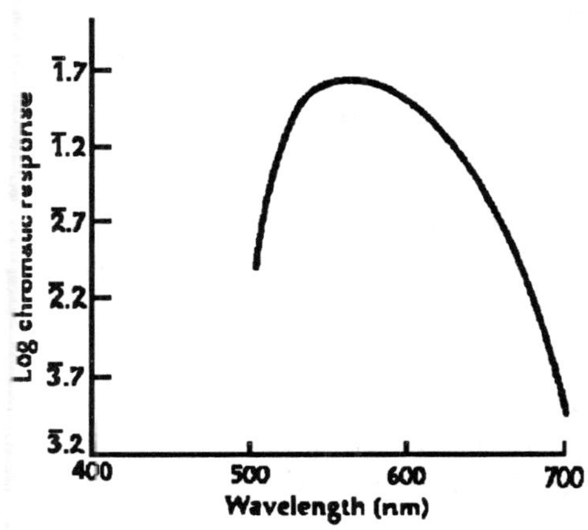

Fig. 9.14. Yellow response. From L.M. Hurvich, *Color Vision* (Sunderland, MA, 1981), p. 57; reprinted by permission of Sinauer Associates, Inc.

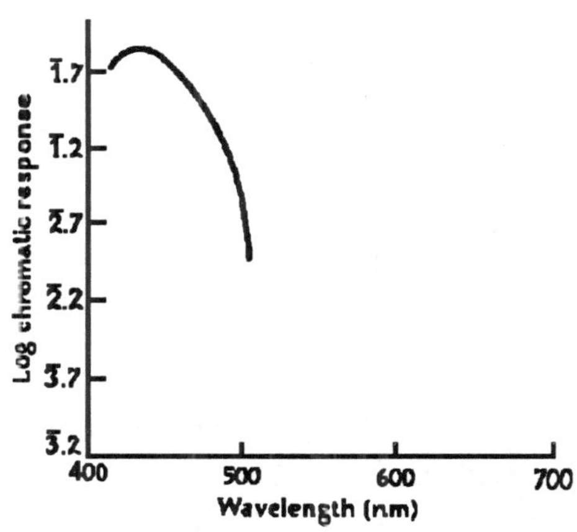

Fig. 9.15. Blue response. From L. M. Hurvich, *Color Vision* (Sunderland, MA, 1981), p. 57; reprinted by permission of Sinauer Associates, Inc.

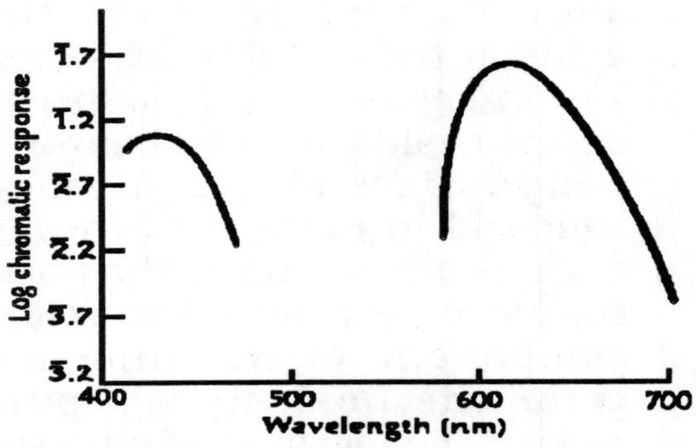

Fig. 9.16. Red response. From L. M. Hurvich, *Color Vision* (Sunderland, MA, 1981), p. 59; reprinted by permission of Sinauer Associates, Inc.

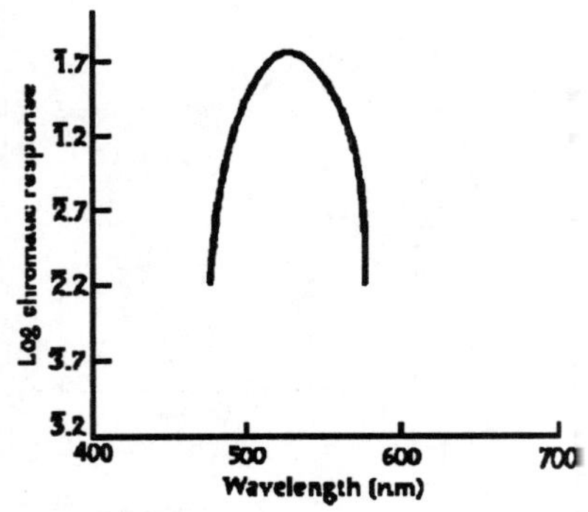

Fig. 9.17. Green response. From L. M. Hurvich, *Color Vision* (Sunderland, MA, 1981), p. 60; reprinted by permission of Sinauer Associates, Inc.

sake, the opposite color-responses by the curves of the opposite signs (Fig. 9.18).

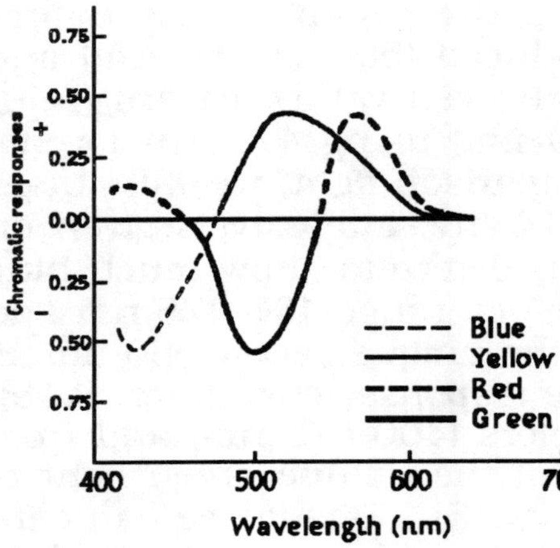

Fig. 9.18. Yellow/blue and red/green responses combined. From L. M. Hurvich, *Color Vision* (Sunderland, MA, 1981), p. 60; reprinted by permission of Sinauer Associates, Inc.

b. color match

Since most colors usually have some white in them, it is necessary to study the *whiteness response* of the eye. This curve can be obtained, for instance, by measuring the light threshold throughout the spectrum, for at the threshold of vision cones do not work and the perception of light is expressed in terms of brightness only. When superimposing this curve with chromatic response curves (Fig. 9.19) in the same graph we can find the total response of the eye to any spectral stimulus. For instance, we see that the stimulus 450 nm will create the response of 0.08 red, 0.44 blue, and .04 white. The unique hues 475 nm, 500 nm, and 580 nm correspond to the intersection of the curves with the axis. Each of these hues can be obtain by adding two properly selected hues. For instance, the unique yellow can be excited by adding 670 nm (yellow-red) and 540 nm (green-yellow) if we manipulate the intensities of these two lines so that the green would cancel red. A color complementary to the one produced by light of 450 nm must have 0.08 green and 0.44 yellow, which, according to Fig. 9.19, gives us about 590 nm.

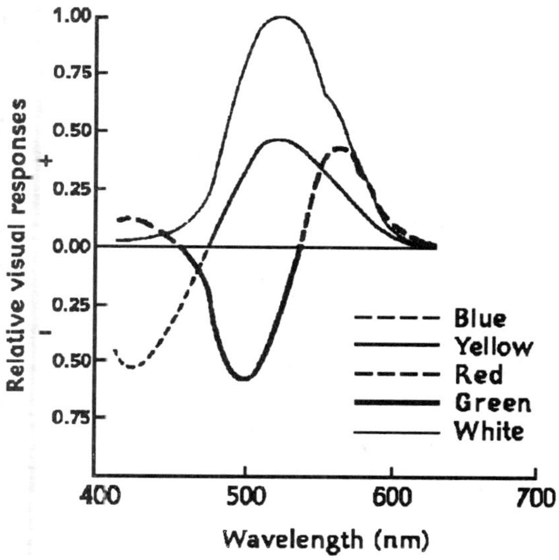

Fig. 9.19. Chromatic and achromatic responses. From L.M. Hurvich, *Color Vision.* (Sunderland, MA, 1981), p. 67; reprinted by permission of Sinauer Associates, Inc.

It is clear from Fig. 9.18 how an arbitrary stimulus can be matched by a sum of three primary hues. The primaries are usually taken from the ranges 400 to 450 nm, 510 to 520 nm, and 630 to 700 nm, however, the choice is the subject of practical considerations rather than of theoretical principles. For instance, we selected 440, 520, and 650 nm. The lines between 650 and 700 nm are matched by varying the intensity of 650 nm, the lines between 400 and 440 nm are matched by varying the intensity of 440 nm. The stimulus in the range between 440 and 520 nm is matched by an appropriate mixture of these two hues; and if the stimulus is between 520 and 650 nm, the mixture of these two will do the job. Matching the hue may be accompanied by matching the brightness if we change the intensities of the components. A perfect match will be achieved only if we also adjust whiteness. If λ is the wavelength to be matched, λ_1, λ_2, and λ_3 are matching wavelengths, a is the amplitude of the compound light, and $\mathbf{a}_k$ are the amplitudes of the components, then

$$a\lambda = a_1\lambda_1 + a_2\lambda_2 + a_3\lambda_3.$$

Sometimes, one of the amplitudes $\mathbf{a}_k$ may be negative, which means a complementary color, or that this wavelength will be added to the test field instead of the investigated field. Recently, the theory of three primary color receptors obtained a direct support. Using microspectrophotometrical methods scientists measured responses of visual pigments in single cones of human retina and discovered three peaks of sensitivity: 450 nm, 525 nm, and 555 nm (Fig. 9.20). They interpreted these peaks as belonging to three classes of cones. Similar experiments with monkey cones gave maxima of 445 nm, 535 nm, and 570 nm.

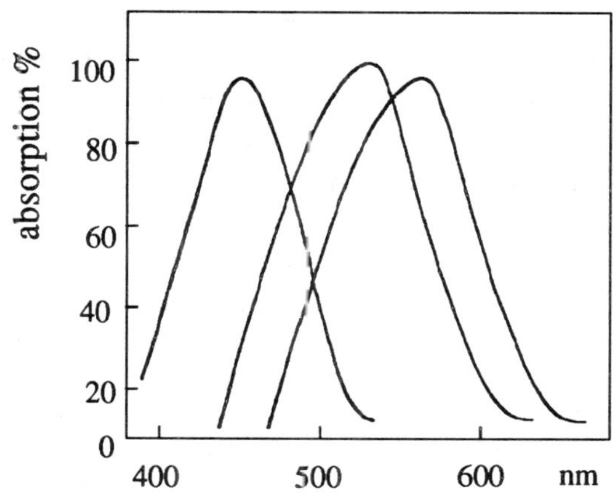

Fig. 9.20. Three classes of cones. The heights of the curves are arbitrary.

review

2. PHYSIOLOGICAL RESEARCH

a. optical potentials

To obtain direct evidence of how the receptors respond to light it is necessary to measure the electrical potentials produced by neural cells. Among the bioelectrical signals the best known is the action potential, which travels large distances and changes only the rate of pulses but not their magnitude. The resting neural cells usually fire at a slow rate. When a stimulus is applied the magnitude of the potential remains the same but the rate of firing the pulses either increases (stimulation) or decreases (inhibition). Not all neurons equally respond to the turning light on and off. For instance, in the frog's eye some neurons produce a short burst of pulses in both cases; others do it only when light is turned on, and the third group stops firing when light is "on" and begins firing again when the light is "off."

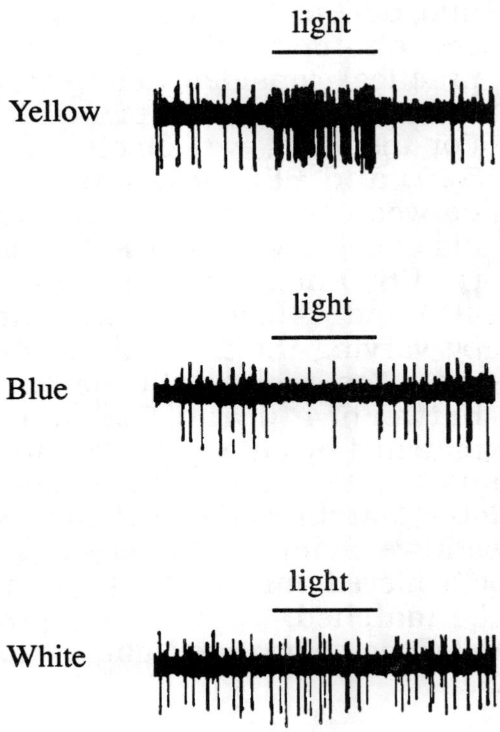

Fig. 9.21.

The local potential travels only a short distance and has a variable magnitude that may depend on the energy of light, which involves the wavelength. Some cells in the lateral geniculate nucleus of a rhesus monkey are excited by yellow light and inhibited by blue light, while white light has little effect (Fig. 9.21). Such cells are called the *opponent color cells*. These cells are responsible for the phenomena of successive contrast or afterimages.

b. afterimages

lab

Teacher. Take a colored rectangle, cover one half of it with white paper and stare for 20-30 sec at the colored part. Then remove the white cover and report what you see.

Dorothy. The covered part appears to be darker than the open one.

Ruth. Perhaps more saturated than darker.

Teacher. Here is another experiment. Look at a small white strip on black background for at least 30 sec and then either shut your eyes or continue to look at it. What do you observe?

Michael. When I shut my eyes the white strip appeared darker and the background lighter.

Mary. I continued to look and saw a similar phenomenon.

Teacher. And now let us try the same thing with a colored rectangle. After staring at it for 30 sec cover it with white paper.
Dorothy. I saw a green rectangle instead of a red one.
David. My rectangle was blue, and the image was yellow.

Teacher. Colored afterimages were discovered in 1742 by Buffon, who called them "accidental colors." When looking for some time at a red square against a white background he began perceiving a faint green color around it, and when he turned his eye away he saw a green square. In the same way, yellow upon white produced a pale blue, blue upon white a pale red, etc. The law of mixing of accidental colors was the same as for the natural ones. For instance, an accidental green color produced by a natural red when projected upon red background produced yellow, and an accidental blue color produced by yellow appeared green upon yellow background. Aepinus investigated the accidental colors seen by a person after he looked at the sun or another very bright object. He found that at first, the image appeared yellow, then green, and finally blue. While looking upon white paper when the effect of sunlight on his eye continued he observed a brown image verging towards red. This corroborated the observation of Lahire: if after looking at the sun one shuts his eye, one sees first the red color, then yellow, and then green and at last blue. **history**

The complementarity of afterimages suggests an involvement of antagonistic neural processes. Imagine, for instance, a cell which is excited by red light, inhibited by green light, and indifferent to white light. If after looking at a red object we turn the eye to a white one, it is equivalent to switching the stimulus off, or an inhibition. Since the latter procedure is associated for this cell with green light, it reacts so as to produce a sensation of green color. Colored afterimages can even be produced by a slow sequence of black-and-white stimuli. This is called the *subjective colors* or "Fechner-Prevost colors." A disk with a white/black pattern making from 5 to 15 revolutions per second produces colored appearances. The best known pattern is of Benham's disk (Fig. 9.22). When rotated clockwise, the outermost lines look bluish, the center greenish, and the innermost lines lines reddish. When rotation is counterclockwise, the order of colors is the opposite: the innermost lines are bluish, and the outermost ones reddish. ***review***

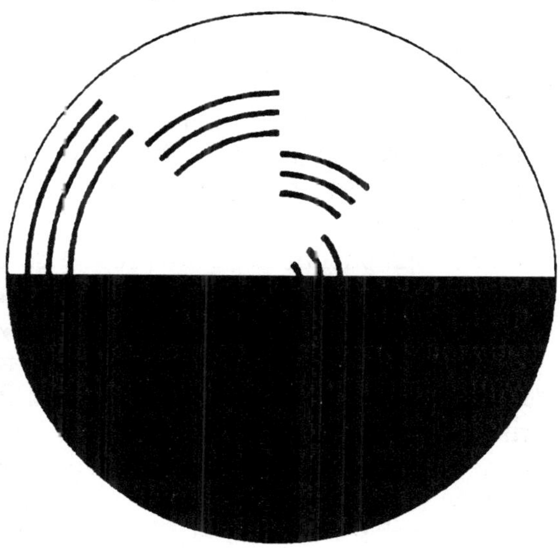

Fig. 9.22. Benham's disk.. From L. M. Hurvich, *Color Vision*. (Sunderland, MA, 1981), p. 190; reprinted by permission of Sinauer Associates, Inc.

c. simultaneous contrast

1. Color shadows

lab

Teacher. Take two small gray squares and place one on a white paper and the other on a black one. Compare the appearances of these squares.

Michael. The one that is set against white paper is darker than the other.

Teacher. Thus, the background (or surround) affects our perception. Let us now try colors. Set two pieces of paper of the same color against different colored background. What do you see? Have only a brief look in order to exclude an afterimage.

Dorothy. Yellow appears reddish against a green surround.

John. A blue strip held against a red surround appeared greenish.

Ruth. I saw a yellow object placed on red paper becoming greenish. It appears from our observations that the additional hue is complementary to that of the surround.

Teacher. Very good! Now let us try projecting bodies using a colored light. Darken the room. Take your desklight and cover the housing's window with a red filter. Hold a white screen at about 50 cm and project your finger on it. Before doing it predict the color of the shadow.

Dorothy. Shadows are always black.

Mary. Wait a second! Since red light will create a red surround for the shadow, the shadow should appear greenish rather than black.

Teacher. Very well, test your hypotheses.

Dorothy. What did I tell you: it's black!

John. Perhaps it's black only because this light is too weak.

Teacher. Perhaps, let me turn on another desklight.

Mary. Wow! It's green! I was right!

David. This is strange: the other light has no filter on. How can an ordinary light produce a green shadow?

history

Teacher. This phenomenon was first described by Georges Buffon (1707- 1788) in July 1742. The sun was near the horizon, very large and very red. Buffon noticed that the shadows of the trees 30 feet from the wall were colored with light bluish green and the shadow of another object three feet from the wall was very green. The appearance lasted five minutes and vanished with the sun. The next morning he found that the rising sun produced shadows of the indigo color. Subsequently, he usually observed blue shadows. He said that everyone could see those colors at sunrise or sunset by simply holding a finger before a sheet of white paper. Beguelin claimed to have observed the blue shadows indoors any time of the day with direct or reflected sunlight, and, in particular, in the places not reached by the light of the blue sky. He also observed a blue shadow from an object illuminated by a candle and by the morning twilight. The 18th century scientists tried to explain the blue shadows in the same way as the color of the sky or of clouds: by reflection, refraction, and absorption of light. We'll discuss the modern explanation after studying the *Mach contrast*.

2. Mach contrast

Teacher. ***Ernst Mach*** (1838-1916), physicist and philosopher, is primarily known for his work in acoustics. He discovered an effect of contrast (Mach contrast) which you are going to reproduce. Hold an index card 1-2 cm over a sheet of white paper and observe the edge of its shadow cast by a desklight. Do you see anything unusual? ***discussion***

Mary. I see a narrow black band at the shadow's border, then there is a gray band less than 1 cm wide, and the rest of the shadow is black.

David. If I am not mistaken, there is also a narrow white band near the black one, on the lighted part of the paper. It is brighter than the lighted part because there is a fine border between them. I would also add that the darkness of the "black band" increases from the white band inside the shadow.

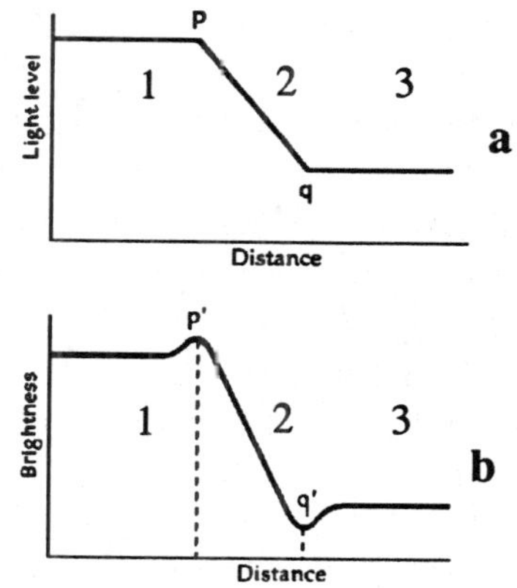

Fig. 9.23 Mach's contrast.. From L.M. Hurvich, *Color Vision.* (Sunderland, MA, 1981), p.165; reprinted by permission of Sinauer Associates, Inc.

Teacher. Your observations are correct. Let **1** be the lighted part, **2** the black band, and **3** the gray band (Fig. 23a), **p** the location of the white band, and **q** the location of the darkest part of the "black band." Let us mark as **p'** and **q'** (Fig. 9.23b) the brightness at these points showing that **p** is brighter and **q** is darker than the surround. Recently, it has been been found that a single ganglion cell responds to illumination of a large part of the retina (receptive field) but the response differs for different parts of this area. For instance, when the center is stimulated the cell fires when light is "on", while at its periphery or surround area the cell fires when light is turned off. Thus, the central part is excitatory and the periphery is inhibitory (Fig. 9.24). If a weak light is falling on the center of the receptive field, there is a small positive response. However, if at the same time the surround is illuminated stronger, this inhibits the cell, and the result is such as if there were

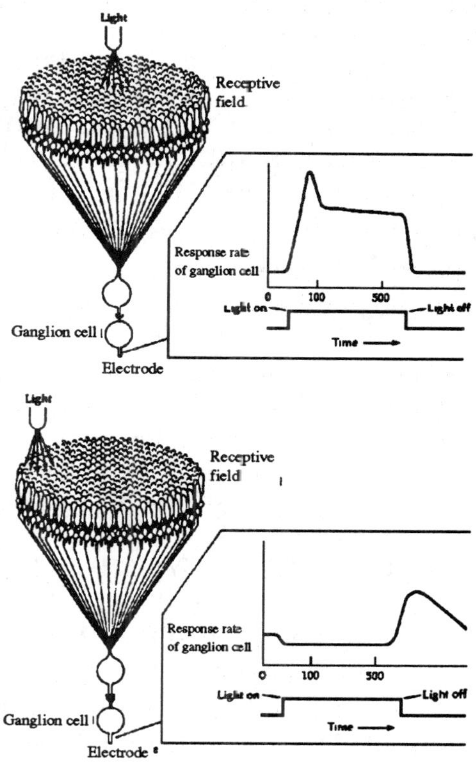

Fig. 9.24. From L.M. Hurvich, *Color Vision.* (Sunderland, MA, 1981), p. 170; reprinted by permission of Sinauer Associates, Inc.

even less light at the center: the gray center appears blacker. This explains an ordinary spatial contrast. To explain the Mach contrast let us assume that the eye moves from the surround **1** to the center **2** (Fig. 9.23). When crossing the border **p** the inhibition of the neural response will decrease, and this edge will appear to the eye brighter than the area **1**. On the other hand, when coming from the center **2** to the surround **3** and crossing the border **q**, the inhibition increases, and this edge will appear to be darker than the area 3.

Ruth. Can we apply this to a colored shadow too? If the shadow is a center, and the lighted area is a surround, does it mean that the surround of the receptive field is inhibited by the color of the lighted area, and as a result, the center, which is stimulated by a dark shadow produces a complementary color?

☞ *good point!*

Teacher. Very good! Cells had been found in some animals the response of which is color-specific. For instance, when the center of a cell from striate cortex of rhesus monkey produces an "on" response to a green spot the surround produces an "off" response to red light (Fig. 9.25). These cells are called *double opponent* cells. They may be involved in simultaneous contrast effect. For instance, if the screen around the shadow is yellow-red ("blue shadows"), the surround is inhibited; moving the eye from the screen to the shadow removes this inhibition, and the cell fires as if the shadow were stimulated with a complementary blue-green color.

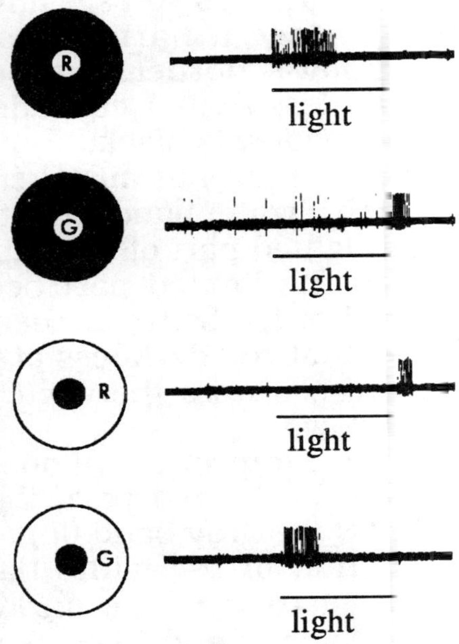

Fig. 9.25.

d. color vision deficiencies

discussion

Michael. I confuse red and green colors, but I have no problems with the blue and yellow. How many defective receptors do I have: one or two?

Teacher. One: like John Dalton, you belong to a group called *dichromats* because they have only two receptors properly functioning. Your red/green system is not activated by light of any wavelength, as shown in Fig. 9.26. Thus, light of, say, 520 nm (the green-yellow region) will produce in your eye a yellow color which is much brighter and less saturated than for the normal eye, because of the extra amount of white light corresponding to the green component.

Dorothy. What is so magic in the red and green colors that quite a few people can't see them?

Teacher. Nothing, because there are dichromats can't see blue and yellow as shown in Fig. 9.27. Moreover, there are a few people called

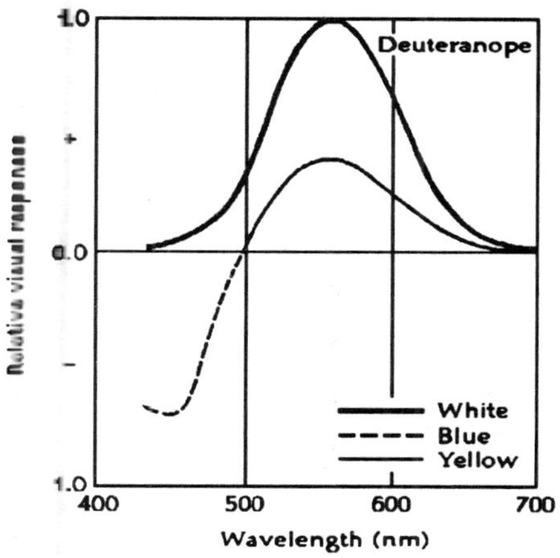

Fig.26. From L.M. Hurvich, *Color Vision* (Sunderland, MA, 1981), p. 243; reprinted by permission of Sinauer Associates, Inc.

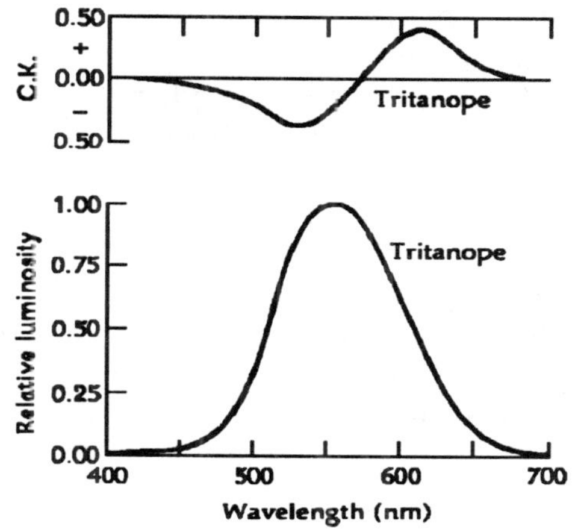

Fig. 27. From L.M. Hurvich, *Color Vision.* (Sunderland, MA, 1981), p. 255; reprinted by permission of Sinauer Associates, Inc.

monochromats who cannot see any color. The world for them looks as the one in a black-and-white TV. Monochromats have two receptors malfunctioning, and the remaining one cannot produce an opponent-type neural signal. There is also another defect of color vision which is common to people with any number of functioning receptors, including the trichromats. It consists of a 10-20 nm shift of the whole spectrum in the direction of either the red light *(deuteranopes)* or violet light *(protanopes)*. A protanope will see, for instance, the unique blue light of 475 nm as blue-red, while a deuteranope will perceive the same light as blue-green. Physiologically, this is due to a shift in the photopigment absorption. As to the physiological cause of the dicromatic and monochromatic deficiencies, it is not clear yet whether they result from defects in specific photopigments or in the neural system.

Michael. Can a dichromat learn to distinguish the missing colors?

Teacher. Yes. Carry two good color filters with you, as a sort of eye-glasses, with the names of the filters clearly marked on them. In your case, a green and a red filters are necessary. If you'll look in turn through each of these filters at a rose bush, you would easily distinguish the colors of leaves and roses.

V. EXPERIMENTS

1. Glass color mixer

(student instruction sheet)

Equipment. A 3x5 in index card, colored paper, a microscopic slide or another clear plate, paper clips. ***lab home***

Procedure. Fold the card so as to make the right angle, tape a red paper strip to one side and a green one to the other side. Press the slide to the fold so as to make 45° with the horizontal surface. When looking at the glass from above (Fig. 9.28) you will see a mixture of two light beams: one is reflected from the vertical surface and then from the glass, and the other is reflected from the horizontal surface and then transmitted through the glass.

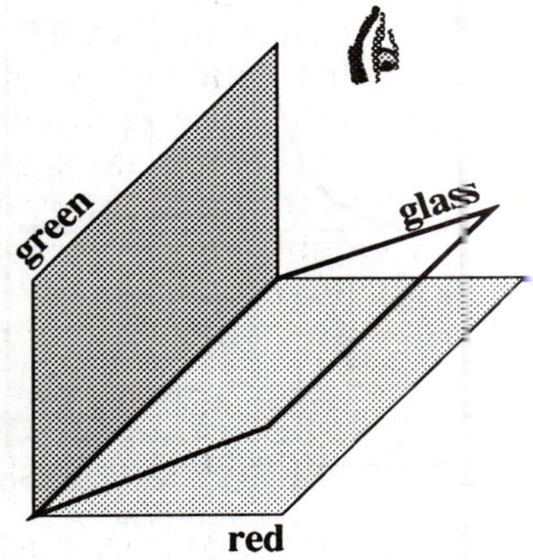

Fig. 9.28.

Investigate.
1. How does the compound color vary when you change the angle which the slide makes with the horizon from 0° to 90°?
2. Which of the two colors prevails in the compound at an arbitrary position of the glass: the one closest to the glass or the farthest from it?
3. Which features of color (intensity, hue, or tint) change with a change of the angle?
4. Is this an additive or subtractive mixture?
5. What color do you expect to see from a combination of red and green colors? Check your prediction. Repeat the experiment with other pairs of colors.

Conclusion. Can you make white by mixing two colors in this experiment?

Teacher's note. Varying the angle of the glass changes the intensity of reflected and refracted light: at a small angle of incidence most light is refracted, near 90° all light is reflected. Thus, if you want one color to prevail, bring the glass close to the corresponding surface. This method of mixing colors is additive

2. Maxwell's tops
(student instruction sheet)

lab

Equipment. Cut a cardboard disk of about 10 cm in diameter and make a hole in the center so as to fit tightly to the shaft of a small DC electrical motor. The motor may be powered by two D-size 1.5v batteries or by a DC adapter 6-9 v, 100 - 300 ma. Cut out two sets of paper circles of different colors, one of 10 cm in diameter and the other of 5 cm, and make a radial cut in each of them (Fig. 9.9). Also make a white and black circles of the smaller diameter.
Procedure. Place two color circles of the larger diameter on the disk so as to obtain two colored sectors Turn the motor on and watch the color obtained through rotation.

Investigate.
1. How does the compound color change when you vary the relative size of the open sectors of the same pair of circles? Why does such a change

affect the color perceived?
2. Try other pairs of colors. Can you obtain white color by using complementary pairs (blue-yellow and red-green)?
3. To determine the compound color, place on the disk three circles of the smaller diameter (two colored and white or black) and adjust their sectors so as to make the same color in both sets of sectors.
4. Try to combine three colors, in particular, red, yellow, and blue; or, red, green, and violet in order to obtain white. For a comparison, use a combination of white and black circles smaller circles.

Conclusion. Have you obtained a white image? an uncolored one? What is the difference between them?

Remember: change only one variable at a time.

Teacher's note. Obtaining gray instead of white is not a failure: what matters is discoloration. "White" differs from "gray" only by certain amount of "black."

3. Color shadows

a. shadows in colored light

(student instruction sheet)

Equipment. Two desklights with housings; the housing has a window 3x3 cm provided with a cardboard track to insert color filters, white screen. Paul Lombardi (South Davis Junior High, Bountiful, UT) uses a single housing with three windows which gives an opportunity to combine two or three beams. To converge the beams he sets two mirrors on lens-holders and places them inside the housing behind the bulb. There are opaque lids to cover the windows which are not in use. For a home experiment, use two candles and make two free-standing cardboard C-shaped screens 15 cm x 20 cm, each with a window covered with a color filter. Provide a window with two cardboard rails to make the replacement of a filter easier. Place the two screens at about 50 cm from a white screen and at an angle to one another. To prevent a candle from overturning make a stand for it from a cardboard 4 cm x 5 cm. In the middle of it make two perpendicular cuts with a razor blade slightly longer than the candle's diameter, unbend the four corners, and insert the candle. ***lab home***

Procedure. Direct two light beams towards the white screen. Intercept one beam with a red filter and another with a blue filter. Observe the shadows of narrow and wide objects.

Investigate whether the color of a shadow depends on:
1. the color of the beam which produces this shadow;
2. the color of the other beam;
3. the distance from the light source to the object, or the distance

from the object to the screen;
4. the color of the body;
5. the color of the screen.
6. If the colored shadows overlap, do the colors mix in the additive or subtractive way?
7. Is it possible to obtain a colored image if one of the two beams is uncolored?

Conclusion. How to obtain a shadow of a specific color, for instance, red?

Remember: change only one variable at a time

Teacher's note. Changing the distance from a light source to a white screen changes the intensity of the image which may affect the color. To use uncolored light remove one of the filters and move this light source back until you see the color of the image clearly. For instance, with blue and red filters the images are orange and bluish-green (complementary colors!), while with a blue filter and an open candle light the image is yellow.

b. blue shadows

(student instruction sheet)

home

Procedure. At sunset or sunrise, within 5-10 minutes of the total disappearance of the sun, you may see that your finger casts a color shadow on a white screen. Describe the color.

Investigate which of the following factors affect the color of the shadow:
1. the object's dimensions or shape;
2 the color of the object;
3. the distance between the object and the screen;
4. the color of the screen;
5. time left to the total disappearance;
6. Can you observe the phenomenon outside?
7. Can you see a similar phenomenon at another time of the day?

Conclusion. When and where can this phenomenon be seen the best?
Remember: change only one variable at a time.

Teacher's note. The color is blue, sometimes greenish-blue. It cannot be seen during the day because sunlight then is not sufficiently colored. You can imitate this experiment with an electric light instead of the sun. Take a cardboard box, open its narrow side, attach a white screen to the opposite side, and fix a pencil in the middle of the box by means of *Tak* . When you shine light from an incandescent bulb inside the box the pencil's shadow appears bluish. You may need to find a proper distances between the light and the box and also between the pencil and the screen.

4. Afterimages
(student instruction sheet)

Equipment. Pieces of colored paper 10 x 10 cm for a background, small color squares 1 cm x 1 cm. *lab home*
Procedure. Observe a small red square on white background for 30 sec without moving the eyes, then move the eyes to white background or cover the square with white paper and look at it. What do you see?

Investigate whether the effect depends on:

1. the color of the square;
2. the color of the background (start with white);
3. the color of the cover;
4. the length of time of observation;
5. on the color of light.

Conclusion. What would you do to make afterimages more noticeable?
Remember: change only one variable at a time

Teacher's note. A colored background changes the color of an afterimage because of a mixture of the two colors. Increasing the observation time makes a brighter image but only up to a limit, which is for many people around 30 sec. To make colored light use a desklamp with a color filter.

5. Benham's disk
(student instruction sheet)

Equipment. A small DC motor powered by two D-size, 1.5v batteries or by a 3v DC adapter, a cardboard disk 5 cm in diameter, a paper circle with the Benham pattern on it. *lab*
Procedure. Install the disk on the shaft of the motor and cover it with the Benham circle. Turn the motor on and watch the circle from above. Describe what you see.

Investigate how the effect changes if:
1 you change the angle at which you look at the disk;
2 you relax your eye (focus at infinity);
3 change the speed of rotation;
4 change the light illuminating the disk (for instance, from daylight to incandescent light, or to sunlight).

Conclusion. How to conduct the experiment to achieve the strongest effect?
Remember: change only one variable at a time

Teacher's note. Colors change from the periphery to the center from reddish to bluish or the vice versa, depending on the direction of rotation. To gradually slow down the rotation press a sheet of paper against the spinning disk.

BIBLIOGRAPHY

R. Boyle, *Experiments and Considerations*, part 3.

J. Priestley, *History of Light*, pp. 438-445, 631-634.

T. Young, "On the Theory of Light and Colours," in his *Miscellaneous Works* 3 vols (London, 1855), I: 140-69 (p.147).

T. Young, "Chromatics [1817]," ibid., pp. 280-283.

T. Young, *A Course of Lectures on Natural Philosophy and the Mechanical Arts*, 2 vols (London, 1807), I:437-40.

D. Brewster, "On a new analysis of a solar light, indicating three primary colours, forming coincident spectra of equal length," *Trans. Roy. Soc. Edinburgh 12* (1834), 123-36.

H. Helmholtz, "On Sir David Brewster's new analysis of solar light. " *Phil. Mag.*, 4 (1852), 401-16; "On the theory of compound colours," ibid., 519-534

H. Helmholtz, *Treatise on Physiological Optics*, 3 vols (New York: Dover, 1962; a facsimile of the 1924 translation of the 1909 German edition), I: 120-165.

H. Grassmann, "On the Theory of Compound Colours," *Phil. Mag.*, 7 (1854), 254-64.

J. Maxwell, "On the theory of colours in relation to color- blindness [1855], in his *Scientific Papers*. 2 vols (Paris, n.d.; a facsimile of the Cambridge edition of 1890), I: 119-125; ; "Experiments on colour as perceived by the eye [1855]," ibid., 126-155; "On the theory of compound colours [1860]," ibid., 410-447.

C. Everitt, "Maxwell, James Clerk" in *Dictionary of Scientific Biography*, ed. by Ch. Gillispie, v. 9 (New York: Charles Scribner's Sons, 1974), pp. 198-202.

K. Brown & G. Wald, "Visual pigments in single rods and cones in the human retina," *Science* 144 (1964), 45-52.

L. M. Hurvich, *Color Vision* (Sunderland, MA: Sinauer Associates Inc. Publishers, 1981).

Adler's *Physiology of the Eye, Clinical Applications* (St. Louis: C. Mosby Co , 1987), chs. 22, 23.

Chapter 10

INTERFERENCE of LIGHT

I. PHENOMENA OF INTERFERENCE

Teacher. The term *phenomena of interference* appeared in the 19th century and was applied to some of the phenomena previously called "periodical colors." This term referred to alternating fringes of different colors seen, for instance, in soap bubbles, around narrow obstacles, etc. The new term implied a specific explanation of these phenomena, namely through the principle of interference of light. However, the phenomena themselves had been known long before the discovery of the principle of interference and were explained differently at the time. We are going to examine one of these phenomena, first experimentally and then theoretically.

1. Colors of thin plates

Preliminary part

investigation

Equipment. Two microscopic slides, paper, scissors.

Procedure Put together two microscopic slides, while supporting them with index fingers from below, use your thumbs to move the upper slide over the lower one with so much friction as to produce colors. The colors are seen better against a dark background. If you succeeded in obtaining colors try to change them.

Experiments

Dorothy. I don't see any colors.

Michael. We obtained some, mostly near the edges.

Mary. I saw the colors even in the middle of the plate in the form of irregular circles and rings. They alternate: green, red, green again, red again, and so on.

David. I saw colored rings near my thumb so long as I pressed it to the glass. When I removed the thumb the rings began contracting, until they completely disappeared.

Ruth. I've noticed a change in colors when I looked at the same place at different angles without changing the pressure.

Formulating a problem

What is the cause of the colors created by two glass plates?

Selecting variables

Teacher. Name the factors which affected the appearance of colors.

Mary. Pressure and the angle of reflection. Thus, we have two variables.

Teacher. Since pressure is not an optical concept, its influence must be indirect. We have to find optical factors affected by pressure, and those will be our variables.

John. Perhaps, pressure changes the **shape** of plates: we know that if surfaces are not parallel, the image is distorted and may even be colored.

Main part

a. shape of the plate

Hypothesis

Michael. Let us suppose that pressure bends the plates into a prism

Test

Ruth. We've bent a single plate (one has to be **careful** not to break the glass), but no colors appeared.

Dorothy. We've also tried it. No luck!

conclusion ➡ The hypothesis is wrong: colors are **not** due to bending of plates. A single plate doesn't create colors.

b. shape of the space between plates

Hypothesis

Mary. Perhaps two plates make an air prism which produces colors.

Test

David. This sounds unlikely, because a prism produces only a **single** spectrum. However, we can test it directly by separating the plates at one end.

Michael. We placed a piece of cardboard between the plates near one edge and pressed the plates together, but we haven't seen any colors, even a single set.

conclusion ➡ The hypothesis is false: the shape of the air gap between the plates is **not** important.

c. substance between plates

Hypothesis

Dorothy. Perhaps the colors are due to the air between the plates

Test

John. Let us remove the air from between the plates. Our group replaced air with water, and observed no change in coloration.

Mary. We did the same and found small bubbles between the plates, which implies a mixture of air and water, but the colors are there nonetheless!

Dorothy. We put some vinegar between the plates and obtained colors, similar to those with water.

conclusion ➡ The hypothesis is false: the air is **not** necessary to produce colors.

d. distance between plates

Hypothesis

Ruth. Let us summarize the facts once more: 1) we need two glasses but not to hold a specific substance between them; and 2) we need pressure but not for bending the plates. We can reconcile the two by assuming that the distance between the plates plays some role. ***good point!***

Test

David. We can obtain a variety of distances between the plates in a single trial by inserting something (for instance, a piece of paper or cardboard) between the plates on one side. ***good point!***

Dorothy. We did it with cardboard and the result was negative.

Mary. We inserted a paper strip, and it worked. The colors appeared, however, only on the side opposite to the paper strip. No colors appeared near the strip however hard we pressed. This means that the distance necessary to produce colors is much smaller than the thickness of paper.

conclusion ➡ | The colors are produced by two glass plates put together only if they have a **very small** distance between them. Pressure is needed to reduce the distance between the plates.

Teacher. We still have the angle of reflection to investigate. Mark a point on the upper plate and watch how the colors will change when you increase the angle of reflection.

e. angle of reflection

Preliminary experiments

Dorothy. When I looked at the plates at a smaller angle I saw a blue ring; when the angle increased the ring became green.

Hypothesis

David. The greater the wavelength the greater the angle of reflection at which we see that color.

Test

Michael. We've seen the red color at a larger angle than the yellow one, and the yellow fringe appeared at a greater angle than the green one.

conclusion ➡ | The angle of reflection increases with the wavelength. ***important***

Teacher. Very good! You've repeated some of the experiments of Robert Hooke (1635-1703) and Isaac Newton (1642-1727). Let us now see how they explained this phenomenon.

2. Hooke's theory

history

A sickly boy who barely survived childhood, Robert enjoyed making mechanical toys. In 1653, he came to Oxford to become Robert Boyle's assistant, involved in making scientific instruments and performing experiments. In recognition of his outstanding mechanical skills, Hooke became curator of instruments and demonstrator of the newly founded Royal Society of London. Every week he had to prepare several major experiments and demonstrate them at the Society's meeting. This greatly stimulated his interest in scientific research and inventions. On the other hand, it prevented him from realizing in full many of his brilliant ideas and dragged him into priority disputes with Huygens, Newton, and others. Hooke invented an air-pump, a spring balance, which eventually made possible the watch and the maritime chronometer, and many other instruments. He was one of the first to systematically use a microscope for biological observations which he described in his book *Micrographia*. He discovered "Hooke's Law" for elastic deformations, influenced Newton in his work on universal gravitation, gave a theory of the origin of fossils, and distinguished himself as an architect.

In his *Micrographia* Hooke described multiple spectra in thin pieces of mica. He found that the order of these colored rings was inverse to that in the primary rainbow, and that by pressing a particular place he could change the color there. By splitting mica further and further he obtained pieces which exhibited a single color. Hooke supposed that colors depend on the thickness of mica and that this is a property of all thin transparent substances. He confirmed this hypothesis through observations of a very thin glass film, a thin film of air or of a liquid between two glass plates, soap bubbles, and finally of the film that covers tempered steel. Hooke thought that light was a periodical sequence of pulses and he explained the colors of thin plates as follows.

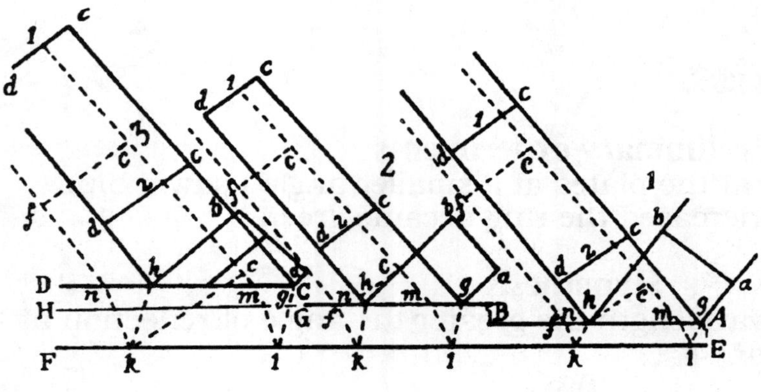

Fig. 10.1. Hooke's theory of thin plates. From Hooke, *Micrographia* (1665)

The pulse **ab** (white light) falls on the font surface of a plate **AB**, which reflects a portion of it as the pulse **cd.** The rest of it is refracted and then reflected by the back surface **EF** as the pulse **ef** (Fig. 10.1). Naturally, the pulse **ef** is weaker than the pulse **cd**. When the combination of two pulses enters the eye it produces a perception of blue color when its weaker part precedes the stronger; and of red color when the stronger part precedes the weaker one. Since the position of the weaker pulse between the two following stronger pulses depends on the thickness of the plate, the latter will determine the resulting color (Hooke, 47-56, 65-7).

3. Newton's theory

history

Newton adopted Hooke's idea that the color of a thin film depends on its thickness. For a film of air between two lenses he was able to find the relation between the radius of a colored ring and the thickness of the film. From his measurements Newton found that the thicknesses that corresponded to the brightest part in consecutive spectra formed an arithmetical progression 1, 3, 5,... while the thicknesses in their darkest parts were 2,4,6,... ... The unit in these progressions was 1/178,000 of an inch, which was the thickness of the film at the brightest part in the spectrum of the first order, which Newton associated with yellow color. He found experimentally the ratio of the thicknesses that corresponded to the red and violet parts of the same spectrum, and using an acoustical-optical analogy, calculated the thickness of the film for every color (*Opticks*, Book II, 193-224). How difficult these quantitative experiments were may be seen from the fact that no one repeated them until the middle of the 19th century. While acknowledging Hooke's experimental results on thin films, Newton rejected his explanation of these colors and offered his own.

In Newton's view, light can periodically change its property of being reflected or refracted (he called this "fits of easy reflection and transmission"). This meant that after entering a refracting medium, at every distance from the surface multiple of 2/178,000 in. light acquired a property to be transmitted further, while in the middle between these distances it could be reflected back. Therefore, light either passes through a film or returns, depending on whether the film's thickness consists of an odd or even number of 1/178,000's of an inch (Fig. 10.2).

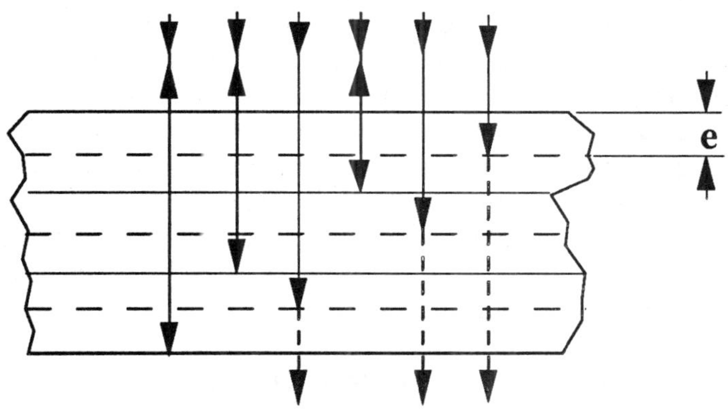

Fig. 10.2. Newton's "fits of easy reflection & transmission".

Newton called the distance between the two consecutive points of easy transmission or easy reflection the *interval of fits*. Its magnitude depended on the color and on the index of refraction of the film. Thus, to some extent the interval of fits resembled the wavelength. Although Newton's concept of periodicity of light differed somewhat from the modern one, he was able to obtain some interesting results with its help. The most important of these was calculating the dimensions of colored rings produced by concave glass mirrors ("colors of thick plates"). Newton's measurements confirmed his theory with great precision (*Opticks*, 278-315).

The first wave explanation of the colors of thin films was provided by Thomas Young.

history

II. PRINCIPLE OF INTERFERENCE OF LIGHT

1. Thomas Young and his discovery

Teacher. Thomas Young was the eldest of ten children in a family of Quakers. His grandfather stimulated his interest in classic literature, and his neighbor, a surveyor, acquainted Thomas with mathematical and physical instruments. Most of his knowledge he obtained by his own efforts. Thomas first distinguished himself in classical languages. At the age of 14, he was more or less versed in Greek, Latin, French, Italian, Hebrew, Persian, and Arabic. While studying botany he decided to make a microscope from a description in a book. For this purpose, he acquired a lathe and for a while abandoned science for learning the art of turning. Then he found in a book some fluxional symbols and did not stop until he mastered an introductory calculus. Classical scholars were greatly impressed with his Greek translations, but Thomas decided to pursue a career in medicine. In 1792, Young began his medical studies in London, continuing them in Edinburgh and finally in Göttingen. He became interested in physics because he believed that a thorough knowledge of that science was important for becoming a good physician. In 1796, while preparing for his medical degree in Göttingen, Young studied the formation of human voice, which led him to acoustics. Upon returning to England in 1797, he found that to practice medicine he had to get a medical degree from an English university. Thus, he enrolled at Cambridge, and since he was already well prepared for the examinations, Young devoted his time to acoustical experiments. While studying beats of sound he discovered in 1799 the principle of interference for sound: two sounds may not only reinforce but also destroy one another. In 1801, he generalized this idea for all sorts of waves, including light. This meant that light added to another light can produce darkness. Young first presented this new concept to the Royal Society of London in November 1801 in his article "On the Theory of Light and Colours." He called it *principle* (or *law*) *of interference.*

discussion

Dorothy. You mean that turning on two lights instead of one can make a room darker?

Teacher. No, the principle of interference is applicable not to every light. Neither two electrical bulbs, nor two candles, nor any other two independent sources of light can produce interference.

John. Can we produce darkness by combining direct light from the sun with sunlight reflected by a mirror?

Teacher. No. Another limitation concerns the angle between interfering rays: it must be close to zero, which does not take place in the experiment you have described.

Mary. How about two mirrors set at almost 180° to one another and reflecting sunlight? Will this work?

Teacher. A similar idea was realized by Augustin Fresnel in 1816. To succeed, this experiment has to fulfill another condition: the two rays must differ very little in the length of their routes. This is possible if the reflection occurs at a very small angle. Besides, for this experiment, the source of light must be very small, and sunlight can be used only after passing through a pinhole.

important

Michael. So many restrictions! It is probably a very rare phenomenon, which would be difficult to observe or reproduce. If so, what is the point in studying it?

Teacher. Indeed, interference of light occurs much less frequently than reflection or refraction of light. It can be observed only under the following conditions (called *conditions of coherence*):

The interfering rays must:

1. originate from the same light source;
2. have the same wavelength (color);
3. have about the same intensity;
4. form a very small angle between themselves;
5. have a very small path difference; and
6. in some cases, come from a luminous body of a small angular size.

important

When these conditions are not fulfilled our eye cannot detect any interference pattern. We study this phenomenon because of its great theoretical and practical value. As to its rarity and difficulty to observe, you simply don't know that some well known phenomena belong to this category. For instance, the colors of soap bubbles, with which you all have played are produced by interference. The same is true about the multicolored spots in pools of water contaminated by gas or mineral oil.

Mary. Do colors we've observed between glass plates also result from interference?

Teacher. Precisely.

Ruth. How can this be? These colors have another explanation!

Teacher. You mean **two** other explanations. There is nothing wrong with advancing a new explanation of the phenomenon which had already been accounted for. The purpose of any new theory is to give a fuller explanation of the given phenomenon and to extend its application to other phenomena as well.

important

Dorothy. But how can the principle of interference explain colors? When light destroys another light, that means darkness rather than colors.

Teacher. This is correct as applied to monochromatic light. According to Young, if two rays of, say, red light meet after traveling different routes, and their path difference at the meeting point is an integer number of the wavelength for red light, we will see a bright red spot. If at another point their path difference equals an odd number of the half-wavelength, this point will appear dark. When two rays of white light interfere, the red component of the first ray will interfere with the red part of the second ray, the blue light from the first ray will interfere with the blue light from the second ray, etc. Since the ratio of the wavelengths for red and blue light is about 1.5, wherever the path difference is one wavelength for red light, it will be three half-wavelength for blue light. This means that at the same point where red light will be reinforced, blue light will be extinguished. As a result, this point will appear not white but reddish. Let us now see how this theory explains the colors of thin films.

2. Young's theory of thin films.

theory

According to Young, the interfering rays are those which are reflected from two parallel surfaces (Fig. 10.3) of the film. The ray **1** is reflected from the first surface and travels the path **SCO**, while the ray **2** first is refracted and then reflected from the second surface following the route **SAEFO.** is P=SA+AE+EF+FO-(SC+CO). Let us draw **AB** perpendicularly to **SC**, and **CD** perpendicularly to **AE**. If the angle **ASC** is very small,

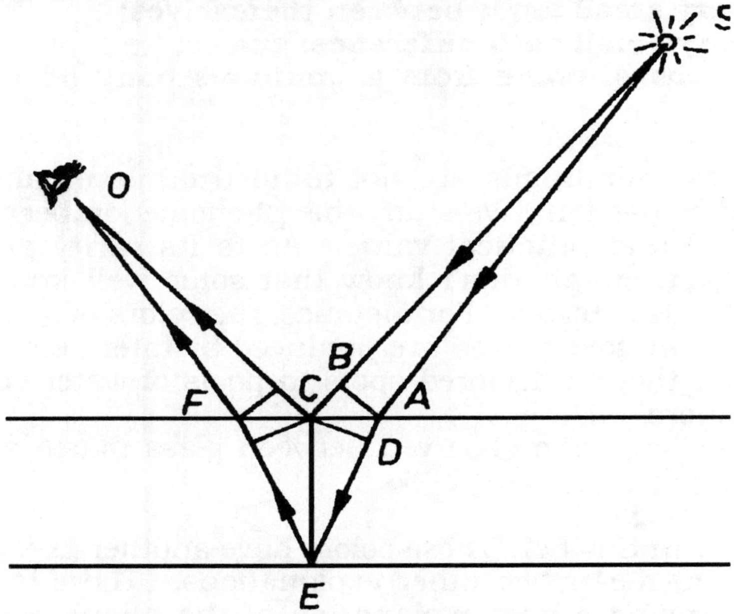

Fig. 10.3. From T. Young ,"Theory of light," *Misc. Works*, v.1, p.169.

SB=SA, and for the time necessary for light to come from **B** to **C** it will also pass from **A** to **D**. Thus, at the points **C** and **E** the ray **2** will be delayed relatively the ray **1** by nDE, where **n** is the film's index of refraction. Similarly, one can find that on the way from points **C** and **E** to the eye **O** the ray **2** will be delayed by the same distance nDE. Thus, the path difference of the two rays is P=2nDE, and since DE=CEcos*r*, we have

$$P=2tn\cos r$$

where **t** is the plate's thickness, and **r** is the angle of refraction. The results calculated with this equation differed from Newton's observations, and to reconcile the two Young supposed that a ray coming from a rarer to a denser medium to be reflected back loses half of a wavelength from its path. Thus, Young's modified equation for the path difference in reflected light is

$$P=2tn\cos r + \lambda/2 \qquad (10.1)$$

The bright fringes will be seen in the directions where P=mλ (λ is the wavelength and **m**=0, 1, 2, ... which is the spectrum's order), or where

$$2tn\cos r_{max} + \lambda/2 = m\lambda \qquad (10.2)$$

Analogously, the minima of light will appear at

math is a friend !

$$2tn\cos r_{min} + \lambda/2 = (2m + 1/2)\,\lambda \qquad (10.3)$$

Thus, if the incident light is white, the rays producing maxima are reflected under different angles for different wavelengths, as the result of which white light is decomposed into fringes of different colors. The spectra of higher orders overlap, and colors mix until they produce white. For this reason, ordinarily, only spectra of the first few orders are seen. Since the number of spectra produced increases with the path difference which, in turn, depends on the film's thickness, only extremely thin films (about 1 micron) produce colored fringes in white light. Thicker plates produce an interference pattern only in more or less monochromatic light, and the thicker the plate, the narrower the range of wavelengths must be.

important

3. Principle of interference and waves of light

discussion

Dorothy. This theory is more complicated than the previous two.
Teacher. Perhaps, but it is better.
John. Why?
Teacher. Hooke's theory showed that a parallel plate can create colors which depend on the plate's thickness. However, it was a qualitative proof, it dealt only with two colors, and it didn't show why only thin plates produce colors. Newton's theory was quantitative, it explained the origin and place of all colors, and it was applicable to two other phenomena. However, it could not explain why thicker plates don't make colors. Young's theory was also quantitative and applicable to all colors. In addition to this, it accounted for the limitation in the thickness of the plates and could be extended to many other phenomena. Thus, we see how each theory improves over its predecessors. Young's theory of thin films was also superseded by others which explained, for instance, the intensity of the interference fringes. However, as far as the location of the fringes is concerned, Young's theory remains a good theory because it provides a precision satisfactory for many purposes by very simple mathematical means.

important

Michael. Perhaps it is a fine theory, but it is very obscure, at least as you presented it. Why did Young compare the path difference to the wavelength? Where did he get this rule from?
Teacher. Young supported the wave hypothesis of light, thus he compared light waves to sound waves. He knew from mechanics that when a point participates in two vibrations the compound amplitude can be either a sum or a difference of amplitudes, depending on

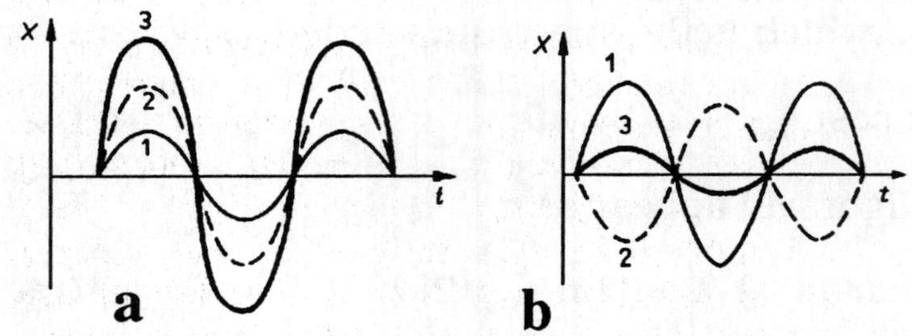

Fig. 10.4. Superposition of waves of the same frequency.

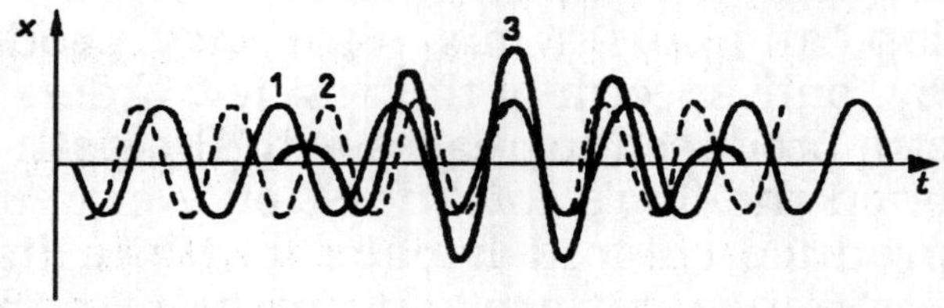

Fig. 10.5. Superposition of waves of different frequency.

whether the vibrations were in the same or the opposite phase. This can be expressed graphically as in Fig. 10.4 (equal frequencies) or in Fig. 10.5 (slightly unequal frequencies). Initially, Young applied such graphical addition to explain beats of sound, or a periodical increase and decrease of loudness (Fig. 10.6) produced by two sounds of slightly different frequencies. He extended this principle to light waves: if two waves bring to to the eye vibrations in the same phase, their amplitudes add; if the phases are opposite, the amplitudes subtract. An addition of amplitudes is supposed to produce a spot brighter than the background, while their subtraction should create a darker spot. The relative phase of the two vibrations depends on the path difference of the two waves. The phases are the same when the path difference is a multiple of the wavelength, and they are opposite when the path difference equals an odd number of the half-wavelength. The concept explaining at which point we should expect an increase or decrease of the intensity of vibrations was also discovered by Young and is called the *principle of superposition of waves*. The principle of superposition is applicable to any two intersecting waves, but the result of such an intersection is not always observable. The principle of interference refers to *observable* phenomena, thus it includes the principle of superposition of waves and the conditions of coherence.

important

David. The idea of adding two waves as expressed in Fig. 10.4 and Fig. 10.5 is so simple that it was probably easy for Young to stumble upon it after he adopted the wave hypothesis. Is that right?

Teacher. Wrong. It took him two years to advance from interference of sound to interference of

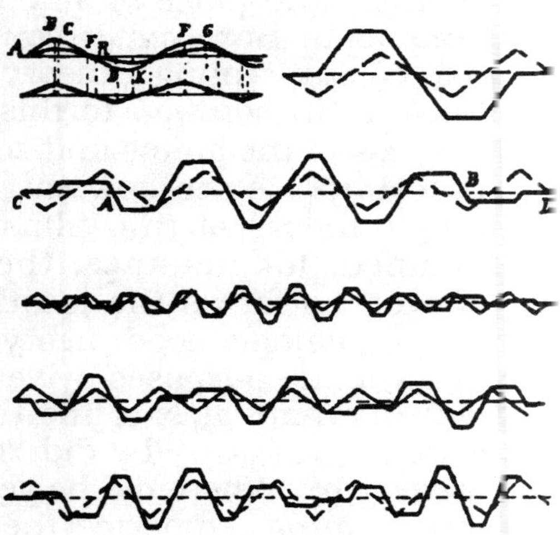

Fig. 10.6. Superposition of waves of unequal frequency. From T. Young, *Misc. Works*, v.1, pl. III.

light. One of the reasons for this can be seen in Fig. 10.6. At the time, mathematicians believed that a sine wave representation of sound (introduced by Newton) was too simplistic and they used more complicated curves. However, such curves are not applicable to light. To produce beats of sound we need two waves of different frequency which produce vibrations of differential frequency changing their intensity in time (it is called *temporal interference*). The eye cannot perceive temporal interference of light because the difference of any two optical frequencies is beyond the range of visible frequencies. The eye can only perceive *spatial interference*, or the one which changes the intensity only in space and not in time. Such interference is produced by waves of identical frequency.

important

Mary. Is that is why you said that red light interferes with red, and blue with blue?

Teacher. Exactly. Of course, you can mix, red light with blue but you won't see any interference pattern.

Dorothy. O.K. So it took Young some time to decide on selecting sine waves to represent light mathematically, but after that everything was obvious, wasn't it?

Teacher. Not at all! For instance, he had to decide whether to use sine waves of finite or infinite length. What would be your choice?

math is a friend !

John. I would probably choose an infinite wave.

Teacher. Wrong. Infinite sinusoidal waves must interfere whatever their path difference, which contradicts experiment. Among other things, Young had yet to discover all the conditions of coherence, and to confirm his theory by a quantitative experiment.

Michael. How was Young's principle of interference received?

Teacher. Most scientists misunderstood it. They were asking (not always explicitly) the same questions you did, however Young didn't bother to answer them in detail. Instead, he relied on analogies, comparing the interference of light with an intersection of waves on the surface of water or with beats of sound. He meant that all these phenomena can be explained by an addition or subtraction of amplitudes of vibrations. His opponents, however, looked at them from an entirely different perspective. In their view, the intersecting water waves did not interfere at all because they preserved their form and velocity; and the beats of sound had a psychological rather than a physical cause. For these reasons, the principle of interference of light was accepted only after 1815, thanks, to a considerable degree, to the efforts of Augustin Fresnel.

interesting!

David. Why didn't Young bring forth new experiments to support his case for interference?

Teacher. He did, and we will repeat some of them in a short while. In 1804, for instance, he offered an experiment (I call it the "screening experiment") which later became very influential in winning support for the principle of interference and the wave theory. The idea was to show that **two** rays of light are necessary for interference, and that blocking one of them destroys the interference pattern. Young placed a wire into the cone of sunlight entering a dark room through a fine

interesting!

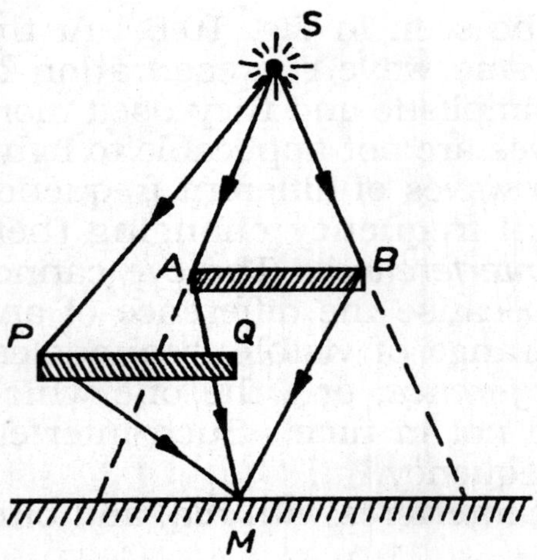

Fig. 10.7. Young's "screening experiment".

pinhole and observed parallel fringes inside the shadow. That was a phenomenon of diffraction known since Grimaldi (see Ch. 11). Young supposed that these fringes resulted from the interference of two rays of light **AM** and **BM** (Fig. 10.7) coming to the screen from the opposite edges of the wire **AB** , and his calculations agreed with experiments (Ch.11). When he inserted a small opaque screen **PQ** in the shadow in order to intercept the ray **AM**, he discovered that the fringes disappeared. Young interpreted this outcome as proving that **two** rays of light are necessary for producing the interference pattern. In his view, this refuted Newton's explanation of periodical colors as resulting from periodical changes in a **single** ray.

III. INTERFEROMETER

1. The idea

In 1816, ***François-Dominique Arago*** (1786-1853), a member of Paris Academy of Sciences, was engaged by the Academy in reviewing Fresnel's early papers. Arago reproduced Fresnel's and Young's experiments on interference of light. While repeating

history

Young's "screening experiment" he happened to have in his hand a glass plate $\mathbf{P_1Q_1}$ (Fig. 10.8) instead of an opaque screen **PQ**. To his surprise, the fringes inside the shadow **CD** of the body **AB** disappeared although the plate did not stop the ray **AM**. Fresnel supposed that the ray **AM** was delayed inside the glass on its way to the meeting point **M**, or, that the path difference of the two rays increased and they met at another point $\mathbf{M_1}$ outside the shadow. There, the background was so bright that the fringe became invisible. To test his hypothesis Fresnel suggested replacing a thick glass plate with an extremely thin one: the fringe reappeared within the shadow at the point $\mathbf{M_2}$. Arago found that the same result could be achieved by means of two thick glass plates $\mathbf{P_1Q_1}$ and $\mathbf{P_2Q_2}$ of a slightly different thickness placed at both sides of the shadow: a small difference in their thickness produced a noticeable shift of fringes. This gave Fresnel and Arago the idea to measure small changes in length or in the index of refraction, because the optical path of a ray depends on both. That is how the first *refractometer-interferometer* was born. Astronomers provided the first task for the new method: to check whether refraction of the air depends on its humidity. That was necessary for calculating the atmospheric

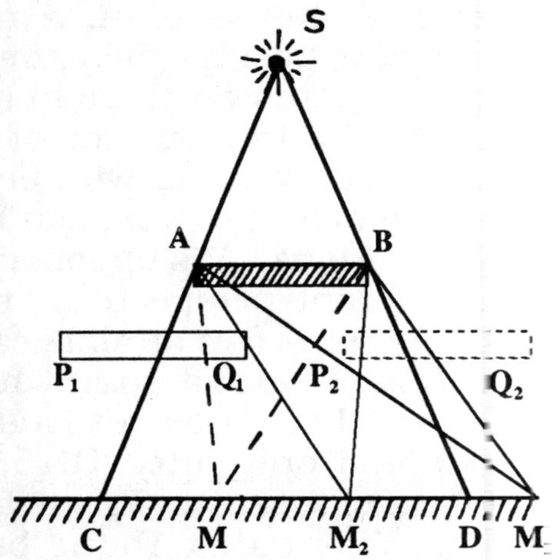

Fig. 10.8. Arago's experiment..

refraction and determining the true positions of heavenly bodies. Interferometers were much improved in the second half of the 19th century, to a considerable extent due to the American physicist Albert Abraham Michelson (1852-1931).

2. Michelson interferometer

The son of a Jewish immigrant from Poland, Albert early displayed courage, inventiveness, and determination. He dreamed of entering the U.S. Naval Academy but was tied with two other applicants, and the place was given to another person. With a letter of support from his senator, Albert intercepted President Grant during his walk and pleaded with him for an extra appointment. He succeeded and eventually became the first American to win the Nobel prize. After graduation he taught at the Academy for a while, and at that time he distinguished himself by measuring the velocity of light (1878). Two years later, he went to Europe for graduate studies, and it was in the laboratory of Helmholtz that he conceived of the idea of a new interferometer. The instrument had to be sensitive enough to discover whether or not the earth moves relative the ether. If "Yes," the "ether wind" must depend on the direction of light relative to the direction of the earth motion, which can be discovered by rotating the interferometer. The result was negative. In 1887, Michelson and Edward Morley repeated the experiment with an improved interferometer, but the result was the same. The negative results of Michelson's experiments played an important role in physics, by preparing the ground for the theory of relativity. Michelson, however, realized that his instrument has also important practical uses. For instance, with the help of his interferometer he found that the Paris standard of meter contained 1,553,163.5 wavelengths of the red line of cadmium. Subsequently, this became the foundation for replacing the metal meter as a standard of length with a much more constant "light meter", based on the wavelength.

The Michelson interferometer consists of two front-surface mirrors **A** and **B** and a beam-splitter **C** (Fig. 10.9). While Michelson used a spectral

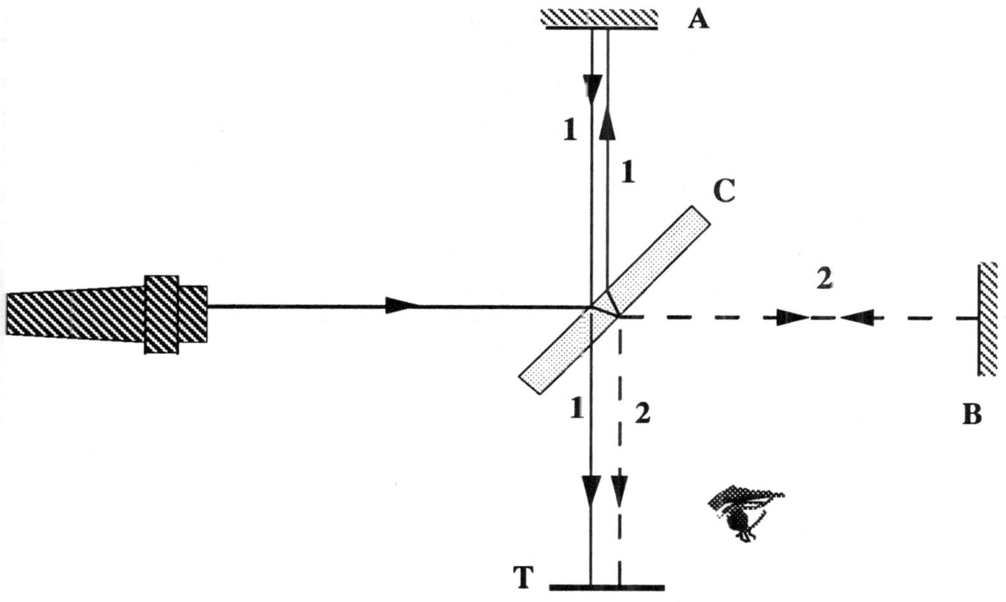

Fig. 10.9. The Michelson interferometer.

tube for a source of light, we will have a laser **L**, which will considerably simplify the procedure. The beam is divided in two parts in the plate **C**. The ray **1** enters the eye after being reflected by the silvered side of the plate **C** and then by the mirror **A**; the ray **2** is transmitted by the plate **C**, reflected by the mirror **B** and again reflected by the silvered side of the plate **C** towards the eye. The two rays interfere in the same way as on a plane-parallel plate (if the mirrors **A** and **B** are perpendicular) and produce a system of concentric bright and dark rings. With white light, one cannot see more than a few rings, but their number increases with the degree of monochromaticity of light. With a laser, one may see many rings, which implies a large path difference and a high monochromaticity of light. When the path difference is very large, it may be difficult to locate the center of the ring system.

The purpose of the interferometer is to measure a **change** in the path difference of two interfering rays, caused by either a movement of a mirror or by a change in the index of refraction of the medium crossed by one or both rays. The index of refraction varies with pressure or temperature. If $\Delta n = n_2 - n_1$ is the change of the index of refraction and **L** is the length that light traverses, the change in the path difference will be $\Delta P = 2L\Delta n$, and, if expressed in the number **N** of wavelengths λ, $\Delta P = N\lambda$. Hence,

$$N = 2L\Delta n / \lambda$$

Thus, when the path difference changes, the fringes move: a shift by N times the width of the fringe corresponds to a path difference change of N wavelengths. The shift is difficult to measure when it is large: one counts how many fringes crossed a chosen mark on the screen. This is easier to do when we see the center of rings, in this case one counts how many new circles appeared from the center (the path difference increased) or how many circles collapsed in the center (the path difference decreased). Since the ray **1** crosses the plate **C** twice (at 45°) while the ray **2** does not cross it at all, the original path difference of the two rays is about 2.8t where **t** is the thickness of the plate **C**. The precision of measurements increases when the initial path difference of the two rays is very small. To achieve this, one has to compensate the extra path of the ray **1** by introducing a proper glass plate (or several plates) into the ray **2**.

IV. EXPERIMENTS

1. Soap bubbles

lab home

Background. The first investigations of soap bubbles were conducted by Boyle, Hooke, and Newton.
Equipment. Plastic cup, detergent, glycerin or corn syrup, cocktail straw.
Procedure. In a plastic cup, prepare a detergent solution, for instance, "Joy" diluted 20 times by volume with a few drops of glycerine. Moisten the walls and the rim with the solution. Make a bubble using a cocktail straw, set it on the rim, and remove the straw.
Investigate.
1. The form of colored fringes and their sequence from the top.
2. Do colors change with time? Watch the same spot on the bubble, for instance, the reflection of a window.
3. Why do colors change? Why do bubbles break? Are the two causes connected?

Conclusion. Can you prove that these colors depend on the thickness of a soap film?
Teacher's note. The soap liquid flows from the top of the bubble to its bottom because of gravity. Thus, the bubble's wall has the same thickness at equal height, and the thickness increases downwards. That is why the colored fringes are horizontal circles. A small bubble has thick walls and does not produce colors.

2. Colors of "mixed plates"

lab home

Background. Young discovered that two glass plates which held between them a mixture of two different substances, water and air, for instance, produced what he called the *colors of mixed plates*. He supposed that these colors resulted from interference of light passing through different media, i.e. water and air, and their path difference was determined by the difference in velocity of light in these two media. After assuming that light travels faster in air than in water, he confirmed this theory by experiment.
Equipment. Glass plates, water, vinegar, corn syrup.
Procedure. Wet two glass plates and press them together.
Investigate.
1. Describe the shape of colored fringes and their sequence when you look at them in reflected light. Repeat the experiment with other liquids.
2. Mark a point on the plate. Watch the color at this point in reflected light, then move the plates without changing their orientation in space so as to see **through** them. Did the color change?
3. Do colors change if you squeeze the plates or look at them at a different angle?

Conclusion. What is the main difference between the colors of thin and mixed plates?
Teacher's note. Those colors are much wider than the colors produced

by the air film. Squeezing the plates should not change these colors.

3. Two-slit interference

lab

Background. This most famous Young's experiment was very briefly described in his *Lectures* (1807). The diagram provided (Fig. 10.10) did not clarify much. Lack of any details about the experimental arrangement led some historians to suggest that it was a thought experiment introduced to explain the principle of interference rather than a real one. Indeed, the hyperbolic lines of maxima and minima in the upper part in Fig. 10.10 remind one of similar lines in another diagram by Young drawn to illustrate the interference of water waves coming from two centers, and which certainly was constructed on the basis of his theory rather than experiment (Fig.10.11) Evidence has been found, however, that Young did perform this experiment, although perhaps not exactly in the way described in physics textbooks, which show a pinhole before the slits and a white screen behind them. He may have used the **direct observation** method, that he had applied in other experiments: holding the slits near the eye and looking through them at a candle flame a few

Fig. 10.10. Two-slit interference. From T. Young, *Lectures*, v.1, Fig. 442.

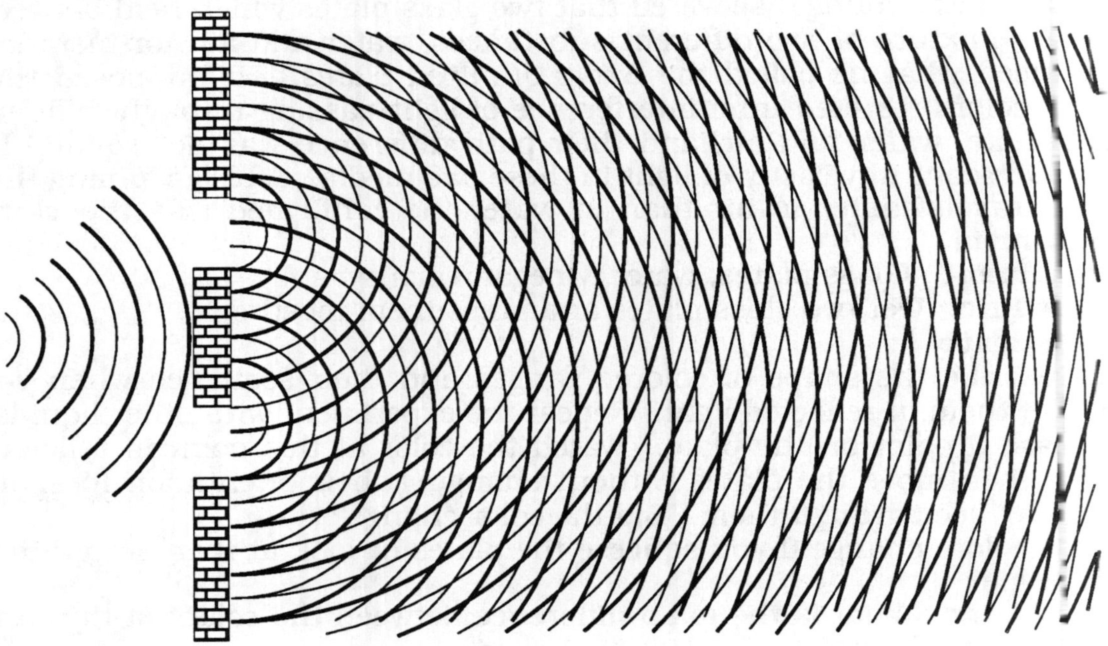

Fig. 10.11. Interference of two water waves . From T. Young, *Lectures*, v.1, Fig. 267.

meters away. It also appears that Young did not realize that the fringes he observed with two slits were very different from those obtained with a single slit (see Ch.11).

Equipment. Two microscopic slides, candle, desklight with a housing, V-slit made out of an index card, pins, razor blades, pieces of wood or cardboard, short rulers, masking tape.

Procedure. To make a V-slit, hold a microscopic slide over a candle flame and deposit soot on the middle part of it. Place the slide **1** (Fig. 10.12) on a table between two supports **2** made of wood or cardboard that are taller than the slide. To prevent the slide from moving while you draw on it, press the supports to it and tape them to the table. Take a fine razor blade and with the aid of a ruler **3** draw a line on the sooty surface. Then, without changing the ruler's position draw another line: a slight change in the position of your hand is sufficient to produce two lines crossing at a very small angle (the intersection point should be visible). At some distance from the two lines draw a single line: you will need it in some experiments where you will compare the phenomena produced by a single slits and by two slits. To make a double slit take two razor blades, separate them with a single layer of masking tape, and tape them together. Using this double razor draw the lines as described above. When a slide with a V-slit or a double slit is ready, wipe the soot at two opposite edges to free a margin about 5 mm wide, and cover it with several layers of masking tape. Place another slide (a clean one) on the top of the first and tape the slides together. This will preserve your slits from destruction. Use a magnifier with a scale to measure the distance between the lines.

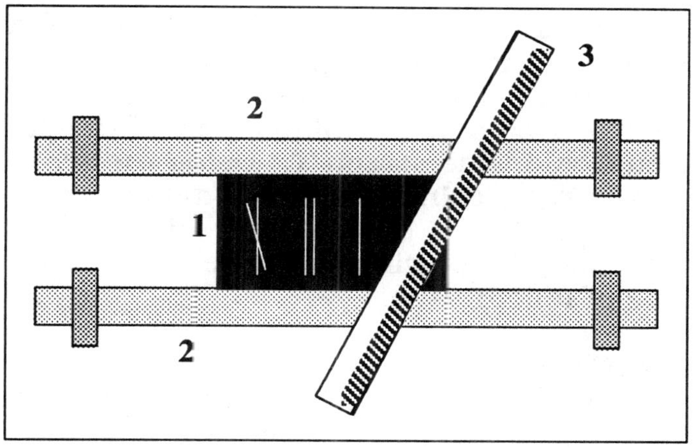

Fig. 10.12. Making double-slits and V-slits (view from above).

Investigate.

1. Look at a distant source of light (a candle flame or a filament of an electric bulb) through a double slit (Fig. 10.13). Describe the fringes.
2. Now look through a single slit. Do you see any difference?
3. Move your eye to and from the vertex and away from it and describe the change. How would you go on establishing the cause of this change?

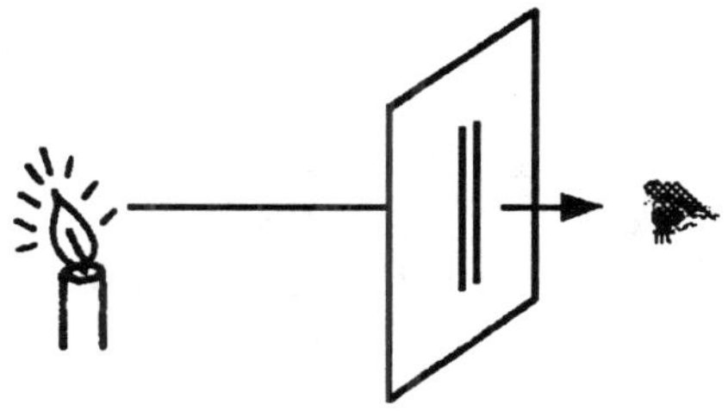
Fig. 10.13. Direct method of observing the two-slit interference.

Conclusion. What is the main difference between the colors produced by a single slit and a double slit?

4. Michelson interferometer

Equipment. Interferometer, laser, air cell, vacuum pump, microscopic slides, ground glass, paper tissue.

lab **Procedure**. Set a laser and the screen **T**, as shown in Fig. 10.9. Direct the laser beam at the middle part of the plate **C** so as to obtain two rows of images on the screen **T**. Using the screws adjust the mirror **B** so that the two rows coincide. At this moment you will see fringes. If you don't see them, repeat the process. It may be helpful to reduce the beam's brightness by placing near the laser's exit window a ground glass or a piece of toilet tissue.

Investigate.
1. Press the base of the instrument with a finger and check whether this affects the fringes. Is there a correlation between the amount of pressure and the shift?
2. Place a microscopic slide perpendicularly to the ray **2** and slowly turn it around the vertical. Describe the change. Suggest a reason for the change and test your hypothesis.
3. Insert several glass plates into the path of the ray **2**. How can you determine whether you increased or reduced the path difference?
4. Measure the index of refraction of the air. Evacuate the cell and place it on the cell support between **C** and **B**. Readjust the mirrors if the fringes disappeared. Mark a dot on the screen, open the valve so as to let the air in **very slowly**. Count the number of rings passing by the dot. Calculate the index of refraction (the length **L** is shown on the cell).

Conclusion. How great a length change can you measure with this instrument?

BIBLIOGRAPHY

R. Hooke, *Micrographia*, pp. 47-67.
I. Newton, *Opticks*, Books II and III.
T. Young, *Misc. Works*, I:140-91, 279-358.
T. Young, *Lectures*, I:464.
G. Peacock, *Thomas Young* (London, 1855)
A. Wood, *Thomas Young Natural Philosopher* (Cambridge, 1954).
N. Kipnis, *History of the Principle of Interference of Light* (Basel: Birkhäuser, 1991), chs. II, III, IV.

Chapter 11

DIFFRACTION of LIGHT

I. EARLY STUDY OF DIFFRACTION

1. Is there light inside shadow?

discussion

Teacher. By definition, the rectilinearity of light means that in a uniform medium the rays of light don't bend. This means that light cannot penetrate shadow. Right? Have we actually tested this?

David. No, we didn't look for light inside a shadow. But isn't this a contradiction in terms? I guess that "shadow" is **defined** as the zone inaccessible to light.

Teacher. You are right, I was not precise. By "shadow" I meant the so-called "geometrical shadow", or the area behind a body limited by the light rays coming from a **point** source and touching the body. Thus, to check whether light bends around corners we must have a point source of light.

Dorothy. Can we use a laser?

Teacher. Yes. I will aim a laser beam at a white screen and place a nail in the path of the beam. Michael, would you please approach the screen and tell us what do you see?

Michael. I see a long shadow of the nail, probably caused by stray light, which appears to be narrowing where the nail crosses the beam with its outlines becoming dim.

Mary. Doesn't this mean that light entered the shadow on both sides of it?

Teacher. It certainly does. Now, I will replace the nail with a pin. What do you see?

John. There is practically no shadow to speak of. Light reaches the middle of the geometrical shadow but it isn't uniform, for there are two narrow dark lines running lengthwise. Where do these lines come from?

Teacher. Forget the dark lines for the moment (we'll discuss them in a short while). Do you still believe that light follows straight lines?

Ruth. Perhaps what we see is an exception? Why don't we try a wider object, such as a pencil?

Teacher. Let's do it.

Dorothy. I moved a pencil all the way from the laser to the screen but there was practically no effect.

Ruth. This may imply that rays bend only around narrow bodies, or, more exactly, a deviation from a straight line is very small and hence it is noticeable only when a body and its shadow are narrow.

good point!

Michael. Sounds interesting. Can this be proven without a laser?

Teacher. Yes, this phenomenon has been known for more than three centuries.

2. Grimaldi

history ***and discussion***

Teacher. That narrowing of the shadow was discovered by Robert Hooke in 1672, when he let sunlight into a dark room and placed in its path a round wooden body. Hooke claimed that it was a new phenomenon rather than the result of reflection or refraction. He interpreted it as the result of bending of light rays inside the shadow, which he called *deflection*. Hooke was unaware that a similar property of light had already been discovered by ***Francesco Maria Grimaldi*** (1618-1663), Professor of Mathematics at a Jesuit Collegium in Bologna. Grimaldi found that when light entered a dark room through a very small opening **AB** (Fig. 11.1) and illuminated a narrow body **EF**, its real shadow **MN** was considerably larger than the geometrical shadow **IL**. On the outside, the shadow was terminated with three colored fringes, red outside and violet inside (Fig. 11.2). These external fringes always followed the body's outlines (Fig 11.3). He also found dark and light fringes inside the shadow of a narrow body which generally ran lengthwise, except for the corners (Fig.11.4). Grimaldi was also greatly surprised to find that the image of a small opening or of a narrow body produced by a sunbeam appeared to be wider than it was supposed to be for geometrical reasons. Grimaldi stated that he discovered a new phenomenon of bending light around corners, different from reflection or refraction, and called it *diffraction*. To him, diffraction consisted of a splitting of each ray passing near an obstacle into a number of rays, scattered into all directions. Grimaldi's discovery of diffraction was published posthumously in 1665 in the first chapter of a large volume on light and colors. Perhaps the size of the book scared readers off because few people knew of Grimaldi's discovery and still fewer understood its importance. An indifferent review of Grimaldi's book published by Henry Oldenburgh ,the Secretary of the Royal Society, inspired neither Hooke nor Newton to read it. Having failed to repeat Grimaldi's experiments,

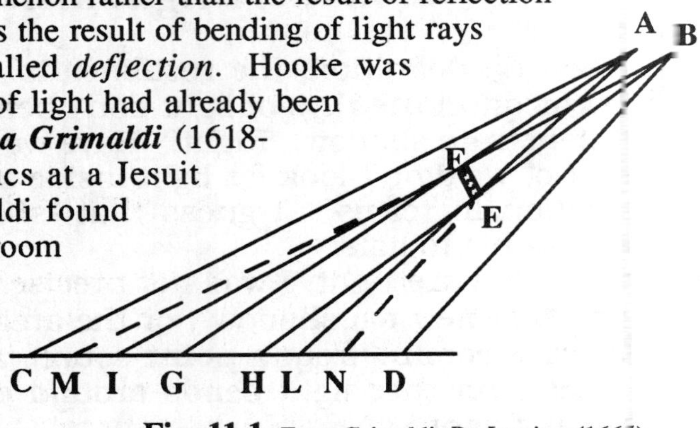

Fig. 11.1. From Grimaldi, *De Lumine* (1665).

Fig. 11.2. From Grimaldi, *De Lumine* (1665).

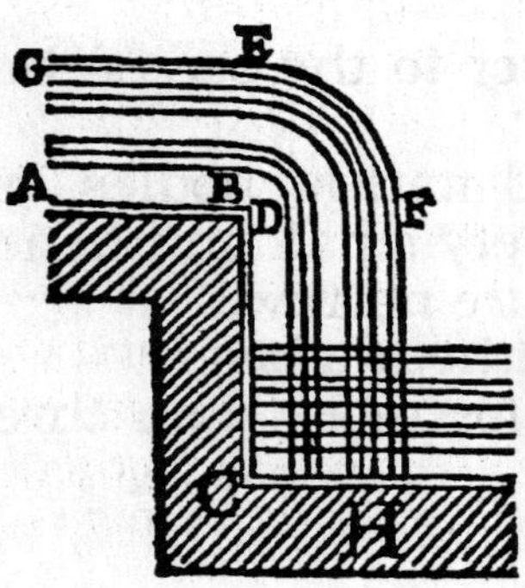

Fig. 11.3. From Grimaldi, *De Lumine* (1665).

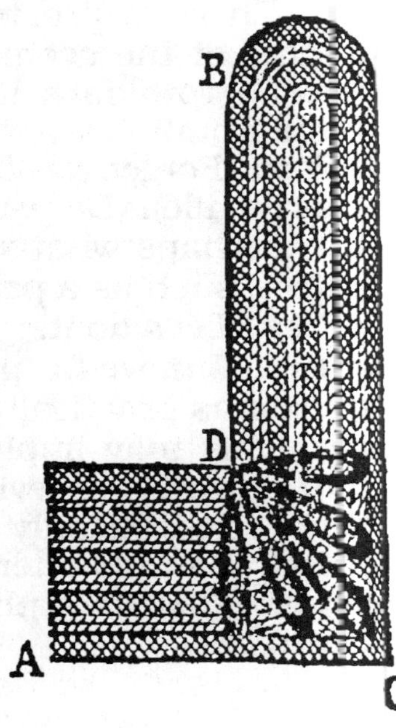

Fig. 11.4. From Grimaldi *De Lumine* (1665).

Huygens even questioned the validity of his discovery.

Dorothy. How come a scientist of Huygens' stature couldn't observe diffraction?

Teacher. Well, it is not clear whether he learned about diffraction from Grimaldi's book, and not from a secondary source. Moreover, even if he did, he might have missed the some details, such, for instance, as the size of the opening in the shutter. You will see for yourself, how important this size is.

David. How did Newton respond to the discovery? He was probably aware of Hooke's work on diffraction, wasn't he?

3. Newton

history and discussion

Teacher. Yes. Newton had heard Hooke's presentation (a very brief one) in 1675, and he read something about Grimaldi's experiments in a secondary source. He agreed with Hooke that all of Grimaldi's phenomena could be attributed to bending of light rays rather than their splitting. Newton believed, however, that rays bent outward rather than inward, and called this phenomenon *inflection*. Initially, Newton stated that the new phenomenon was actually a refraction produced by a change in the density of the ether adjacent to a body. After 1687, he adopted a new explanation of the bending of light: a body repels light rays passing by, the stronger the closer they approach the body (Fig.11.5).

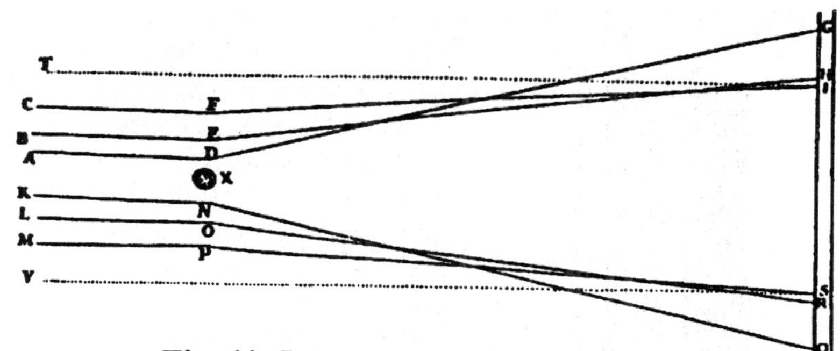

Fig. 11. 5. Trajectories of diffracted rays. From Newton, *Opticks* (1704), Bk.III.

Newton was the first to have performed quantitative diffraction experiments: he measured positions of fringes and tried to express them by a mathematical law. He was also the first to use a monochromatic light to study diffraction. Some of his measurements were very precise and physicists used them as late as the early 19th century for testing new theories. Other observations, however, were erroneous or incomplete. For instance, he never mentioned the internal fringes described by Grimaldi. Actually, since he experimented primarily with a hair, he could not see on the screen multiple internal fringes. However, it is difficult to miss a light band which runs along the hair in the middle of the shadow. It is possible that Newton chose to ignore this phenomenon because it didn't fit into his theory. Newton was the first to have found the following phenomenon: when a glass mirror with a silvered second surface was illuminated by sunlight through a hole in an opaque screen, the light reflected back to the hole displayed colored rings around it. Newton named the rings the *colors of thick plates* and explained them as follows. The rays that fall on the first surface are diffracted in different directions by particles of dust. Some of the diffracted rays travel towards the second surface where they are reflected back and are again diffracted by dust.

Newton's experiments stimulated a considerable interest in diffraction. In 1723,

Giacomo Filippo Maraldi (1665-1729), an astronomer and a member of Paris Academy of Sciences, rediscovered the internal fringes (Fig. 11.6). He found that the interval between them decreases with the thickness of the body, which explains why these fringes cannot be seen in the shadow of a hair. He was the first to use a magni-

Fig. 11.6. **Internal fringes.** From Maraldi, *Mem. Acad. Sci. Paris*, 1723.

fier to observe diffraction fringes on a screen. In 1723, he made a very important discovery by observing a bright spot at the center of the shadow of a small disk or a ball. In 1755, ***Michel Ferdinand d'Albert d'Ailly, Duc de Chaulnes*** (1714-1769), subsequently a general of the French Army and a governor of Picardie, confirmed Newton's explanation of the colors of thick plates and discovered another phenomenon similar to it. He shined a beam of sunlight on a wooden frame with a piece of gauze stretched out on it, with a metal mirror behind it, and observed colored rings

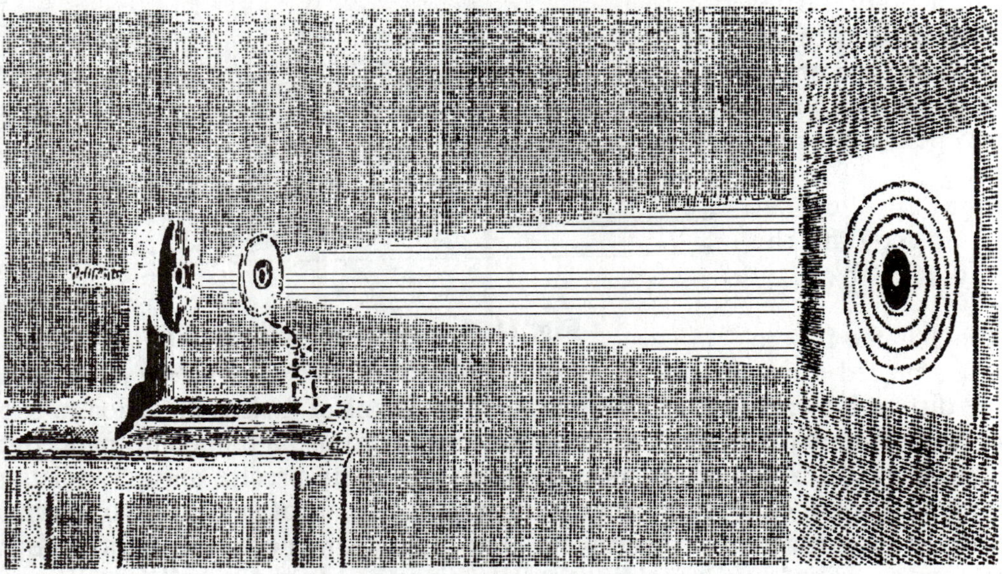

Fig. 11.7. From Duc de Chaulnes, *Mem. Acad. Sci. Paris*, 1755.

(Fig. 11.7). In his view, the fibers of the fabric diffracted light in the same way as particles of dust in Newton's experiments.

David. Why should the results of diffraction on a hair be different from those on nails and other wider objects?

Teacher. If you want a theoretical explanation of the difference, you have to wait until we discuss the wave theory of diffraction. If you doubt the existence of such a difference, see for yourself. We are going to repeat a number of historical experiments, including the experiments with various long and narrow bodies.

4. Experiments with sunlight

lab

Equipment. It is easier to run these experiments when the sun is comparatively low in the sky. Cover the windows with cardboard sheets. Cut with a razor blade several square openings of about 1 cm and cover them with aluminum foil. Make a small hole in the foil with a tip of a fine sewing needle (have one hole per group). Prepare several narrow objects: metal and cardboard strips 3-10 mm wide, nails of about 1-2 mm in diameter, wires 0.1-0.5 mm in diameter, and a V-slit. Use a floor stand with a slider to which you attach the larger objects using masking tape or reusable adhesive ("Tak"). Wires and hairs are to be mounted on a slide frame which is taped to the slider (Fig. 11.8)

Procedure. Bring the object into the sunbeam close to the hole and so as to have the whole object illuminated. Place a white screen about 30 cm from the object and examine its shadow. Move the screen back and forth and watch whether the shadow changes, both inside and outside (remember: the range of distances should be large!). The most difficult part will be to see colored fringes outside the shadow. With a naked eye you should see them at about 3 m from the object. If you cannot move your screen far enough, use a magnifier (x5 - x10), it certainly helps. Make a drawing of what you see.

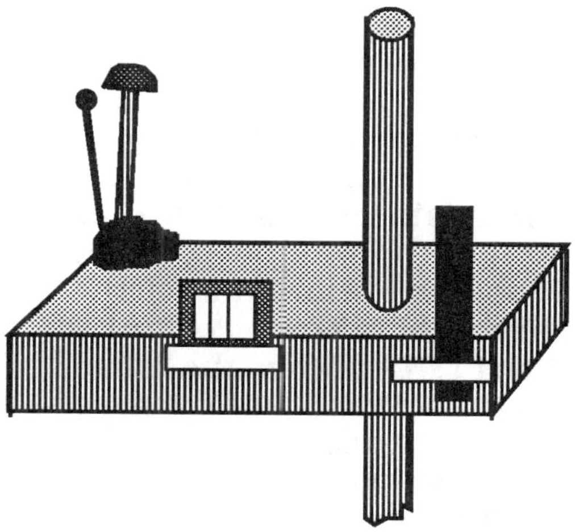

Fig. 11.8. Diffraction experiments with sunlight.

Purpose. Investigate diffraction on the following objects.

a. narrow body (Grimaldi, Newton)

Equipment. Use the stand with a slider, that carries nails, wires, hair, narrow paper and metal strips as shown in Fig. 11.8.

Procedure. In these experiments your primary concern should be with the coloration of external fringes and the number (or the width) of internal fringes.

Investigate.

1. Under what circumstances the external fringes are colored? Show in a diagram the sequence of colors.
2. How do the internal fringes change with the distance between the object and the screen (what should you keep constant)?
3. Try to establish a qualitative relation between the number of dark inner fringes and the width of the body (what must be constant?).

b. aperture (Newton)

Equipment. A V-slit is made of two pieces of an index card 10 cm long

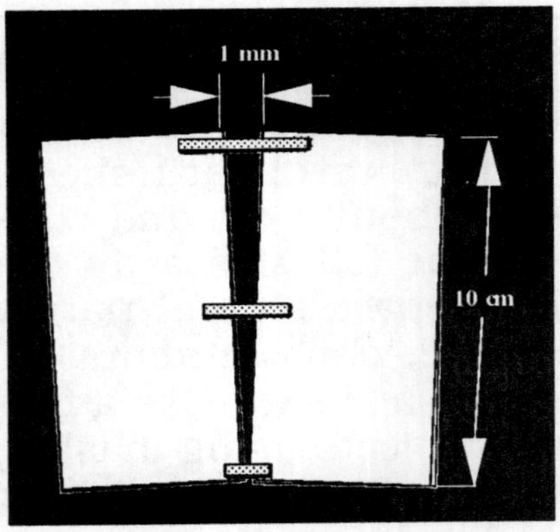

Fig. 11.9. V-slit.

taped together at the ends and in the middle so that on one end they touch one another and on the other end they are 1 mm apart (Fig. 11.9)

Procedure. Place the slit 1 m from the pinhole and shine light on it so as to include the apex.

Investigate.

1. What is the sequence of colors?
2. Is the number of internal fringes the same at different distances from the apex?
3. What is the shape of fringes at different distances from the apex?

c. "thick plate"

1. Newton's experiment.

Equipment. Concave glass mirror, pieces of nylon stocking or some other fabrics. For these experiments you need a larger opening (5-10 mm) in the window covering.

Procedure. Hold the mirror against the window and reflect sunlight back through the opening (Fig.11.10). Move the mirror back and forth until you see colored rings. If you don't see any, try reflecting light by different parts of the mirror and make note about the conditions of the parts which do produce rings.

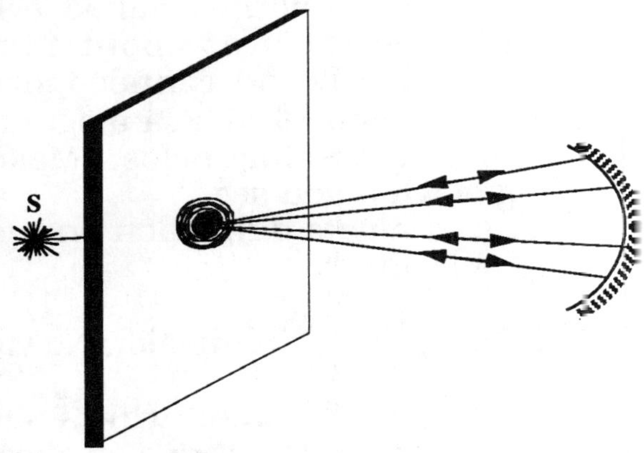

Fig. 11.10. Newton's experiment with colors of "thick plates".

Investigate.

1. Test Newton's hypothesis that particles on the mirror surface are responsible for the phenomenon. Dust the mirror with a chalk powder or other fine powder. Describe the rings observed.
2. Breath on the mirror where you see the sun spot. Why do the rings appear and vanish after a short time?
3. If the rings are produced by particles on the mirror surface, what is essential about these particles: their diameter or the distance between them? How can you test your hypothesis?

2. Duc de Chaulnes' experiment.

Equipment. A concave mirror, pieces of a pantihose and other fabrics stretched over a slide frame.

Procedure. The frame is held close to the mirror. The rest of the experiment is conducted similarly to the previous experiment.

Investigate.
1. Do the colors or the size of the rings depend on the fiber's thickness or on the distances between fibers?

d. ball (Maraldi)

Equipment. Pins with well-rounded heads.
Procedure. Place the pinhead into the sunbeam and watch the center of the shadow. Can you see any light there?
Investigate.
1. Does the shape or brightness of the light spot change if you move the screen to or from the pin?
2. Does the diameter of the pinhead affect the result?

5. Is diffraction a new phenomenon?

investigation

Ruth. O.K. We've completed the experiments and found the claim of Grimaldi and others about light entering the shadow to be correct. But how could they prove it was a new phenomenon?
Teacher. This time, I will let you experiment first and tell about their results afterwards. We are going to resolve the following problem.

Problem

Prove that diffraction is a new phenomenon, different from reflection or refraction.
Michael. What does bending of light have to do with reflection or refraction?
Teacher. Bending of light is an **interpretation** of the phenomenon rather than a fact. What you see is simply light entering a zone supposedly forbidden for it. This can be explained in different ways, and some scientists contended that it was light reflected by the edges of either the aperture or the body, or both. Our purpose is to prove that this is not the case. How would you eliminate reflection?

Hypothesis 1

Mary. A black body reflects much less than a polished one, thus if the material of both the diffracting body and the aperture is black there should be no light inside the shadow.

Test 1

John. If we take a polished nail or a pin and blacken a part of it with ink or paint, we can compare diffraction from both parts.
Dorothy. Our group found no difference between the two.
Michael. So did ours.

Test 2

David. It is difficult to get a sturdy black material in which to make a pinhole. On the other hand, only the edges of the hole are supposed to reflect. Thus, perhaps we can burn a hole in an index card with the tip of a red-hot pin?

Michael. I held a needle over a candle flame with pliers and then pressed it lightly against a card. The holes produced had black edges, and some of them were quite small. The result was the same.

Hypothesis 2

Ruth. The amount of reflected light depends not only on color but also on the size of the reflecting surface. If the diffracting body has a sharp edge, it should not produce fringes.

Test 3

John. Let us try diffraction on a razor blade.

Dorothy. We observed a razor blade producing the same phenomenon as a pencil, although the latter has a much larger reflecting surface.

Teacher. Congratulations! You follow Hooke without knowing it. Now, with reflection out, how can we test the role of refraction? I mean refraction by the surrounding medium.

Hypothesis 3

Ruth. The refraction of air depends on its temperature. Thus, heating a diffracting body may change the diffraction pattern.

Test 4

David. I've used the tip of a soldering gun as a diffracting body and found that the fringes didn't move when it became hot.

general conclusion ➡ **Diffraction is not due to reflection or refraction of light.**

II. WAVE THEORIES OF DIFFRACTION

1. Diffraction and theories of light

history

Diffraction became a major battlefield in the debate on the nature of light. Newton's chief argument against the wave hypothesis was that light does not bend around obstacles as sound does, when we hear it behind a building. In his view, the deviation of light from its rectilinear path observed in diffraction phenomena was too small to support the wave nature of light. Euler, the champion of the wave theory of light, responded to this that hearing a sound behind a building has nothing to do with it bending around corners: sound simply penetrates the walls of the building. Strangely, no one tried to test Euler's statement by direct experiment. While in modern textbooks diffraction is considered to be unquestionably a wave phenomenon, it was not so in the 18th century. Newton's theory of an interaction between light and bodies explained some phenomena of diffraction, and although it was merely a qualitative theory, the "undulationists" had nothing to offer even at the qualitative level. The situation began to change early in the 19th century due to the efforts of Thomas Young and Augustin Fresnel.

2. T. Young

a. diffraction by a hair

history

In 1802, Young discovered a new method of observing diffraction. Instead of projecting the diffraction fringes on a white screen as Grimaldi did, he observed them directly by placing the eye behind a small obstacle and looking "through" it at a small distant source of light (a candle flame). With this method (we will call it "direct observation") he was able to identify without difficulty the internal fringes of a hair missed by Newton. Young explained the colors produced by a hair by an interference of two rays of light diffracted at the opposite sides **A** and **B** of the hair (Fig. 11.11). If **M** is a bright point at a screen, **AM** and **BM** are the interfering rays. If AB«AM, **BC** is perpendicular to both **AM** and **BM**, and θ is a very small angle. Then, the path difference will be P=AM-BM=AC. If b=BN, x=MN, d=AB, then $\theta \approx \sin\theta \approx \tan\theta$, $P = d\sin\theta \approx d\theta$, $x = b\tan\theta \approx b\theta$, and $P = xd/b$. Since bright fringes correspond to $P = m\lambda$ ($m = 0, 1, 2, \ldots$), $xd/b = m\lambda$, and the angular radius of the $\mathbf{m}^{th}$ bright fringe will be

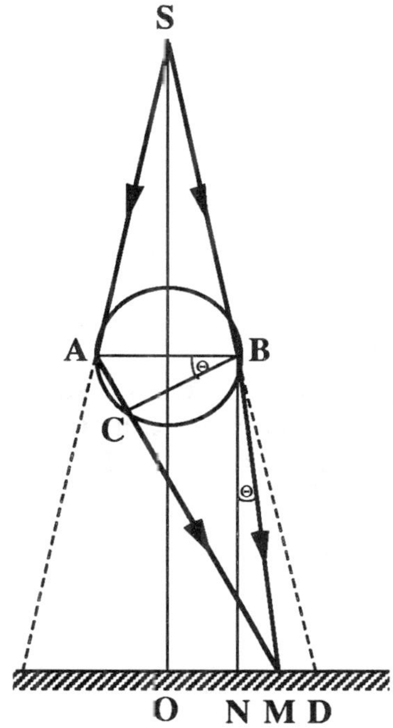

Fig. 11.11. Young's explanation of internal fringes.

$$\theta = m\lambda/d \text{ (rad)} \tag{11.1}$$

Subsequently, Young realized that he could improve the agreement of this equation with his measurements (there were discrepancies for low **m**), by assuming that the inflected ray loses one half of the wavelength, which changed the equation to

$$\theta = (m+1/2)\lambda/d \tag{11.2}$$

Young used this equation to measure the wavelength.

b. diffraction by many particles

One of the best achievements of Young's theory was its explanation of colored rings seen around the sun behind a cloud, or produced by a thin layer of wool illuminated from behind. He supposed that these rings are of the same nature as the fringes produced by a hair: both result from interference of light diffracted by the edges of a single fiber (or water globule). Young asserted that the more equal are the globules or fibers, the sharper and brighter are the rings. He confirmed this theory in experiments with various fibers, globules, and particles using a device he called *eriometer*. Young's eriometer consisted of a piece of a cardboard or metal **A** with a small hole **O** (Fig. 11.12) in the middle. The hole was a center of two circumferences the diameters of which were 1/2 inch and 1 inch, each circumference being pierced through so as to make 8 tiny holes. Young held an object **C** near his eye **E** and looked through the hole in the plate **A** at the candle flame **D**. He moved the plate **A** back and forth until a chosen colored ring coincided with

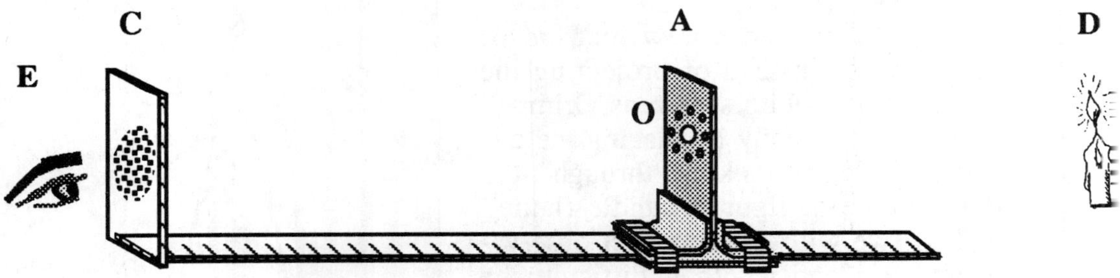

Fig.11.12. Eriometer.

either circle. After measuring the distance **b** between the object **C** and the plate **A** and the circle's radius **R**, he obtained the angular radius θ of a ring as θ=R/b. By comparing this result with eq.(11.2) he calculated the diameter of particles (or fiber's thickness) **d**:

$$d=(m+1/2)\,\lambda b/R \qquad (11.3)$$

3. Fresnel

history

a. "a man of genius"

Like Young, Augustin Fresnel (1788-1827) entered optics with an ambitious plan to establish the truth of the wave hypothesis of light, and he began with diffraction. He was born in the family of an architect at the North of France. As a child, he was good in math and drawing and hated languages. His health was always poor, and if other kids accepted him to play with them it was because he was very inventive. They called him "a man of genius." One of his discoveries was, for instance, determining the best dimensions of a pipe for a blow-gun. In 1806, he graduated from the *École Polytechnique* (Polytechnical School) in Paris where he received a thorough knowledge of mathematics and chemistry. Then he continued his education at the School of Roads and Bridges and became an engineer. Building roads bored Fresnel, and he was looking for a better application for his brain. He tried his hand in inventing, first in hydraulics then in chemistry, but without much success. In 1814, he turned to optics. His duties did not leave him much time for science until he obtained an unexpected break. In the spring of 1815, Napoleon returned from Elba. Fresnel, who hated Napoleon, went to the South of France to join the royalist army. By the time he reached the army he was so sick that he was turned back. However, when Napoleon seized power, Fresnel was dismissed from his job for disloyalty and exiled to his native village. It was there that he made his first discoveries in diffraction. Lacking any scientific instruments, he made a good use of his inventiveness. For instance, when he needed a short-focus lens, he made it out of a drop of honey covering a small hole in a copper sheet. To measure very small distances he invented a micrometer, which consisted of a thread and a wooden frame.

interesting!

In October 1815, Fresnel rediscovered the principle of interference and applied it to diffraction on a narrow body. Fresnel's theory was the same as Young's, however the response to it was more favorable. In 1817, he entered the contest announced by Paris Academy of Sciences for the best theory of diffraction supported by exact experiments. While working on this, Fresnel discovered a new idea which helped him build a more general and exact wave theory of diffraction than Young's. Fresnel called it the *Huygens Principle*, but later it was renamed the *Huygens-Fresnel Principle* because it differs from the concept introduced by Huygens himself. According to this principle, the wavefront can be divided into small parts **Am**, $\mathbf{mm_1}$, $\mathbf{m_1M}$, **Mn**, $\mathbf{Nn_1}$ each of which becomes a source of secondary waves. The intensity of light **I** at the point of observation **P** (Fig. 11.13) is the result of interference of all secondary waves reaching this point. When the number of zones increases to infinity, their width tends to zero, and the compound sum is transformed into a definite integral over the open part of the wave front. Fresnel applied this integral to a number of phenomena of diffraction and obtained a very good agreement with his observations. This impressed his judges so much that, despite their opposition to the wave hypothesis, they awarded the prize to Fresnel. Fresnel's saying was "Nature is not afraid of mathematics." He realized, however, that many people are. Thus, he developed a simplified version of his theory, called the *zone theory*. This is a beautiful theory which is simple enough for high school students.

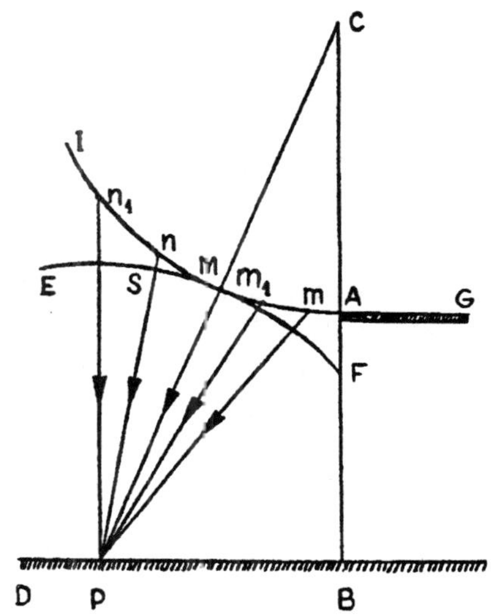

Fig. 11.13. Huygens-Fresnel Principle. After A. Fresnel. *Oeuvres*, v.1, p.174

b. the zone theory

The wave front is divided into *zones*. A zone is a part of the wave front with the origin at **S** cut off by two spheres with the center at the point of observation **O**, the radii of which differ by a half-wavelength (Fig. 11.14). If the radius of the first wave is b+λ/2, the radius of **m**th wave will be

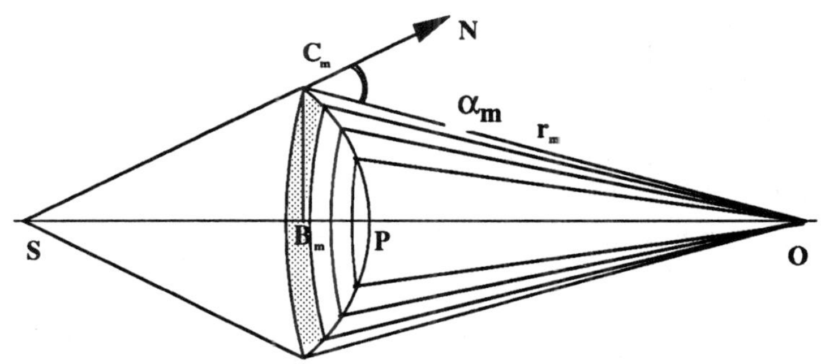

Fig. 11.14. Fresnel's zones.

$$r_m = b + m\lambda/2; \qquad (11.4)$$

math is a friend!

From the triangles $\mathbf{SC_mB_m}$ and $\mathbf{OC_mB_m}$ one can determine the radius ρ_m of the **m**th zone $\mathbf{C_mB_m}$ approximately as

$$\rho_m = \sqrt{\frac{m\lambda ab}{a+b}} \qquad (11.5)$$

where a=SP is the radius of the wave, and b=PO is the minimum distance of the wave from the observation point. Since a path difference of one half-wavelength means an inversion of the phase, the total amplitude of vibrations **A** at the point **O** created by all zones is

$$A=A_1-A_2+A_3-A_4+\dots; \qquad (11.6)$$

The amplitude of vibrations created at the point of observation **O** by a wave from a given zone diminishes with its area and with an increase of its distance $\mathbf{r_m}$ and the angle α_m. Fresnel studied only cases of small angles of diffraction, for which all zones have approximately the same area. This leaves only $\mathbf{r_m}$ and α_m as variables, both of which increase with the zone number. For this reason, $A_1>A_2>A_3>\dots$, and equation (11.6) may be rewritten as follows:

$$A=\frac{A_1}{2}+\left(\frac{A_1}{2}-A_2+\frac{A_3}{2}\right)+\left(\frac{A_3}{2}-A_4+\frac{A_5}{2}\right)+\left(\frac{A_5}{2}-\frac{A_6}{2}\right)+\frac{A_6}{2}\dots; \qquad (11.7)$$

Assuming that $A_k=(A_{k-1}+A_{k+1})/2$, we obtain

$$A=\frac{A_1}{2}\pm\frac{A_m}{2}; \qquad (11.8)$$

where "+" and "-" correspond, respectively, to an odd and even number of open zones. This means that an odd number of zones produces at **O** a maximum of intensity, while an even number of zones creates a minimum.

important

Thus, if only the first zone is open, the total amplitude is A_1, while the whole wave produces only $A_1/2$. When less than the whole first zone is open there will be a bright spot at any distance from the aperture. This leads us to the wave explanation of the rectilinear propagation of light: **almost all energy carried by a light wave is propagated within a very narrow channel connecting the source of light and the eye**. For a century and a half, the rectilinearity of light had been the prime argument against the wave theory. However, by the time Fresnel offered this theory, the debate on the nature of light began to focus on other things.

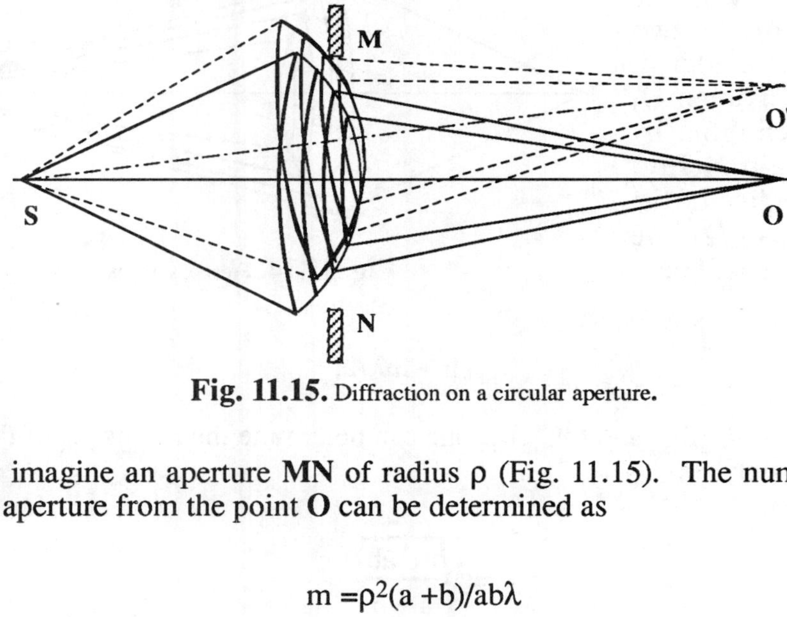

Fig. 11.15. Diffraction on a circular aperture.

Let us now imagine an aperture **MN** of radius ρ (Fig. 11.15). The number of zones **m** seen at this aperture from the point **O** can be determined as

$$m=\rho^2(a+b)/ab\lambda \qquad (11.9)$$

Thus, when the distance from the aperture to the eye increases, the center of the hole will appear alternatively bright and dark. In white light, the central spot is colored, and the colors vary according to the distance **b**. An off-center point **O'** creates another set of zones, in which the number of open half-zones in the upper and lower parts of the aperture is different. For this reason, the intensity of light at **O'** will be different from that at O, but it will be the same at all points equidistant from **O.** As a result,we have alternating bright and dark (or colored) concentric rings. If the aperture is rectangular, we will see parallel dark and bright straight lines (for convenience sake, Fresnel's zones will be made rectangular). In both cases, the number of fringes, as well as whether the central fringe will be dark or bright depends, according to eq. 11.09, on the distances **a** and **b** and the width of the aperture 2ρ.

If instead of a hole we have a a disk (or a ball) covering the first **m** zones (Fig. 11.16), at the shadow's center the total amplitude will depend solely on the amplitude of the first open zone,

$$A = \frac{A_{m+1}}{2} \qquad (11.10)$$

which means that whatever the diameter, the center of the shadow will be bright.

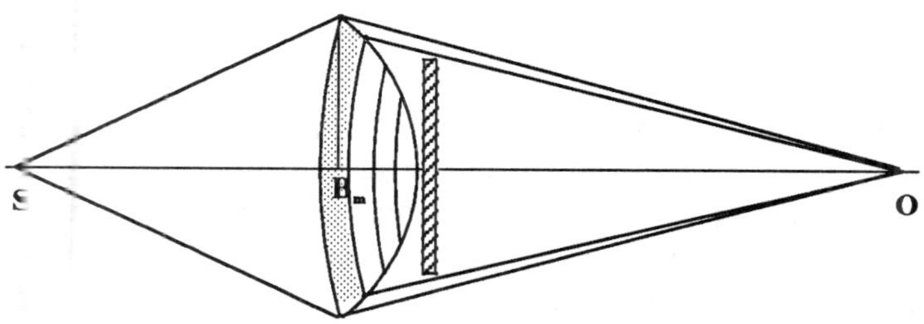

Fig. 11.16. Diffraction on a disk.

In his memoir submitted to the contest on diffraction, Fresnel did not discuss the case of a disk. ***Simon-Denis Poisson*** (1781-1840), a renown mathematician and one of the judges in the contest, applied Fresnel's integral to a disk and found that there must be a bright spot in the center of the shadow. With Maraldi's experiment forgotten, Poisson considered this result to be physically impossible, which was a strong objection to Fresnel's theory. However, François Arago, another judge, performed an experiment together with Fresnel and demonstrated that there was no mistake. It was Poisson's criticism that inspired Fresnel to invent his "zone theory," which made everything crystal clear and easy to grasp.

math is a friend

Fig. 11.17. Zone plate.

We have shown that by covering almost all of the wave front and leaving open only the first zone we actually increase the image's brightness. Therefore, **a small hole plays the role of a converging lens.** It is possible to collect even more light by covering all the even zones, for in this case the total amplitude is

wow!

$$A=A_1+A_3+A_5+.... \qquad (11.11)$$

Thus, a plate (*zone plate*) made so as to let through light of the odd zones and absorb that of the even zones, will concentrate light like a lens (Fig. 11.17).

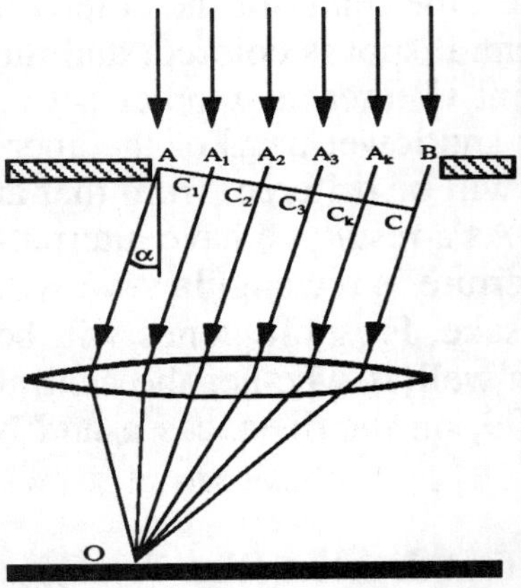

Fig. 11.18. Diffraction on a rectangular slit.

Applying Fresnel's zones is simpler for diffraction in parallel rays. For instance, when a plane wave falls on a a parallel slit **AB** (Fig.11.18), light will diffract in all possible directions. Let us select the rays which diffract at an angle α with incident rays and which are collected by a lens **L** at the point **O**. Since refraction in a lens does not change the path difference of diffracted rays, the total path difference of the two rays coming from points **A** and $\mathbf{A_k}$ of the slit **AB** equals $\mathbf{A_kC_k}$, where $\mathbf{AC_k}$. is perpendicular to $\mathbf{A_kC_k}$. If we select the points $\mathbf{A_1}$, $\mathbf{A_2}$, $\mathbf{A_3}$,.. so that $A_1C_1=\lambda/2$; $A_2C_2=2\lambda/2$; $A_3C_3=3\lambda/2$..., then $\mathbf{A_1A_2}$, $\mathbf{A_2A_3}$, $\mathbf{A_3A_4}$...can be treated as consecutive Fresnel zones. Whether the point **O** will be dark or bright will depend on the slit **AB** containing an even or an odd number of zones. The total number of zones seen at the angle α can be obtained by dividing the path difference BC=asinα of the extreme rays (**a** is the width of the slit) by the half-wavelength. Thus, diffraction minima (dark fringes), will be observed where the number of zones is even, which means

$$a\sin\alpha_{min}=m\lambda; \qquad (11.12)$$

math is a friend !

where m=1, 2, 3... Incidentally, diffraction maxima differ in their intensity. If, for instance, the slit contains three zones, two of them cancel the effect of one another, and the remaining zone produces the amplitude of vibrations equal one third of the maximal amplitude that is produced when there is only a single zone. Similarly, in the case of five zones, four would cancel one another, and the remaining zone would produce only one fifth of the maximal amplitude. The number of zones increases when the point of observation moves away from the center of the slit. For this reason, the central diffraction maximum is the brightest, while the side maxima, left and right, lose in their intensity the more the farther they are from the center.

Two parallel slits will produce a system of dark and bright fringes that depends both on the width of each slit **a** and the distance between them AA_1=**d** (Fig. 11.19). The direction at which a single slit produces dark fringes, according to equation (11.12) will be also the direction of minima for two slits. However, in addition to the diffraction pattern produced by interference of light from different zones of the same slit, we will have interference of light from different slits. The latter can be considered as taking place between the rays originating from such points of the two slits the distance between which is always **d** (the *corresponding rays*). The path difference of such rays will be $A_1C=d\sin\theta_{min}$. Thus there will be additional minima in the directions θ_{min} at which the rays of one slit destroy the corresponding rays of the other slit, such that

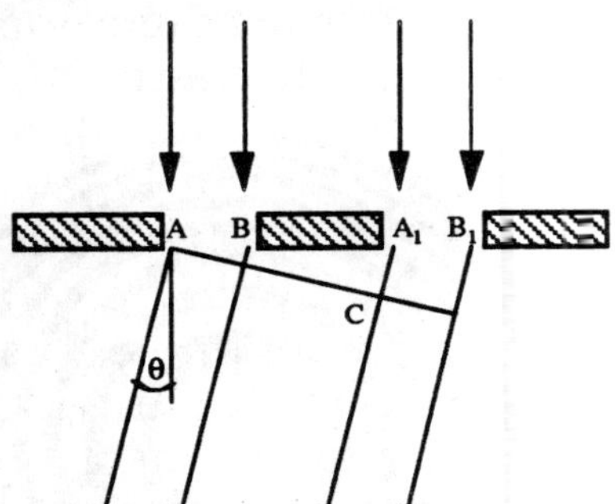

Fig. 11.19. Diffraction on two slits.

$$d\sin\theta_{min}=(m+1/2)\lambda; \qquad (11.13)$$

where m=0,1, 2,... On the other hand, there will be directions θ_{max} at which the corresponding rays from both slits reinforce each other and produce bright fringes, or,

$$d\sin\theta_{max}=m\lambda \tag{11.14}$$

Since d>a, according to eqs. 11.12-11.14, within the central maximum produced by a single slit (or between the two minima corresponding to m=±1) there will be several maxima produced by interference of light from two slits (Fig. 11.20) The same is true when we have a set of many parallel slits, or a *diffraction grating*, only in this case the maxima are much brighter because in such a direction the resulting amplitude of vibrations equals the amplitude produced by each slit multiplied by their number.

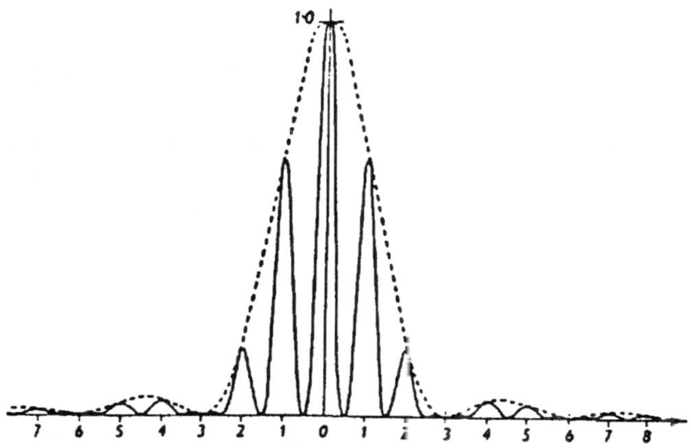

Fig. 11.20. Intensity of light distribution produced by two slits.

III. DIRECT OBSERVATION OF DIFFRACTION

1. Hair (Young)

Equipment. Nails, pins, paper clips from 0.5 mm to 2 mm thick; copper wires of different diameter from 0.5 mm to 48 gauge; hairs, slide frames for mounting hairs and wires, a candle, a tape measure, a meterstick.

Teacher. In the following experiments you will study the external and internal diffraction fringes by looking through an object at a source of light of a small angular size. Investigate how the fringes' width and coloration vary with the size of an object, its shape, material, position, etc. We'll begin with Young's experiment with a hair. Mount a hair on a slide frame and holding it near the eye look at a distant candle flame. What do you see? ***lab***

Dorothy. Nothing of interest.

John. I see colored fringes on both sides of the flame, several spectra on each side, red and green are visible the best.

Teacher. If some of you cannot see the fringes, move the frame across the eye so as to bring the hair exactly opposite the pupil.

Mary. I've just tried the same experiment with a paper clip instead of a hair. I saw a yellowish narrow band perpendicular to the paper clip but no colors. Why does a hair produce color fringes and a paper clip doesn't?

Michael. Perhaps metals cannot do it. I've tried a nail, and it doesn't work either.

David. I've mounted on a slide frame a few copper wires and found that a thin wire produced colors but the thick ones didn't.

Ruth. I know: it all depends on the object's thickness and not on its material.

John. If this hypothesis is true, a fine thread pulled out of a fabric should produce colored fringes as well. Indeed, it works! Ruth was right!

good point!

Dorothy. If a single thread creates colors, shouldn't the whole piece of the fabric do the same? Let's try it.

Michael. It works, but I saw two perpendicular rows of colored fringes instead of one. Strange, isn't it?

David. Not really. If you look closely at the fabric, you'll see the fibers making a neat rectangular lattice. Since a single fiber produces fringes parallel to it, perpendicular fibers should make perpendicular fringes.

Ruth. But why do many parallel fibers create only a single set of fringes?

John. I mounted on a slide frame a few thin copper wires so that they were all parallel and close to one another. I obtained, however, the same set of colors as with a single wire, only brighter.

Michael. We taped to a glass plate two hairs that were not parallel. What we've found was that each hair produced its own row of fringes which made an angle with one another.

Mary. Probably the spectra produced by parallel wires or hairs coincide, which makes fringes brighter.

Dorothy. Why should they coincide? Can't two parallel wires make two rows of fringes parallel to one another but located at different places?

David. Such two rows of fringes couldn't be spaced vertically because of the whole wire only the part equal to the size of the pupil participates in creating colors. I discovered this by placing behind the wire a small aperture: although only a small part of the wire was visible through the aperture, the fringes were the same as those produced by the whole wire. On the other hand, the two sets of fringes cannot be spaced horizontally (within the same row) either, because, according to equation (11.2), identical wires produce fringes equally distant from the flame.

Teacher. Can you now explain why a hair or a thin wire produce colored fringes while nails, paper clips, and pins don't?

Mary. We see that in the spectrum of each order the red fringe is seen farther away from the object's shadow than the green one, which agrees with equation (11.2). The same equation tells us that the thicker the object, the closer the fringes to the shadow and to one another. Perhaps, when an object is too thick the fringes of different orders overlap, and we see no colors.

good point!

Teacher. Can you prove that the same fringe is visible at the same angle whatever the position of an observer?

John. A given fringe (for instance, the red of the first order) is seen farther from the shadow when you move away from the candle, which hints at the possibility that the ratio of the two distances (x/b) is constant. To test it, we have to measure the distance of the first red fringe to the shadow **x** and the distance between the frame and the light source **b**. For a better precision, we can measure the distance 2x between two red fringes of the first order on different sides of the flame and divide it in two. Two persons are needed to do this: one watches the fringes and directs the other one where to put two white paper riders on a meterstick held near the light. Then the partners exchange their roles, each making two measurements.

Teacher. Is it possible to use Young's experiment to measure the wavelength of light, for instance, of red light?

Ruth. We can use the same equation (11.2). We already know how to measure the angular distance of the fringe from the shadow ($\theta = x/b$). To measure the diameter of the hair **d** we need a microscope with a reticle. To improve the precision we can repeat the experiment with the red fringe of the second order and take the average wavelength.

Teacher. I think you are now ready to do another Young's experiment.

2. Eriometer (Young)

lab ***lab***

Equipment. An eriometer, a microscope with a reticle, lycopodium powder, a blood smear, a flock of wool, cotton puffs. The eriometer consists of three parts: an object, the measuring plate, and a ruler (or tape). An object is placed between two microscopic slides taped together. To reproduce Young's original measuring plate take an index card, draw two concentric circles, the radii of which are, for instance, 1cm and 2cm; make the central hole 3 mm in diameter and 8 peripheral holes in each circle 1 mm in diameter. Install the object in a holder at the beginning of a ruler with the millimeter scale (or attach it to the ruler with a "Tak"). Place the measuring plate in a sliding holder that can be moved along the ruler.

Teacher Hold the object **C** near the eye (Fig.11.12) and look through it and through the central hole **O** of the plate **A** at a source of light **D** of a small angular size (a candle flame or an electric bulb). You will see colored rings around the light. To measure the angular radii of these rings, move the sliding holder until a chosen color ring coincides with the perforated circle on the plate **A**. Try the red rings of the first and second orders. A candle flame should be sufficiently close to the eye (30-50 cm). To obtain a bright and distinct image, you may have to move sideways both the object (to find a better spot on it) and the plate **A**.

Dorothy. Why do we need two concentric circles at the plate **A**?

Teacher. To cover a large range of fiber thickness without moving the plate **A** too far. Young wanted farmers to use his device to measure the thickness of wool produced by different sheep.

discussion

John. How do we know that Young was right to apply his theory of diffraction on a hair in this experiment. Not only the objects here are different (a flock of wool or a smear of lycopodium, for instance), but even the shape of fringes ?

David. We know that each fiber produces fringes parallel to it, thus a number of identical fibers oriented in all directions will produce arcs say, of red color, of the same radius, which form a red circle.

Teacher. Very good! The key word is "identical." If fibers differ in thickness, they will produce arcs of different colors which overlap and destroy any coloration.

good point!

Mary. We've tried various fabrics instead of yarn and found that they produce straight parallel fringes arranged in two perpendicular rows rather than circles. The only fabric that made circular fringes was nylon stockings. Why is it so?

Ruth As seen under a magnifier, in most fabrics fibers form a rectangular lattice. That is why the result is the same as with a mesh. In nylon stockings, the fibers are curved in different directions which makes them work rather like a yarn than like other fabrics.

Michael. But why do particles produce colored circles like wool fibers?

good point!

Dorothy. Not always! Lycopodium does it, but not a chalk powder or my face powder.

Teacher. We've already noted that when seen through a small hole a fiber creates the same fringe pattern as when seen unobstructed and we've found that only thin fibers produce colors. Thus, the length of a fiber appears to be of no importance, and cutting it into small pieces should not change the result. But these "small pieces' are nothing else as particles. To produce colored circles the particles or fibers must be: 1) small, 2) identical, and 3) randomly located. Each fiber from the stockings create the same fringe pattern, and when the fringes of the same color coincide the resulting fringe becomes brighter. But if the fibers (or powder particles) differ from one another, they create circles of different radius for light of the same color. As a result, the circles of different colors overlap and produce white. As to the last condition, it can be proven, for instance, by demonstrating that changing distances between fibers doesn't affect the result. Do you know how to do it?

John. I increased the distances by stretching a stocking, but the circles didn't change.

David. I mounted a piece of a stocking on a slide frame. First, I observed circles by holding the frame perpendicularly to the visual line. Then I tilted the frame so as to make a sharp angle with this line. The circles remained the same although in the latter case the distance between the two next fibers, as projected onto the original plane of the frame, certainly decreased.

Ruth It is now clear to me why Young used the equation derived for a single hair in the experiment with a flock of wool: it is a single fiber which is responsible for the angular size of the fringes, the multitude of fibers only improves the fringes' brightness.

Teacher. Very good! Moreover, Young extended this theory to **transpar-**

ent particles as well. In his view, colored rings around the sun or the moon hidden behind a cloud are due to diffraction of light on water drops in the cloud.

3. Fresnel's method

lab

Dorothy. Is Young's method of viewing diffraction fringes directly better than Grimaldi's observation of the fringes on a white screen?

Teacher. In certain cases, yes. For instance, you saw more fringes inside a hair's shadow even in a weak candle light, and you didn't have to move the object following the sun's movement. Your were able to perceive a weaker light because this light entered the eye directly without losing much of its intensity while being reflected from a white screen. Fresnel improved the direct method still further by adding a lens before the eye. We will use Fresnel's procedure to solve a practical problem: to recreate classical diffraction experiments when sunlight is not available. Upon the completion of the experiment we will discuss your results.

Equipment.

a. Take as a source of light the "light-house" described above, which consists of a base-mounted light fixture surrounded with a square cardboard house and covered with a cardboard lid. The electrical bulb must be **clear,** of 40 or 60 watts. In one of the walls make a square window of 1x1 cm at the level of the filament. Take a piece of aluminum foil of 3x4 cm, place it on cardboard, punch a tiny hole in it with the tip of a straight pin (if the pinhole is large, you won't see much of diffraction), and tape the foil to a slide frame and the latter to the window. For home observations use any desk lamp with a housing open on one side: cut a cardboard lid slightly larger than the opening, make a window in its center, and tape the lid to the housing.

b. For a narrow object use human hair and wires taped to a slide frame; straight pins with spherical heads, paper clips, nails of 1 to 4 mm in diameter, toothpick, strips of aluminum and of an index card 2 to 4 mm wide, both rectangular and triangular. For an aperture, drill a circular hole of 1 to 2 mm in diameter in a thin aluminum sheet and polish its edges, or punch a hole with a straight pin in an index card. Make aV-slit of two pieces of an index card taped together.

c. To carry objects use a lens-holder with a flat top (Fig. 11.21): tape the strips and slide frames to the top's side and stick pins or nails into a ball of "Tak" (reusable adhesive). Use another lens-holder to support a lens with the focal distances between 2 and 5 cm.

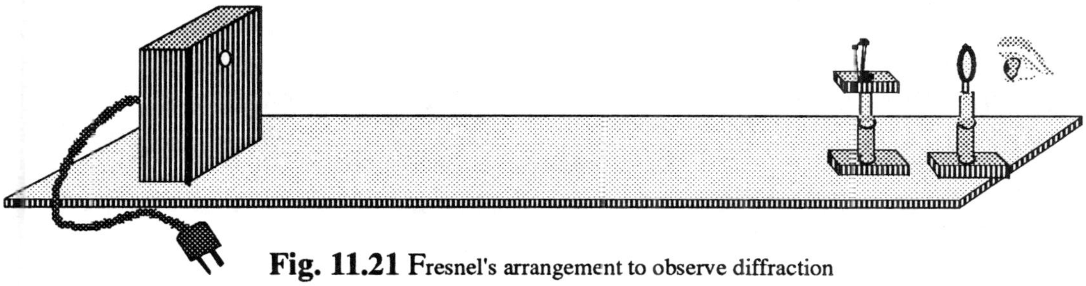

Fig. 11.21 Fresnel's arrangement to observe diffraction

lab

Procedure.
Position yourself at the narrow side of the table and place the "light house" 1.5-2 m from the eye. Move your eye and turn the "light house" until you see the pinhole very bright. Place the lens very close to the eye and adjust it and the eye so as to see a lighted **gray** (not yellow) circle covered with dots, specks, and lines: this is the image of the inside of your eye. Set the object-holder about 30 cm from the lens and adjust it so as to obtain a clear shadow of a pin crossing the circle: if you see parallel lines running lengthwise inside the shadow, your apparatus works fine. In each experiment draw the picture of what you see (use colors if necessary) and compare your diagrams to the original ones. We will check whether such *variables* as the **width** of a body and its **distance** from the lens, so important in the Grimaldi's procedure, retain their significance when using Fresnel's method.

a. narrow body (Grimaldi, Newton, Maraldi)

Watch the fringes produced by long and narrow bodies, such as paper or metal strips, nails, hairs, wires, paper clips. How does the distance between the fringes depend on the body's width and its distance from the lens? Can you make the fringes colored? **Tip**:a nail should be moved much farther from the lens (increase the distance between the light source and the lens), and if this doesn't help, try a smaller pinhole.

b. V-slit (Grimaldi, Newton)

Attach a V-slit to the holder's top vertically and move the eye before the lens up and down so as to look through different parts of the slit. Does the number of fringes depend on the width of that part of the slit you are looking through? Does this number change when you move the slit to and from the eye? Can you produce colored fringes?

c. ball (Maraldi)

This is one of the most impressive optical phenomena illustrating the limitations of the everyday experience. Move the holder carrying a pin with a well rounded head away from the lens and watch the center of the head's shadow. Can you see there a small bright spot? Does the spot change when you move the pin to or from to he lens? Does the spot vary with the size of the head or its shape?

d. small aperture (Fresnel & Arago)

This experiment deserves special attention because of its historical importance for the wave theory. Observe colored rings inside the hole. Investigate how their number varies with the distance between the hole and the lens. Can you obtain at the center a dark spot ? a colored spot? Does this color change with the distance? Do the rings change with the diameter of the aperture?

Teacher. Let us discuss the results starting with a narrow body.

Mary. When investigating the *distance* as a variable, we found that the number of fringes inside a pin's shadow decreased to two when the distance increased, and we adopted this result as a *hypothesis*.

Michael. We *tested* this hypothesis with nails and paper clips and confirmed it. Big nails didn't show any inner fringes, only outer fringes.

Dorothy. We've tried various paper strips: no success. *discussion*

John. Apparently, the *conclusion* is that the number of visible inner fringes diminishes as the distance between the body and the lens increases. This agrees with the results of similar experiments with sunlight. What I cannot understand, however, is why no fringes can be seen at small distances?

David. I guess I know: at small distances the intervals between the fringes become two small to see them separately. Can this be explained in the zone theory? *good point!*

Teacher. Yes, but before answering this question, let us summarize your findings about the second variable, the body's *width*.

Dorothy. It was clear from the previous experiments that only very narrow bodies deserve attention. Thus, we used a straight pin and a small nail. However, the nail didn't show any fringes.

Ruth. When we ran into a similar problem, we recalled that the fringe's width increases with the distance and moved the holder 1 m from the lens: there we observed the fringes in both the pin and the nail. In the pin's shadow, the fringes were wider than in the nail's, thus we supposed that the thinner the obstacle the wider the inner fringes. We tested this hypothesis with a paper clip and a bigger nail: the results were the same.

John. For our test, we've selected a triangular paper strip of 2 mm at the base and 5 cm tall. At its tip, the inner fringes were wide but towards the base they became more and more condensed until coincided. Thus, the hypothesis is correct. *good point!*

Teacher. According to our theory, the central fringe must be always bright. The first dark fringe will be where the first open zone of one part of the wave (say, to the left from the body) is odd, while on the right side the first open zone is even, next to the odd one. It means that the path difference of rays from the edges of the body at the first dark fringe will be $\lambda/2$. Using considerations similar to those on p. 217, one can show that the width of an inner fringe is proportional to the distance from the lens (the light source is far away) and inversely proportional to the body's width. *math is a friend!*

Michael. With a V-slit, the result was different from that with a triangular strip: apparently the same fringes run along the slit becoming wider where the slit is wider. The central fringe could be bright or white depending on the distance from the lens. How come that a dark fringe appears against the center of the slit?

Teacher. The two halves of the slit act identically, thus if one half contains an even number of zones, they will destroy light of each other. To calculate the number of zones, take the path difference for the rays coming from the slit's center and one edge and divide it by $\lambda/2$.

Dorothy. What if the path difference is less than $\lambda/2$?

David. This may happen only at a large distance, and the result will be a bright fringe. Incidentally, if we move the slit farther away, the path difference becomes even smaller. This means that at large distances there is only a single zone within the slit, and the central fringe will stay bright whatever the distance.

Teacher. Very good! How about the shadow of a round pinhead?

good point!

John. Maraldi was right: we did observe a tiny light spot in the center of the shadow. However, a larger head did not produce this effect. Why?

Ruth. According to the theory, the intensity of light at the center depends on the first open zone. However, zones of higher orders contribute less light than the zones of low orders. A large head covers many zones, and the light coming from the rest is too weak to excite the eye.

Teacher. Excellent! Did a circular aperture act similarly to a sphere?

good point!

Mary. Not at all! The central spot was either dark or colored, red and green were especially clear. There were also several concentric rings dark in the proximity of the lens, and colored, at larger distances.

David. Well, we've seen colored fringes with other objects too, although they appeared only at large distances and were fainter.

Michael. Is it correct to say that at small distances fringes of different colors fully overlap and produce an impression of gray?

Teacher. Yes. And who knows why was the center differently colored at different distances?

Ruth. The number of zones depends on both the wavelength and the distance (eq. 11.9).

Teacher. Good! As you see, the zone theory is a powerful tool for explaining diffraction phenomena.

Dorothy. How about regular reflection? You had promised us to explain in the wave theory why an uneven surface can reflect as a mirror.

Teacher. The ideal surface reflects in a single direction because in all other directions the rays reflected by different particles of the surface destroy one another through interference. If the height of the "hills" is less than $\lambda/4$, even at normal incidence the rays reflected by neighboring particles will have the path difference less than $\lambda/2$, and no light reflected at the angle of incidence will be destroyed.

BIBLIOGRAPHY

T. Young, *Misc. Works*, I: 171-72, 179-84, 346-51.

J. Buchwald, *The Rise of the Wave Theory of Light* (University of Chicago Press: Chicago, 1989), pp. 111-202.

N. Kipnis, *History of the Principle of Interference*, pp. 102-118, 218-224.

Chapter 12

POLARIZATION of LIGHT

I. DOUBLE REFRACTION

1. Polarizing sunglasses

investigation

Preliminary observations

Dorothy. Michael, your sunglasses are rather light, do they work well?

Michael. They surely do: these are polarizing sunglasses.

Dorothy. But how it be that your light sunglasses act the same as my dark ones?

John. They don't act the same. I have here both types, and you can try and see that polarizing glasses cut a lot of light from the blue sky and from a water surface but they don't make the ground as dark as the ordinary ones.

Mary. Indeed. How can this be? I know that ordinary sunglasses absorb much of the incident light, which reduces the amount of reflected light (because of the conservation law). That is why such glasses appear dark to an outside viewer. But polarizing glasses must work in a different way because they are not as dark.

David. Maybe they aren't as dark as some absorption glasses but they certainly absorb some light. The main difference must be not in absorption.

Ruth. Some glasses transmit less because they reflect a great deal: they are semi-transparent mirrors. However, polarizing sunglasses don't look like mirrors from outside. But if they neither absorb nor reflect an excess of light, how do they cut the glare?

Teacher. To get a clue to this problem let us do the following experiment. Michael, put your polarizing glasses on! John, hold your polarizing glasses before your eye and look at Michael's eyes! What do you see?

John. I see his eyes, not as well, of course, as without glasses.

Teacher. Now, John, rotate your glasses by 90° in their plane. What do you now see?

John. Wow! Now, I don't see his eyes. How can this be?

Dorothy. I tried the same trick with ordinary glasses but it didn't work: I saw Michael's eyes equally dim in both positions of my glasses.

Formulating a problem

How do polarizing glasses work?

Hypothesis.

David. Apparently, the orientation of polarizing glasses in the plane perpendicular to the line of vision is important in cutting the glare.

Test.

Michael. Let us watch various shiny objects through polarizing glasses while rotating them around the visual line.

Teacher. Good idea! Since only a few of you have polarizing glasses, I'll give you pieces of the same type of a polarizing film as applied in your glasses, that is placed between two plates of glass or plastic. We'll call such a device *polaroid.*

Mary. We've found that by rotating a polaroid we made the sky appear brighter and darker. The pavement didn't change its brightness, however.

John. The images in the pool changed their brightness twice every full revolution.

Dorothy. We've observed the same with car windows but not with the car fenders.

conclusion ➡ There is a specific orientation of a polarizer relative to the visual line at which it transmits more light reflected from certain objects than in the perpendicular orientation.

Michael. Well, we've found that there is a material whose capacity of transmitting light depends on its orientation in space. Are there other materials of this sort?

David. It's not just the material that produces the effect! Have you noticed that neither the pavement nor some parts of car bodies changed their appearances? There should be something different about the light reflected, for instance, by car windows or car bodies.

Teacher. You are right: both the material and the properties of reflected light are essential for this phenomenon. That some materials refract light differently from others was first discovered in a transparent calcite known as *Iceland spar.*

2. Iceland spar

a. Bartholin's experiments

lab

Teacher. In 1669, ***Erasmus Bartholin*** (1625-1698), Professor of Mathematics at the University of Copenhagen, found that an Iceland spar made all objects seen through it double. Let us repeat some of his experiments. Take an Iceland crystal and view different objects through it; if you see them double, investigate the factors affecting the degree of their separation.

Dorothy. I see only the edges of bodies double. Letters, however, do appear double, but it is difficult to see how much they are separated

David. Among the objects we've tried pencil dots were the best, because the distance between them is larger than the dots themselves. The distance between the images increased with the thickness of the crystal.

Ruth. While rotating the crystal relative to the vertical axis we've seen that one image stood still and the other moved around it. Besides, one image was brighter than the other.

John. When I looked along a diagonal the two images almost coincided, perhaps even exactly.

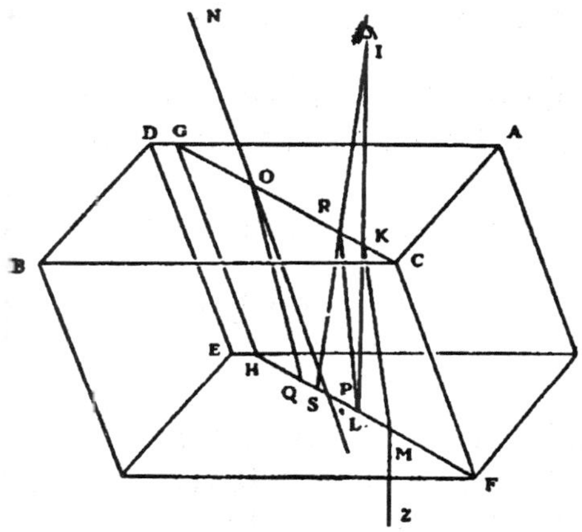

Fig.12.1. From Huygens, *On Light*

history and discussion

Teacher. You have obtained Bartholin's principal results. The direction in which the images coincide is called the *optic axis* (CG in Fig. 12.1). Bartholin concluded that the two images resulted from a different refraction of light. He called the ray that forms the immobile image *ordinary*, because it obeys the law of refraction, and the one that produces the moving image *extraordinary*, because the index of refraction of this ray depends on the direction of incident light.

b. Huygens' theory

In 1678, Huygens presented a detailed account of the properties of Iceland spar, which he published in 1690 (*On Light*, ch. V). He found a specific plane **CFHG** in the crystal, called *principal section*, which retained the incident ray and the two refracted rays for different angles of incident. This plane is parallel to the plane formed by the edge **CF** and the line **CG** bisecting the obtuse angle **ACB**. In the principal section, the incident ray **IK** that is perpendicular to **CG** produces the ordinary ray **KL** and the extraordinary ray **KM**, which form the angle 6°40' between them. Huygens suggested that the extraordinary ray behaves so strangely because its velocity depends on the direction of its propagation in the Iceland spar, while the velocity of the ordinary ray is constant. Since the index of refraction of a medium depends on the velocity of light in this medium, the index of refraction of the extraordinary ray must depend on its direction. According to Huygens, such a phenomenon can be explained in the wave theory by assuming the existence of two types of light waves, spherical and spheroidal. He represented the ordinary light by a spherical wave and the extraordinary light by a spheroidal wave. In the plane of the drawing, the ordinary wave front is represented by a circle and the extraordinary one by an ellipse (Fig. 12.2). The radius drawn from the center C to a point at a wavefront shows the velocity of light in this direction. We see thus that the ordinary wave has a constant velocity, while the velocity of the extraordinary wave depends on the direction of propagation. The two wave fronts touch in points **B** and **S** , which means they have the same velocity in the direction BS, or, in other words, **BS** is the direction of the optical axis. To describe the propagation of the extraordinary wave Huygens uses the same principle as for the ordinary wave, with that difference, though, that the secondary waves are semi-spheroids **SVT** rather than semi-

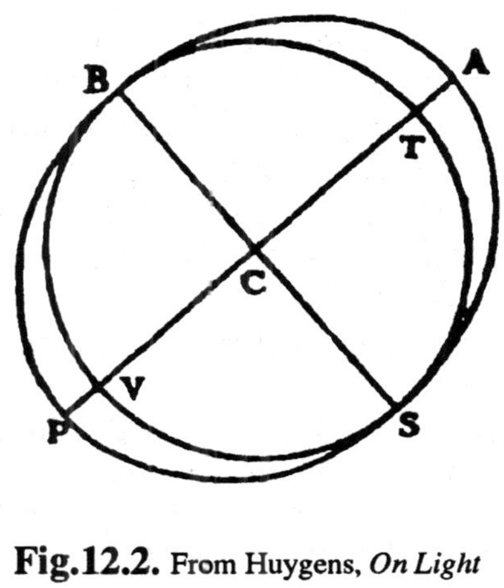

Fig.12.2. From Huygens, *On Light*

spheres (Fig. 12.3). If the incident wave front **AB** is a plane, the extraordinary wavefront **NQ** remains a plane parallel to AB. However, NQ is shifted to the left relative the direction of propagation of light **RA**, which means that the extraordinary ray AV deviates from this direction. On the basis of his theory, Huygens calculated the index of refraction of the extraordinary ray for different directions of incidence and stated (without giving any numerical data) that his theoretical predictions fully agreed with experiment.

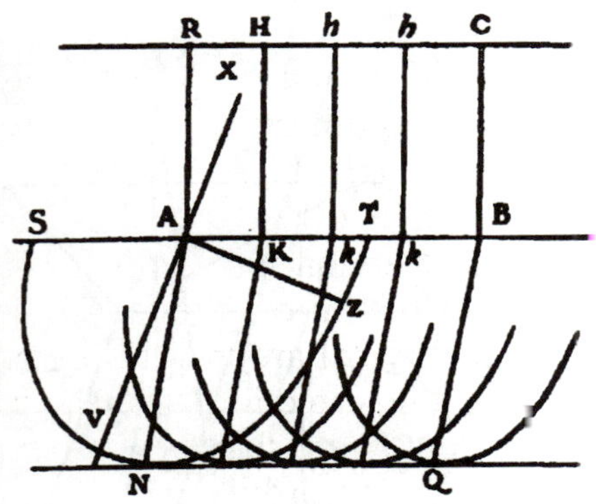

Fig.12.3. From Huygens, *On Light*

Dorothy. This is all interesting, but what does it have to do with polarizing glasses?

Teacher. I am coming to that. Once, Huygens looked at an object through two Iceland crystals and found to his surprise that the number of images varied, depending on the relative position of the crystals. Let us repeat this experiment.

c. Huygens' experiment

lab

Equipment. Two Iceland crystals, index card, scotch tape.

Procedure. Cut a piece of an index card of the size of a side of the crystal, make a 2 mm hole in the middle of it, and attach it to the crystal with scotch tape. Look through the crystal at a light source or a brightly lit surface (the cover must be on the farther side from the eye), and you will see two holes. Place another Iceland spar closer to the eye and slowly rotate it around the line of vision.

John. In most cases, we've observed four images making a parallelogram. The rotation changed their brightness.

Ruth. Apparently the images located on the same diagonal make a pair because they change their brightness at the same time. When one pair is getting brighter the other goes dimmer and the vice versa, and these changes occur four times per revolution.

Mary. We have seen two images, apparently when the principal sections of the crystals were parallel or perpendicular to one another.

history *and discussion*

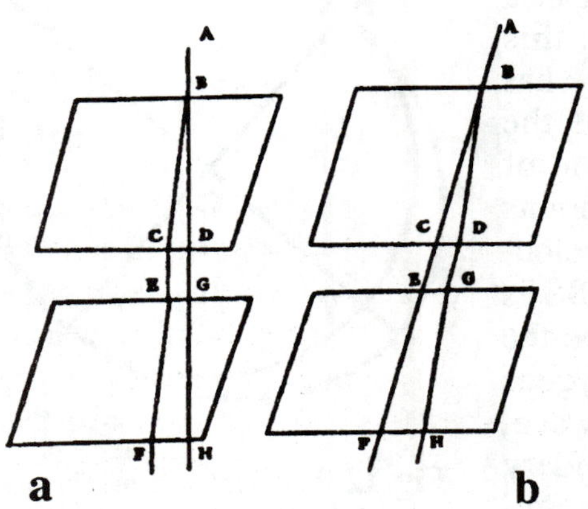

Fig.12.4. From Huygens, *On Light*

Teacher. According to Huygens, two images instead of four appear when the rays emerging from the first crystal lose the ability to split into two other rays in the second crystal. In particular, when the principal sections of the two crystals are parallel, the ordinary ray is transformed into the ordinary one, and the extraordinary ray into another extraordinary one (Fig. 12.4a). On the other hand, when the principal sections are perpendicular, the ray that was the extraordinary in the first crystal becomes the ordinary ray in the second one, and vice versa (Fig.12.4b). Huygens concluded that both the ordinary and extraordinary rays had a new property absent in natural light, but he could not specify it. (*On Light*, 92-94)

It was Newton who made the first attempt to explain this property. In his view, the transformation of the extraordinary ray into the ordinary one solely by turning the second crystal around the direction of light implies an asymmetry in a light ray relative to this direction. He attributed to a ray two pairs of different sides whose positions relative to the principal plane determine whether the ray will be refracted ordinarily or extraordinarily. (*Opticks*, 358-60)

Dorothy. How can a line have "sides?"

Teacher. I've already mentioned that when speaking of rectilinearity of light not all scientists compared a light ray to a geometrical straight line: in the Middle Ages, some of them attributed to a ray transverse dimensions (the *physical ray*). Since such a hypothesis complicated the explanation of many phenomena, scientists tried to avoid it. However, Newton realized that one **cannot** apply the concept of a light ray to Huygens' experiment without assuming that the ray has *sides*, or, that it resembles a rectangular rod with unequal sides (such as a meterstick) rather than a straight line. Imagine a meterstick being pulled through a lattice made of parallel wires the distance between which **d** is larger that the thickness **a** of the meterstick but much smaller than its width **b** (Fig. 12.5). The meterstick will pass easily between two wires only when its narrow side is parallel to the wires or forms a small angle with them. When this angle increases the wires give way, but their friction will slow the passage of the meterstick. Finally, when the angle approaches 90° the friction is too great to let the meterstick through. This may serve as an analogy to the transfer of light by the second crystal, assuming that the amount of emerging light corresponds to the amount of energy of the moving meterstick transmitted through the lattice.

history ***and discussion***

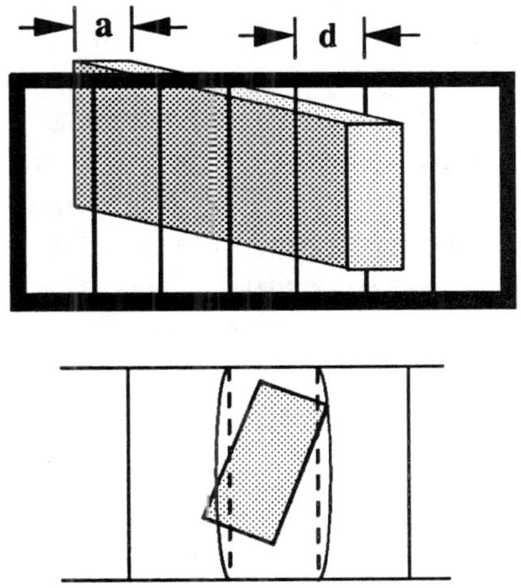

Fig.12.5. A corpuscular model of polarization.

Neither Huygens' theory of double refraction (or *birefringence*) nor Newton's concept of *sides* received much support, and by the end of the eighteenth century both had been practically forgotten. Their revival began in 1802 with experiments of ***William Wollaston*** (1766-1828). Wollaston invented a new method of measuring the index of refraction and applied it to a variety of, substances, including Iceland spar. He knew nothing about double refraction and was surprised to find the index of refraction of Iceland spar depending on the direction of incident light. When he mentioned this fact to his friend Thomas Young the latter referred him to Huygens' theory. Wollaston found his measurements in agreement with Huygens' theory and pronounced this theory to be true. ***Pierre-Simon Laplace*** (1749-1827), a leading French mathematical physicist and a supporter of the emission hypothesis, didn't like Wollaston's conclusion about Huygens' theory, and was looking for someone able to verify it. This role was left for a young military engineer ***Étienne-Louis Malus*** (1775-1812).

II. POLARIZATION BY REFLECTION AND REFRACTION

1. Malus and his discovery

history ***and discussion***

Teacher. In 1794, a new institution was founded in Paris named *École Polytechnique* (Polytechnical School). Students entered this school at the age of 16 after passing very difficult oral exams and studied there for two years. The main emphasis was on mathematics (among the professors were Monge, Lagrange, Poisson, Lacroix, and Legendre). After graduation, a student could join the army (artillery) or to enroll an engineering school, either civilian or military. Malus was in the first class of the *École Polytechnique* and then became a military engineer. In 1798-1801, he took part in Napoleon expedition to Egypt and Syria (1798-1801), during which he almost died from the plague. It was during his recovery there that he started to think about the nature of light and other optical subjects. Subsequently, Malus served in Strasbourg but he frequently visited Paris where he contributed two optical papers to the Academy of Sciences. Laplace highly regarded these articles and decided that Malus possessed all the necessary skills as experimenter and mathematician to solve the problem of double refraction. Thus, he suggested that the Academy made this topic the subject of its next biannual mathematical contest in 1808. The program of the contest required to build a mathematical theory of double refraction and support it by exact experiment. Malus entered the contest, made a number of exact measurements of the index of refraction of the extraordinary ray in several substances, and found them to confirm Huygens' theory of double refraction. This did not mean an advocacy of the wave hypothesis, for Malus (and Laplace too) found a way to reinterpret Huygens' theory of Iceland crystal so as to base it on the properties of light particles. Following Newton's idea of a transverse asymmetry of light, Malus attributed to light particles two different poles **a** and **b** which were perpendicular to one another and to the direction of propagation **c** (Fig.12.6). He called this asymmetry of light *polarization.* According to him, the ordinary and extraordinary rays were polarized in perpendicular directions.

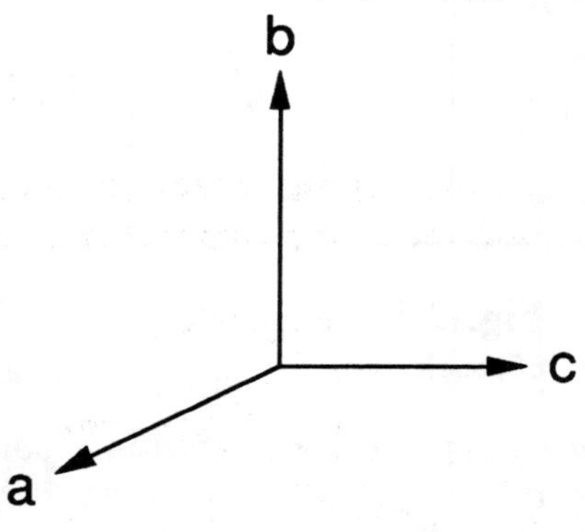

Fig.12.6. Malus' poles.

Dorothy. We don't believe any longer in particles of light having poles, right? If so, what does this *polarization* mean to us?

Teacher. It is simply a word to denote that light has different properties in two directions perpendicular to one another and to the direction of propagation, I've already described its model in the emission theory. Its wave model is similar if we imagine a vibrating string passing between the wires of a lattice: the vibrations will be transmitted from one side of the lattice to another if their plane is parallel to the threads or makes an acute angle with them. The greater the angle the more energy is absorbed by the lattice, which means less light coming through, and no light at all will pass when the plane of vibrations is perpendicular to the threads (Fig. 12.7).

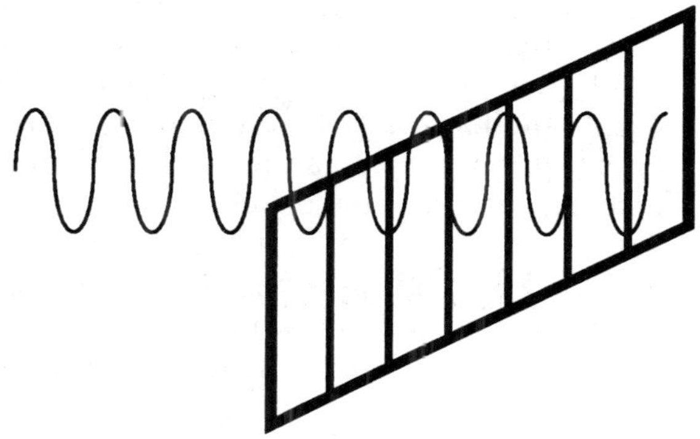

Fig.12.7. A wave model of polarization.

Michael. What does it mean that light is polarized in a specific direction?

Teacher It means a preferential orientation of the poles of light rays (or particles). When speaking of polarized light, we will call *polarizer* and *analyzer* the instruments used to, respectively, produce polarized light and uncover it. The same device can serve as either a polarizer or an analyzer. Prior to Malus, the only known polarizer/analyzer was a double refracting crystal. That changed one evening when, according to Arago, Malus took a look at the image of the setting sun in the windows of Luxembourg Palace through an Iceland crystal. He discovered that the brightness of the two images was different and it changed when he rotated the crystal around the visual line. Malus concluded that since the light reflected by glass behaved similarly to the light transmitted by an Iceland crystal, it had to be polarized. Malus found that other substances, including marble and water (but not metals), can also polarize light, and the effect is more pronounced at a specific angle of reflection, such as 52°45' for water and 56°30' for glass. He tried another glass plate as an analyzer and observed that the image he saw in it changed its brightness with the angle between the two glasses, being the darkest when the two were perpendicular On the basis of such observations, Malus formulated the following law for the intensity of reflected (or transmitted) polarized light ("Malus' law"):

$$I=I_0\cos^2\alpha; \qquad (12.1)$$

where I_0 is the intensity of the incident light, **I** is the intensity of the emergent light, and α is the angle between the planes of polarization of light of the two mirrors, or between two planes of reflection of light (Fig.12.8). In Malus' view, each ray has its specific orientation of the pole **a** in space, and polarization refers to a statistical distribution of these orientations. When light is natural, the poles are randomly oriented; if light is fully polarized, they have the same orientation, Malus called *partially polarized* the type of

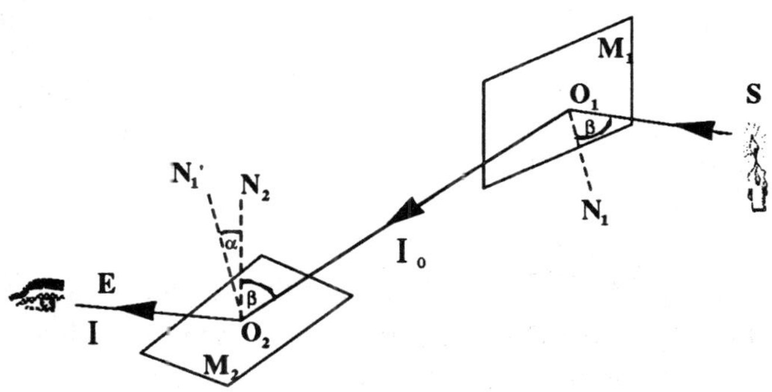

Fig.12.8. Polarizing and analysing light by reflection.

light that shows a preferential orientation of poles in a specific plane: in his view, it was a mixture of fully polarized and unpolarized light. He believed, that a polarizing device selects of all randomly distributed rays only those that have a specific orientation, which implied that the remaining rays will no longer retain a random distribution. For instance, at the boundary of two media the incident light is either reflected or refracted, thus if the reflected portion is polarized, so must be the refracted one, and the two planes of polarization must be perpendicular. Malus' experiment confirmed that the refracted light is partially polarized in the plane perpendicular to that of reflected light. He also found that the degree of polarization of refracted light can be increased by transmitting it through a number of parallel glass plates (Fig.12.9)

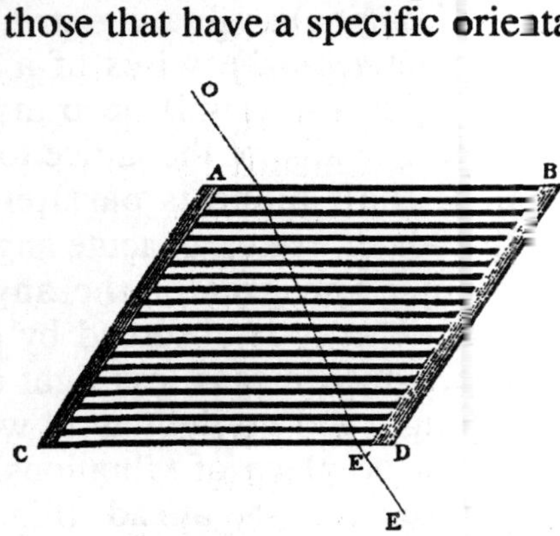

Fig.12.9. Polarization of refracted light. From Brewster, *Phil. Trans.*,1814, 239–6.

2. Brewster

history
and
discussion

Teacher. Although Malus was certain that each substance has its peculiar angle of full polarization, he could not derive any rule for it. This was accomplished in 1815 by ***David Brewster*** (1781-1868) who discovered that the angle of the full polarization β (later, it became known as *Brewster's angle*) for a specific substance depends on its index of refraction **n** as follows:

$$\tan\beta = n; \qquad (12.2)$$

Brewster was born in Scotland in a teacher's family. He attended the University of Edinburgh and graduated as a divinity student. However, although he had a license to preach in the Church of Scotland, he was never ordained a minister. His first interest in optics was in improving microscopes and telescopes, but he became famous for his investigations of double refraction and polarization. A war between Britain and France delayed the news about Malus's discovery, and Brewster's first results in this field (published in 1813) repeated those of French physicists. Brewster's real strength was in experiment, his theories were unsuccessful. Among his best achievements, apart from Brewster's angle, were discoveries of artificial double refraction (1814-15) and of biaxial crystals (1818). During his most productive years, Brewster could not obtain a teaching position (there were few of them in Britain), and he made his income by writing popular books and articles and editing scientific journals and encyclopedias. He was involved in a number of scientific controversies, either on the priority of his discoveries, or on the nature of light and colors. He remained a staunch opponent of the wave theory of light even when it became generally accepted (since 1830s).

Mary. How could he object the theory that was accepted by all scientists?

Teacher. Well, first of all, he had a personal stake in the emission hypothesis, which was the basis of his own theory of biaxial crystals. Second, although the wave theory was accepted as providing a better explanation of double refraction and polarization, there were other phenomena unexplained in this theory, even in the 1850s.

3. Repeating Malus' experiments

lab

Teacher. Let us go outside and repeat Malus's original experiment with light reflected by windows. You will use an Iceland spar as an analyzer: first try it as is, then cover one side of it with paper in which you make a small (2-3 mm) hole. If you turn the opposite side to the eye, you will see two holes instead of one, and you should observe the polarizer through both images of the hole. After completing this experiment you may do other experiments involving reflection and refraction. In addition to an Iceland spar, use a polaroid, and other analyzers. If indoors, use a window or a desklight as a light source; outdoors, watch the image of the sun or the sky. Devise experiments to verify Malus' law.

Michael. We observed a big difference in brightness between two images of a second floor window seen through an Iceland spar. This means that a window can polarize sunlight not only at sunset. The angle was around 50°, so I am not sure whether it was a full polarization.

Mary. We were watching the window reflecting clouds. At some positions of an Iceland crystal one image was dark and the other was bright. When we turned the crystal by 90°, the bright image became dark and the dark image changed to light.

Dorothy. I observed a piece of glass lying on the ground using a polaroid as an analyzer. Every time I turned it by 90° the image of the sky in the glass changed from bright to dark or vice versa.

John. I used an Iceland spar as an analyzer and watched water in a pool reflecting clouds. It didn't work at first, but eventually I found a position from which the reflected light appeared to be quite polarized, for one image was much darker than the other. This occurred at four different positions of an Iceland spar.

David I've tried a glass mirror as an analyzer. First, I found a position where I saw the reflected light at about 56° which is the angle of full polarization. I measured the angle by means of a sighting device provided with a protractor. Then, I turned 90° to my left and, holding a glass plate before me, found in it an image of the polarizing mirror. I started turning the analyzer around the horizontal axis and noticed that the image was the brightest when the analyzer was horizontal and the darkest when the analyzer was vertical . Finally, while holding the analyzer vertically, I moved it to the right to increase the angle of reflection. The image first darkened then lightened, the darkest being at the angle of full polarization. It means that given the same angle of incidence, two glass plates reflect the most light when they are parallel, and the least, when they are perpendicular. Since the angle between the two mirrors is the same as the one between their planes of polarization, the experiment conforms to Malus' law.

Ruth. I've tried to combine a polarizer based on refraction in a glass plate using a polaroid analyzer. One plate showed some difference in brightness of the sky when I turned the analyzer 90°. Five glass plates put together produced a greater difference, and ten plates made the image quite dark.

discussion

Teacher. Very good! You've learned how to polarize and analyze light and you even confirmed Malus' law. He never described how he derived this law or tested it. Since there was no means then for measuring the intensity of light for an arbitrary angle of reflection, he only could do what you did: comparing the orientation of an analyzer at which one sees the image the darkest or the brightest. Since a maximum occurs when the planes of polarization of the polarizer and the analyzer are parallel, and the minimum is perceived when the planes are perpendicular, it would have been natural to suggest the involvement of cosα. According to Biot, Malus probably selected the simplest function containing cosα and providing positive values, which is $\cos^2\alpha$. Subsequently, the truth of Malus' law was established with high precision.

Michael. If glass polarizes light at about 56°, why was I able to see polarization at other angles?

Teacher. Malus emphasized that some polarization occurs at any angle of reflection, but at one angle the reflected light is fully polarized. The greater the degree of polarization the greater the difference in brightness of the two images seen when turning the analyzer by 90°.

David. I wonder why Malus' discovery was called accidental. If you look through an Iceland spar at various objects, such as windows, water, cars, and others, it is difficult to miss a difference in the brightness of the two images. And it was this difference which reminded Malus of the Huygens experiment with two crystals.

Teacher. You are right: for a person well familiar with the properties of two beams polarized by an Iceland spar, it is easy to see the analogy between them and the light reflected by windows (forget about cars!). As ***John Herschel*** (1792-1871), an English astronomer and physicist, said, "*the seeds of great discoveries are everywhere present and floating around us, but they fall in vain on the unprepared mind, and germinate only where previous inquiry has elaborated the soil for their reception and awakened the attention to a perception of their value.*"("Light," 505) On the other hand, Malus did **not** expect light to be polarized by reflection; from that perspective, the discovery was accidental. There are other aspects as well. For instance, if you didn't place a screen with a small hole before an Iceland spar, you would not see the effect. Thus, to succeed Malus needed a proper experimental procedure, but designing such a procedure is to some extent a matter of luck.

important

Dorothy. The Iceland spar is not easy to use: if the hole is too small, you don't see much through it, and if it is large, the two images overlap. Why didn't Malus use a polaroid?

Teacher You are right: in some cases, an Iceland crystal is not very convenient to use because the two polarized beams are difficult to separate from one another. However, there was nothing better available at Malus' time. Later, different means were devised to get rid of one of the two refracted rays. One of the simplest was to use a crystal (tourmaline, for instance) which absorbs one of the rays. The polaroid film used in sunglasses has the same property, although it is made not of a single large crystal but of many elongated microcrystals imbedded in an organic substance so that they all have the same direction.

John. With a polaroid film one can detect even a small degree of polarization. Could physicists do it before the invention of polaroid?

Teacher. Yes, they had found a way to do it using the phenomenon of *chromatic polarization.*

III. CHROMATIC POLARIZATION

1. Arago

history

Teacher. A premature death from tuberculosis in 1812 terminated Malus' brilliant but short career in optics and left a rich new field of polarization to other investigators. At the moment, the general attention was attracted to the phenomenon of *chromatic polarization,* discovered in 1811 by ***François-Dominique Arago*** (1786-1853). As a child, he received a traditional classical education, but he dreamed of becoming an officer in the army. Once, he saw a very young officer and learned from him that students of the *École Polytechnique* obtain a promotion soon after the graduation. Arago abandoned his literary studies and began preparing himself for the entrance exams that were very demanding. In mathematics, he went beyond the secondary school textbooks and studied works of Euler, Lagrange, Laplace, and Poisson. He passed the exams with distinction and was enrolled at the *École Polytechnique* in 1803. Upon graduation in 1805, Arago intended to join the artillery but was talked into taking a position at the Observatory of the Paris Academy of Sciences. He was involved in two optical projects measuring refraction of gases, together with ***Jean-Baptiste Biot*** (1774-1862), and verifying whether the speed of light coming from heavenly bodies depends on their movement. In 1806, Biot and Arago were sent by the Academy of Sciences to Spain to continue measuring the earth meridian. Soon Biot returned to Paris, and Arago continued to do the work alone. The assignment was difficult in itself and it was especially endangered by highway bandits and the war situation between Spain and France. In 1808, Arago was arrested under suspicion of being a French spy. He managed to escape to Algiers. Disguised, with a false passport Arago embarked a merchant ship heading for Marseilles, however, on its way the ship was intercepted by a Spanish corsair, and the passengers were detained. Three months later, the ruler of Algiers threatened to start a war if the Spanish government would not release the ship. Arago was set free and again boarded a ship for Marseilles. The second attempt was no more successful than the first: a severe storm forced them to return to the African coast. During that season there were no sea communications with Algiers, and Arago decided on a very dangerous journey by land joining an Arab caravan. To save his life he faked an intention to convert to Islam. Finally, after some more adventures in Algiers, Arago arrived at Marseilles in July 1809. When the news that Arago is alive and preserved the results of measurements reached the Academy of Sciences, it considered filling the vacancy in the section of astronomy. The Academy had 40 positions in five sections, and each position was for life. The membership carried great prestige and some salary as well, consequently the elections had been very competitive. Some very good scientists had been waiting for decades for a vacancy. Although Arago was only 23 years old and had only two papers to his credit, the academicians decided that his ability to handle so difficult an assignment made him a good candidate and decided to delay the elections until his return. In September1809, Arago was elected to the Academy, and he fully justified the credit given to him. Arago is most known for his work in optics and electromagnetism. Since 1830, he was an active member of the Parliament expressing democratic and liberal views. After the February revolution of 1848, he became a member of a provisional government, serving as a Minister of War and Navy.

Arago's road to fame began in 1811 with his discovery of colors of crystalline plates. While experimenting with Newton's rings produced in thin films by a polarized light, he decided to check whether a thin film of mica produces the same phenomenon as a pair of lenses or glass plates pressed together. While looking at a thin plate of mica through an Iceland crystal, Arago accidentally placed the mica against the blue sky. To his surprise, he found the two images colored.

David. Is the "blue sky" a figure of speech or a physical factor?

Teacher. You have an opportunity to discover this yourself. Go outside, repeat Arago's original experiment and then try to modify it.

2. Colors of mica

Preliminary part

investigation

Experiments

Dorothy. First, I couldn't see any colors at all, so I made the hole in the screen somewhat larger and then I noticed a slight coloration but only when I looked at some parts of the sky.

Michael. We didn't see any colors against clouds or in the part of the sky opposite the sun.

Ruth. Apparently, the colors are seen only in the direction perpendicular to that to the sun. When they do appear, the colors of the two images appear to be complementary, such as greenish and purplish, for instance. The colors changed when I rotated the crystal. Sometimes, I was not sure about the color, but I noticed that one image appeared darker than the other, and when I turned the crystal by 90° the darker one became the brighter of the two.

John. We've got much brighter colors by tilting the mica plate to the line of vision.

David. A thick plate of mica (thicker than 1 mm) didn't show any colors, and neither did very thin plates. While working with a plate of a proper thickness we've discovered that colors changed when we rotated the plate in its own plane. Since the rest was kept the same the cause could be in the plate's thickness, because the crystal was not uniform.

Teacher. O.K. Let us now summarize the results of your experiments in order to formulate a problem and determine the factors that affect the appearance of colors and, thus, could be the variables.

Formulating a problem

To find the cause of colors in mica.

Selecting variables

Mary. The significant factors affecting the colors of mica are of two kinds, some of them refer to mica (its orientation and thickness) while others point to an external cause, apparently connected with the condition of the sky.

Selecting a procedure

David. To have two distinct images in an Iceland spar we need a screen with a hole on the way of light. However, since our mica crystals are not uniform, I suggest attaching this screen to the mica rather than to the Iceland spar: by doing this we will always look through the same region of the mica (Fig. 12.10)

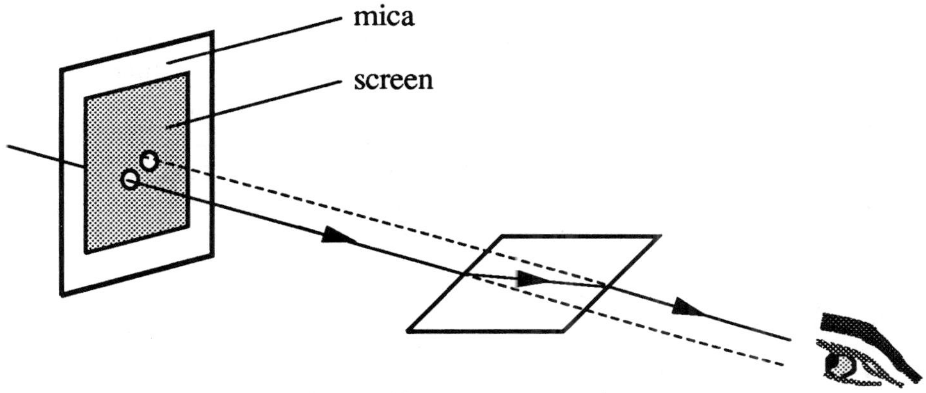

Fig.12.10. Arago's experiment..

Main part

a. the external cause

Preliminary experiments

Teacher. What can be different in different parts of the sky that could affect colors?

John. How about the color of light? A blue light from a clear sky works, while white light coming from clouds doesn't.

Michael. Isn't it true, though, that not all areas of the blue sky produce colors?

Ruth. You're right, of course. Listen, I noticed that the two images in an Iceland crystal were of different brightness. Can't we suppose, as Malus did, that light from blue sky resembles the one emerging from a double refracting crystal, which means it is polarized?

Mary. Perhaps, and if so, the polarization must be partial, because I never saw very dark images.

Hypothesis

The colors of mica are produced by a partially polarized light.

Test

David. Since a window glass partially polarizes the light from the sky (even if overcast), let us look at a mica plate set against a window.

Dorothy. I looked at a window at a large angle, and saw wonderful colors, much better than when I looked at the sky.

Michael. We couldn't find a properly oriented window, so we laid down a glass plate on the ground and watched the reflection of the sky. The colors were very distinct.

Ruth. First, I made a bundle of ten glass plates, taped them together and used them as a polarizer. When holding the bundle in my left hand (it made about 30° with the horizon) and the Iceland spar with mica in my right hand I looked at the sky through both of them and saw nice colors.

conclusion ➡ The colors of mica are produced by polarized light

David. We see that Arago discovered, among other things, that light from the blue sky is partially polarized. Apparently, the degree of polarization reaches a maximum in the direction perpendicular to that of the sun.

b. orientation of mica

Preliminary experiments

Teacher. We've already found that the orientation of mica affects the colors. To investigate this in detail we need a more convenient polarizer-analyzer system. Take two square polaroids and tape them to the opposite sides of a book, or a box, or a table so that the polaroids were parallel, about 10 - 20 cm apart, and have their planes of polarization crossed (perpendicular).

John. Before stopping to use the Iceland spar as an analyzer I decided to check whether the two images are indeed complementary. The slightly polarized light from the sky is not very good for this experiment, so I used a polaroid to make light fully polarized. I also enlarged the hole in the screen so as to make the two images overlap in half. Finally, I selected a piece of mica of apparently the same thickness. With this arrangement, I observed that whatever the colors of the two images, the overlapping part was white, proving that the colors were complementary.

Dorothy. I placed a crystal of mica between the polaroids parallel to them and saw a reddish-purplish color. Then I started slowly turning it around a vertical axis, and the color changed in the following sequence: blue, green, yellow, red, and so on.

Michael. I turned the mica around a horizontal axis and observed a sequence of colors similar to that of Dorothy.

Ruth. We rotated the mica in its own plane, which is difficult to do without giving it a tilt, so we pressed it to the analyzer. We observed the same color (red-purple) but its intensity changed, and in two positions of the mica the image became uncolored. We marked these two directions in the vertical plane and found them to be perpendicular.

Teacher. The existence of such two directions implies that, unlike Iceland spar, mica has two optic axes, and along each of them the velocity of both refracted rays is the same. The existence of two optic axes is not an unusual phenomenon: Brewster discovered in 1818 that most double refracting crystals are biaxial. It became clear that Huygens' theory could cover only a limited class of crystals. This problem was solved by Fresnel in 1821-23.

At this point, given the time constraint and the simplicity of our instruments, we have reached the limit for our investigation of the orientation. As to the thickness, we will resume studying its role with other substances that are more uniform than our samples of mica.

3. Biot and the emission theory

history

Teacher. In 1812-1816, Brewster's main rival in the field of polarization and double refraction was Biot. He was a very thorough investigator intent on deriving quantitative laws of phenomena. Although some of the phenomena he studied were discovered by others, this did not diminish the importance of Biot's contribution. Like Malus, Biot entered the *École Polytechnique* soon after its foundation. After graduation, he taught mathematics and physics first in province and then in Paris. In 1803, he became a member of the Academy of Sciences (mathematics section). Late in his life, for his contributions to the history of ancient astronomy and his writings, he was elected to two other French academies devoted to antiquities and literature. Biot was the leading champion of the emission theory of his time, and he devoted much effort to reintroduce Newton's quantitative method into physics and revive his theory of fits. His first interest in polarization was excited by Arago's discovery of colors of mica. In1812-13, Biot. found that a relation between the colors of thin crystalline plates and their thicknesses was very similar to that discovered by Newton for thin films of air or water, and that the ratio of the thicknesses of Newton's and Biot's plates which displayed the same color was constant for all colors. Thus, after determining this ratio experimentally for one color, aided by Newton's table (Newton, *Opticks*, 233), Biot correctly predicted the color of various plates given their thickness (between 0.03 mm and 0.45 mm). To Biot, that was another confirmation of Newton's theory of fits of easy reflection and easy transmission. To explain chromatic polarization, Biot supposed that when polarized ray penetrated a thin crystalline plate, its plane of polarization oscillated between the initial one and another plane which made the angle **2i** with it (**i** was the angle between the planes of polarization of the polarizer and of the crystal). The oscillations took place at equal intervals, which were different for light of different colors. This theory satisfactorily explained a number of phenomena but later was superseded by Fresnel's theory (1821).

At the same year 1811, Arago discovered that sometimes colors could be produced by thicker plates: a quartz plate 6 mm thick cut perpendicularly to its optic axis exhibited colors similar to those of mica. However, the similarity was not complete, because when Arago rotated the quartz plate in its own plane only the brightness of the transmitted colored light changed but not its hue. Two years later, Biot studied the colors of quartz in detail. He found that when the plate was perpendicular to the beam its color depended on the plate's thickness. While rotating the analyzer from right to left Biot observed a succession of all spectral colors and supposed that quartz polarized light of different colors in different planes. He explained this as follows: while passing a quartz plate the light particles rotated their poles from right to left and the angular speed of this rotation depended on color, This theory suggested that the angle of rotation is proportional to the plate's thickness and depends on the wavelength (or, the interval of fits) if light is monochromatic. Biot proved these hypotheses by exact experiments. He also found a different variety of quartz, which produced the same phenomenon with the analyzer rotating from left to right. In 1815, Biot discovered that some liquids, such as oil of turpentine and citric acid, can also rotate the plane of polarization, the former to the right, and the latter to the left. Later, he found that sugar syrup does the same and suggested measuring the angle of rotation of the plan of polarization in human urine as a test for diabetes.

Dorothy. This "rotation of the plane of polarization" staff literally makes me dizzy. Is there a simpler way to understand it?

Teacher. Let us do an experiment, then the subject will become clear.

4. Experimenting with optical rotation

lab

Teacher. Let us experiment with corn syrup. Use small rectangular containers that we've utilized in experiments on refraction. They are convenient when one needs to measure the thickness of the optically active medium. We will investigate the variables studied by Biot the thickness of a medium and the color of light. The experimental arrangement consists of two crossed polaroids set at a sufficient distance from one another to place the container between them. Attach a red filter to the outer surface of the polarizer. What do you see?

Dorothy. The visual field is dark because the polaroids are perpendicular.

Teacher. What do you see when you insert the container?

Michael. The syrup appears red.

Teacher. Why do you see light now?

Mary. You said that after passing an active medium light rotates its plane of polarization. This implies that the angle between this plane and the plane of polarization of the analyzer is other than 90°, thus some light must come through.

Teacher. Fine. How can we measure the angle of rotation of the plane of polarization?

David. Suppose that the original position of the plane of polarization of light is **P** and the final position is **P'** (Fig. 12.11). It is clear from this diagram that if we rotate the analyzer **A** to a position **A'** where we regain darkness, its angle of rotation α will be equal to the angle we are seeking.

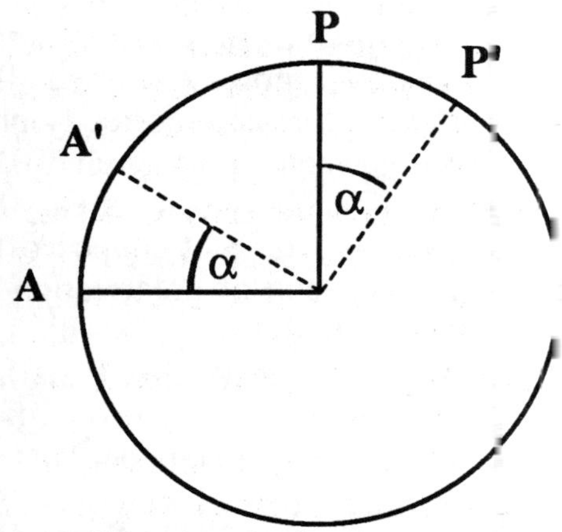

Fig.12.11. Measuring the optical rotation (the idea)

Teacher. Let us determine how the angle of rotation depends on the thickness for a specific color, and then how this angle depends on color. To reduce the range of transmission of your color filters tape several identical filters together. Take a circular analyzer and mark on its circumference the position of its plane of polarization. Make a transparency photocopy of a protractor, whose inner diameter is 5 mm larger than the diameter of the analyzer (make sure the plastic does not polarize light). Make a cardboard ring of 2-3 mm thick and 5 mm wide, whose inner diameter is the same as that of the analyzer. Fix both the scale and the ring to a glass plate with "Tak" and place the analyzer inside the ring (Fig. 12.12).

Michael. With a red filter, one container placed lengthwise (6 cm) produced the rotation of 30°, and two containers aligned lengthwise (12 cm) rotated the plane of polarization by 60°.

John. I've tried a larger container of 10 cm long (also with a red filter) and found the angle of 48°.

Ruth. With a blue filter, one small container produced 60° of rotation, while two such containers aligned made the angle of 120°. It looks like the angle of rotation is proportional to the thickness of the liquid, although the constant of proportionality differs for different colors.

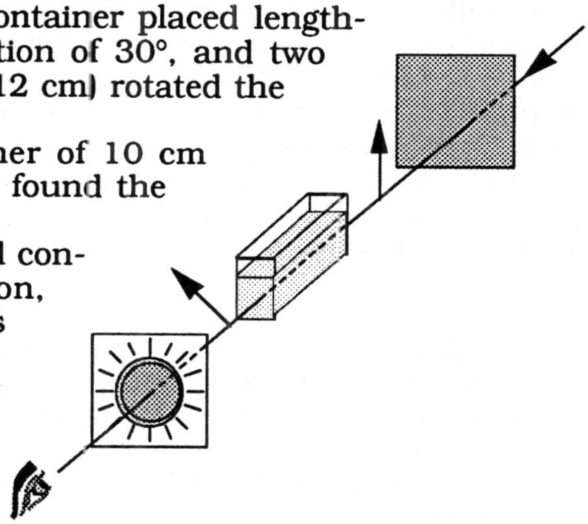

Fig.12.12. Measuring the optical rotation (the experiment)

Mary. We experimented with a small container and three different filters. We obtained 32° for a red filter, 47° for a green filter, and 61° for a blue one. This is consistent with the previous results and shows that the angle of rotation of the plane of polarization increases when the wavelength of light diminishes.

David. This means that when analyzer stood at 32° it eliminated red light, at 47° it absorbed green light, and at 61° the blue one. If so, what can we expect to see in white light while rotating the analyzer? I guess, that at these three positions we would see, respectively the remains of the spectrum, that is, blue-green, blue-red or purple, and yellowish-red.

Dorothy. Let me check your prediction. Yes, it comes quite close for these angles.

John Our group became interested in your remark about measuring the concentration of sugar. We diluted the syrup two and four times with water and found that the angle of rotation decreased 1.9 times with the 50% solution and 3.8 times with 25% solution. We checked that water does not have any effect on polarization. Thus, apparently the angle of rotation is proportional to the concentration of the active substance (sugar).

Teacher. Your results are similar to Biot's, which is very good, taking into account the simplicity of your equipment.

5. Artificial double refraction

history and discussion

Teacher. We have studied a variety of materials that polarize the transmitted light. What do they all have in common?
Dorothy. Some of them are crystals (Iceland spar, quartz, mica), others are liquids. I can't see what these two groups have in common.

lab home

Ruth. These two groups don't exhaust all polarizing materials. I placed a piece of a clear scotch tape between the crossed polarizers, and it exhibited colors like mica.
Michael. We've tried some wrapping cellophane, and it did the same when several layers were used.
Teacher. Did you notice any directional properties in these materials?
Ruth. My material was certainly double refracting, because when I rotated it the colors changed.
Michael. So it was in my case too.
Mary. I understand that crystals are asymmetrical, you can see it. Probably, that is why they transmit light with different speed in different directions. But how about plastics or liquids?
David. Perhaps the active liquids have asymmetrical molecules. As to plastics, isn't possible that they made asymmetric during their manufacturing, for instance, stretched in one direction.
Teacher. Excellent! You understand that the cause of double refraction is an asymmetry in the physical properties of a body, which can be achieved in various ways. Why don't we produce an artificial asymmetry and see whether it leads to double refraction.
Dorothy. I stretched cellophane wrapping and observed colors near its edges.
Michael. We've observed the same when stretching plastic lunch bags.
Mary. I've bent a plastic ruler, and it displayed colors.
Teacher. Very good! That mechanical deformations create double refraction was discovered by ***Thomas Johann Seebeck*** (1770-1831) and Brewster (1815). Glass was the first material utilized for this purpose (Fig. 12.13) Glass, however, required a considerable force, and Brewster was glad to find that a jelly, which is easily deformed, shows signs of double refraction caused by pressure or induration (12.14) I want you to do this experiment at home. Take a packet of a clear unflavored gelatin. Put it in boiling water (take one half of the amount of water recommended per packet) and stir until completely dissolved. Pour the liquid in a plastic container and place it in a refrigerator. When the jelly is set cut a rectangular piece out of it and place it between the crossed polaroids.

lab home

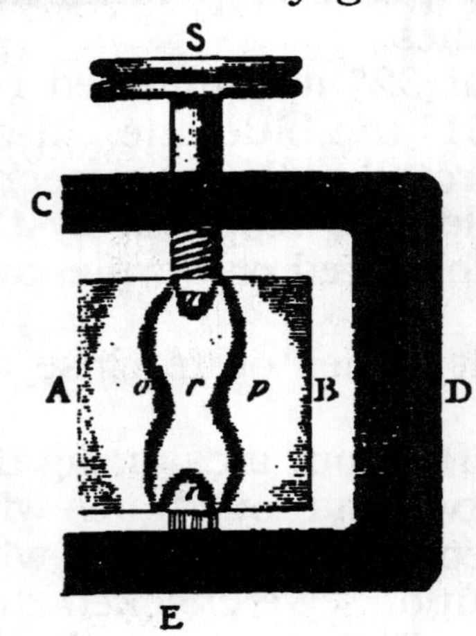

Fig.12.13. Double refraction in glass produced by a compression. From D. Brewster, *Phil. Trans.*, 1816

Watch the image while pressing or squeezing the jelly and try to relate the figures produced to the type of deformation. You have to remember that a coloration is not the only sign of double refraction, because it is observed only for a small range of thicknesses. A more general sign, true for every thickness, is a lightening of the dark visual field after inserting the sample between the polaroids.

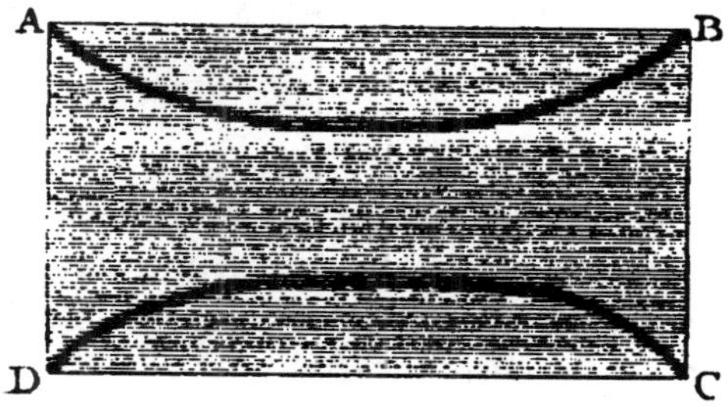

Fig.12.14. Double refraction in jelly produced by induration. From D. Brewster, *Phil. Trans., 1816*

history and discussion

Teacher. Good! Now, is it possible to discover a deformation that is invisible to a naked eye? Such deformations are produced, for instance, during manufacturing, either by mechanical forces or by a heating/cooling of the object.

Ruth. It is possible if the body retains the deformation. You place it between the crossed polaroids, and if you see a coloration only in some parts of it, it is probably the result of a deformation. I've checked that with cellophane.

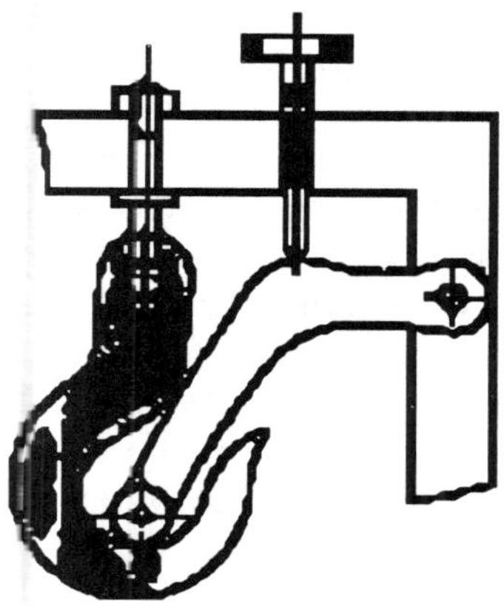

Fig. 12.15. Studying mechanical deformations by means of polarized light.

Teacher. Fine. This idea found a practical application. If one makes a clear plastic model of a working device and studies the colors displayed in polarized light, one can get a qualitative picture of the distribution of stresses in the device, which is important for improving its design (Fig. 12.15). Brewster and Seebeck also discovered in 1814 that glass can be made double refracting by heating it (Fig. 12.16). I want you to repeat this experiment at home. Heat a piece of glass over a candle flame and insert it between the crossed polaroids. Check whether it retains the double refraction after cooling off.

lab home

David. I remember hearing a crackling sound when I put an ice cube in water. This implies that the ice had some residual stresses acquired probably during solidifying. Is it possible that ice has a double refracting capability?

Teacher. I want all of you to check this at home. Compare slices of ice of different thickness. To make ice more clear polish it with your fingers.

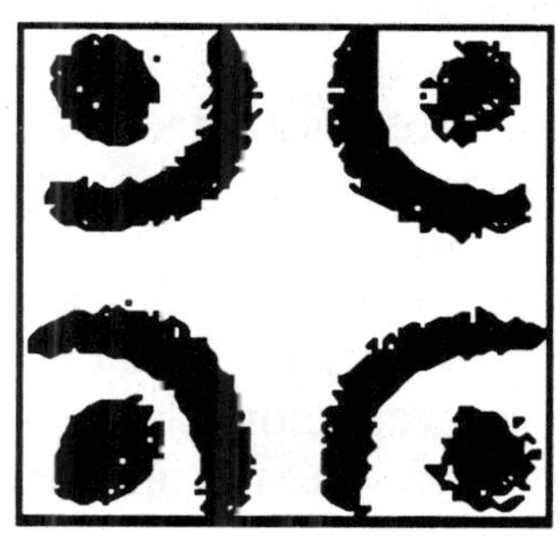

Fig.12.16. Double refraction in glass produced by heating.From T.Seebeck, *Journ. für Chem. u. Physik,* 12 (1814).

IV. WAVE THEORIES

1. Chromatic polarization

history

In 1814, Young suggested that Biot's experiments could be explained through interference of the ordinary and extraordinary waves. He was unaware then that light waves polarized in perpendicular directions cannot interfere. This was demonstrated by Augustin Fresnel and François Arago in 1816. In the two-slit experiment, they placed two crystals before the slits and found that the interference fringes appeared when the two beams were polarized in the same direction and disappeared when they were polarized in perpendicular directions. Thus, they discovered a new condition of coherence: two independent rays which are polarized in perpendicular directions cannot interfere. They also found that two rays polarized perpendicularly to one another could interfere if they originate from the same polarized ray and are reduced to the same plane. Fresnel expressed a light wave by a vector whose length represented the amplitude of light vibrations and the angle this vector formed with a selected axis displayed the phase of these vibrations. He inferred from the experiments on interference that the light vector must be perpendicular to the direction of propagation of light, or, in other words, that light wave is transversal. Mathematically, it means that an arbitrary light vector can be resolved into two (not three!) components that are perpendicular to one another and to the direction of propagation of light. For instance, let **P** be a polaroid with the axis of polarization **x**, (Fig.12.17), and the incident light wave have the amplitude **A** and is polarized in the direction **k**, which forms an angle α with the direction **x**, Then, according to Fresnel, this wave can be replaced with a sum of two waves of the same amplitude but polarized in perpendicular directions. If **x** is one of these directions, **y** being the other, the polaroid will transmit light of the amplitude a=Acosα. By squaring both sides of this equation we obtain Malus' law, because A^2 and a^2 are the intensities $\mathbf{I_0}$ and **I** of, respectively, incident and emergent light.

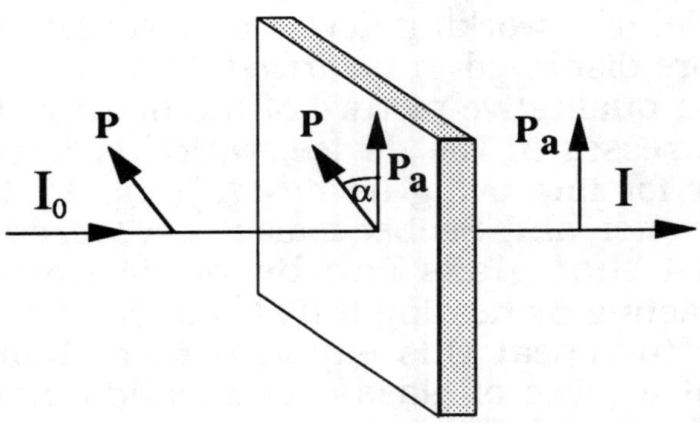

Fig.12.17. Wave explanation of chromatic polarization of light.

theory

Colors of crystalline plates result, according to Fresnel, from an interference of two polarized waves that originated from the same polarized wave, acquired perpendicular polarizations and a path difference when passing a plate, and finally were reduced to the same plane of polarization in an analyzer. For instance, let **P** (Fig. 12.16) be a uniaxial crystal of the thickness **d** cut parallel to its optic axis **O**, and let **x** and **y** be the respective directions of polarization in this crystal of the ordinary and extraordinary rays. Let the incident wave be of the amplitude **A** and polarized at the angle α to the axis **x**. When entering the crystal the incident light is divided in two waves whose vibrations are directed along the axes **x** and **y** and can be expressed as x_0=acosωt and y_0=bsinωt, where a=Acosα and b=Asinα. These waves travel along the same direction **O** but with different velocities $v_e=c/n_e$ and $v_o=c/n_o$. As the result, when exiting the crystal the two waves will have the phase difference

$$\theta = \frac{2\pi d(n_o - n_e)}{\lambda} \tag{12.3}$$

and the projections of the light vector on the two axes will be, for instance, x=acosωt and y=bcos(ωt−θ). To find the trajectory of the end of this vector we have to eliminate time from our equations, which can be done as follows:

$$y=b(coswtcos\theta + sin\omega t sin\theta); \tag{12.4}$$

$$cos\omega t = x/a; \tag{12.5}$$

$$sin\omega t sin\theta = y/b - xcos\theta/a; \tag{12.6}$$

Square both sides of equation (12.5) and multiply them by $\sin^2\theta$:

$$\cos^2\omega t \sin^2\theta = x^2\sin^2\theta/a^2; \tag{12.7}$$

After squaring both sides of equation (12.6) and adding to equation (12.7) we obtain

$$\frac{x^2}{a^2} + \frac{y^2}{b^2} - \frac{2xy}{ab}\cos\theta = \sin^2\theta \tag{12.8}$$

This is an ellipse **ABCD** (Fig. 12.18), the shape and orientation of which depend on the phase difference θ .

For instance, if θ=(2m+1)π/2, we have

$$\frac{x^2}{a^2} + \frac{y^2}{b^2} = 1; \tag{12.9}$$

This is an ellipse **A'B'C'D'** with its axes oriented along **x** and **y,** and with its semi-axes **a** and **b** representing the amplitudes of vibrations. A rotation of the ellipse will thus mean a change in the amplitude (and intensity) of emergent light. Such polarization is called *elliptical.* If a=b, we have a *circular polarization.*

math is a friend !

If θ=mπ (m=0, 1, 2,...), we have

$$y = \pm\frac{b}{a}x; \tag{12.10}$$

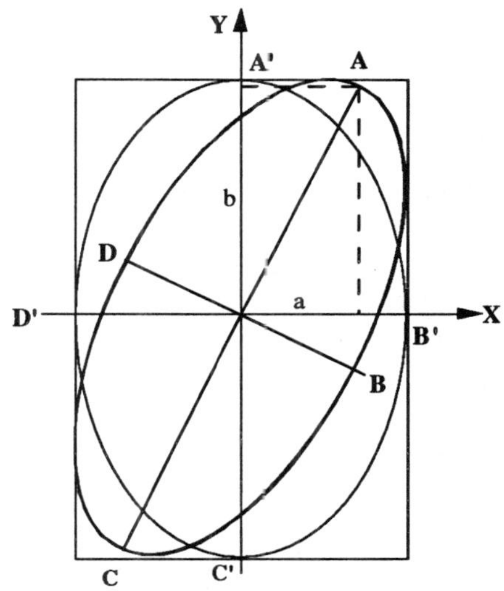

Fig.12.18. Elliptical polarization.of light..

which is the equation of a straight line. This means that after exiting the plate light remains linearly polarized.

discussion

Dorothy I understand the math but not the physical meaning of this elliptical (or circular) polarization. Does it mean that we see colored ellipses?

Teacher. No. Imagine we can slow down optical vibrations so much that we are able to watch the end of the optical vector. If light is linearly polarized, it will move back and forth along the same line; if light is circularly polarized, it will move along a circumference, either clockwise or counterclockwise.

John. How do we discover that a light is polarized elliptically or circularly? Can we use a polaroid or an Iceland spar?

Teacher. Let us have an analyzer whose direction of polarization makes an angle α' with the axis **x**. The two waves will produce along this direction vibrations with amplitudes $a'=a\cos\alpha'$ and $b'=b\sin\alpha'$ and phase difference θ. Their sum will have the amplitude A', and the intensity of emergent light will be $I'=A'^2$ such that

$$A'^2 = a'^2 + b'^2 - 2ab\cos\theta; \qquad (12.11)$$

Thus, for an elliptically polarized light if we rotate the analyzer, changing the angle α', the total intensity will change between a non-zero minimum and a maximum. If light is circularly polarized, the amplitude of the emergent light will not change during the rotation. Thus, a circularly polarized light can be confused with natural light, while an elliptical polarization can be perceived as a partial linear polarization. One can avoid this confusion by transforming an elliptically polarized light into a linearly polarized light. To achieve this we have to change the phase difference by $(4m+1)\pi/2$, which corresponds to the path difference of $(m+1/4)\lambda$. To achieve this, we need a crystal plate of such a thickness **d** that

$$(n_o - n_e)d = (m+1/4)\lambda; \qquad (12.12)$$

The thinnest of such plates (m=0) is called a *quarter-wave plate.*

Michael. How does this theory explain colors of mica?

Teacher. For a plate of a constant thickness the phase difference θ depends only on the wavelength λ. For each position of the analyzer, light of different colors will have different amplitudes, which will result in one color dominating over others.

Mary. If we keep the analyzer steady and rotate the plate in its own plane, neither the phase difference θ nor the angle α' change. If so, why does the field of view change its brightness?

Teacher. Keeping these two factors permanent removes the coloration. However, the rotation of the plate is equivalent to a rotation of the axes **x** and **y**, which produces the same effect as a rotation of the ellipse in Fig. 12.18: the projections of the ellipse' s semi-axes change, and so does the intensity of transmitted light.

Ruth. We see that a linearly polarized light is quite an ordinary phenomenon because it is produced at every reflection of light by a smooth surface. Where can we find an elliptically polarized light?

Teacher. Fresnel discovered in 1817 that when a linearly polarized light experiences two consecutive internal reflections at Brewster's angle it becomes circularly polarized when the plane of incidence forms an angle of 45° with the plane of polarization. A circularly polarized light can be also obtained from a linearly polarized light transmitted through a quarter-wave plate whose direction of polarization makes 45° with the axes of the crystal. Light reflected by metals acquires an elliptical polarization.

2. Double refraction

Like Huygens, Fresnel's connected the variation of the velocity of light in different directions inside a crystal with a change of elasticity of the ether. His problem was more difficult, however, for he had to account for both uniaxial and biaxial crystals. To model the distribution of velocities of light in a crystal, Fresnel used the *ellipsoid of elasticity* with three unequal perpendicular principal axes. For every direction of incident light there are two refracted waves with different velocities (Fig. 12.19). To determine them, it is necessary to draw a plane through the center of the ellipsoid perpendicularly to the direction of propagation of light: this plane cuts the ellipsoid along an ellipse, whose axes give the directions of polarization of the two waves and the magnitudes of their velocities. Among all elliptical sections two are circular, and the perpendiculars to their planes are the optical axes of the crystal. In an uniaxial crystal the two optical axes coincide. The electromagnetic theory of light did not change the essence of this theory, it only gave a different cause for changing the velocity of light in different directions: different elasticity was replaced with different dielectric constant of the material.

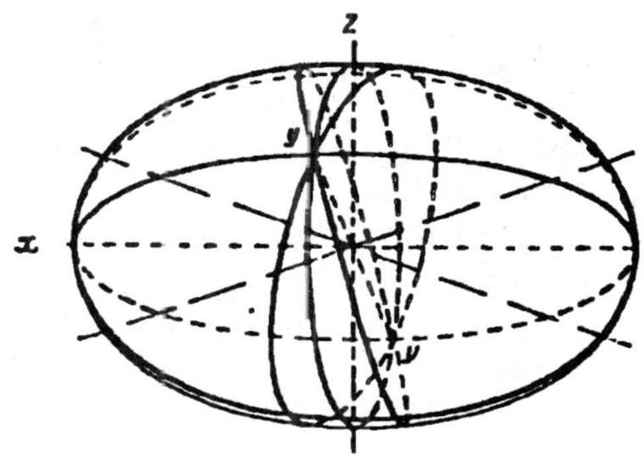

Fig.12.19. Wave model of double refraction.

Thus to explain chromatic polarization and double refraction Fresnel assumed a complete transversality of light waves. This idea initially occurred to him in 1816 after the discovery of the non-interference of light polarized in perpendicular planes. **history** However, it took Fresnel five years to adopt this concept because he could not overcome two difficulties in establishing a mechanical foundation for such a theory (optics was thought of then as a part of mechanics). First, it was known that mechanical waves when crossing a boundary of two media produce both longitudinal and transversal refracted waves. Secondly, the ether was treated then as a fluid, but a fluid could not transmit transversal waves. Thus, Fresnel had to attribute to the ether the properties of a solid body. The idea was perceived as wild, and even such champions of the wave theory as Young and Arago refused to adopt it. Fresnel's alternative consisted of either adopting transversal light waves together with the solid ether, or abandoning a comprehensive quantitative explanation of chromatic polarization and double refraction. He opted for the former. Later, electromagnetic waves replaced the mechanical ones, but the core of Fresnel's theory - the transversality of light waves - remained intact.

BIBLIOGRAPHY

C. Huygens, *Treatise on Light*, ch. V.
D. Brewster, "On the polarization of light by oblique transmission," *Phil. Trans.* 1814, 187-218, 219-230.
D. Brewster, "On the laws which regulate the polarization of light by reflexion from transparent bodies," *Phil. Trans.* 1815, 125-159.
D. Brewster, "On new properties of heat," *Phil. Trans.*, 1816, 46-114; "On the communication of the structure of doubly refracting crystals to glass, etc.... by mechanical compression and dilatation," ibid., 156-178.
K. Petersen, "Malus, Étienne-Louis," *Dictionary of Scientific Biography* Ed. by C. Gillispie, v.9 (C. Scribner's Sons, 1970), pp. 72-74.
J.Z. Buchwald, *The Rise of the Wave Theory of Light* (Chicago, 1989), pp. 203-290.
R. Wood, *Physical Optics* (Dover, 1961), 329-387.
R. Longhurst, *Geometrical and Physical Optics* (Longman, 1973), pp. 533-586.

Index